Dan Corbett

AN INTRODUCTION TO EXPERT SYSTEMS
The Development and Implementation of Rule-Based Expert Systems

AN INTRODUCTION TO EXPERT SYSTEMS

The Development and Implementation of Rule-Based Expert Systems

James P. Ignizio

The University of Houston

McGraw-Hill, Inc.

New York   St. Louis   San Francisco   Auckland   Bogotá   Caracas
Hamburg   Lisbon   London   Madrid   Mexico   Milan   Montreal
New Delhi   Paris   San Juan   São Paulo   Singapore   Sydney   Tokyo   Toronto

This book was set in Times Roman by Publication Services.
The editors were Eric M. Munson and Jack Maisel;
the production supervisor was Friederich W. Schulte.
The cover was designed by Carla Bauer.
R. R. Donnelley & Sons Company was printer and binder.

AN INTRODUCTION TO EXPERT SYSTEMS
The Development and Implementation of Rule-Based Expert Systems

1 2 3 4 5 6 7 8 9 0 DOC DOC 9 5 4 3 2 1 0

P/N 031712-7
PART OF
ISBN 0-07-909785-5

Library of Congress Cataloging-in-Publication Data

Ignizio, James P.
   An introduction to expert systems: the development and
   implementation of rule-based expert systems / James P. Ignizio.
      p.   cm.
   Includes index.
   ISBN 0-07-909785-5 (set)
   1. Expert systems (Computer science)   2. System design.
I. Title.
QA76.76.E95I387   1991
006.3'3—dc20   90-6343

TRADEMARKS
80286, 80386, and 80486 are registered trademarks of Intel
Apple, Hypercard, Mac II, and Macintosh are registered trademarks of Apple Computer
Corporation
Arity Compiler and Arity Interpreter are registered trademarks of Arity Corporation
ART is a registered trademark of Inference, Inc.
Compaq is a registered trademark of Compaq Computer Corp.
dBase, dBase II, dBase III, and dBase III PLUS are trademarks of Ashton-Tate, Inc.
DEC is a registered trademark of Digital Equipment Corp.
EXSYS is a trademark of EXSYS, Inc.
Goldworks is a registered trademark of Gold Hill Computers, Inc.
Guru is a registered trademark of **mdbs,** Inc.
IBM and IBM PC are registered trademarks of International Business Machine Corp.
IntelliCorp is a trademark of IntelliCorp, Inc.
KEE is a registered trademark of IntelliCorp, Inc.
KEEconnection, IntelliScope, SimKit, KEEtutor, and PC-Host are trademarks of
IntelliCorp, Inc.
KES is a trademark of Software Architecture and Engineering, Inc.

ABOUT THE AUTHOR

James P. Ignizio is Professor of Industrial Engineering and Chairman, Department of Industrial Engineering, at the University of Houston. He received his Ph.D. degree in Industrial Engineering and Operations Research from the Virginia Polytechnic Institute. His work in artificial intelligence began in 1962 in the development of a framework for heuristic programming and in the applications of heuristic programming to problems of combinatorics in engineering design.

Professor Ignizio is the author of five textbooks and numerous papers, including more than sixty referred papers in various international journals. Most recently, he served as Guest Editor for the Special Issue on Expert Systems and Operations Research of the *International Journal of Computers and Operations Research*. Professor Ignizio is a member of the American Association for Artificial Intelligence, the International Neural Network Society, the Operations Research Society of America, the Operational Research Society of Britain, and the Institute for Industrial Engineers. In 1980, he was awarded the First Hartford Prize by the U.S. National Safety Council.

CONTENTS

PREFACE

There has been, in the past decade, a virtual explosion of interest in the field known as expert systems (or, alternatively, as knowledge-based systems). Appearing from seemingly out of nowhere, expert systems has quickly evolved from an academic notion into a proven and highly marketable product—one that offers an efficient and effective approach to the solution of an exceptionally wide array of important, real-world problems. In particular, expert systems provides a powerful and flexible means for obtaining the solution to a variety of problems that often cannot be dealt with by other, more orthodox methods. Typically, such problems have always been considered too large and too "messy" for solution by conventional approaches. In this regard, expert systems may be considered to represent a potent new instrument for addition to one's decision-making "tool kit."

By the end of the 1980s, the implementation of at least 2000 expert systems had already been documented within the corporate world alone. That number, however, was most likely only the tip of the iceberg since at least as many, if not more, implementations have occurred in the military sector, within local, state, and federal government, and in a host of other classified, proprietary, or otherwise undocumented areas. And it is expected that the number of expert systems implementations will increase at an even faster rate in the 1990s and beyond. In fact, it is estimated that, in the United States alone, revenues from expert systems should top $6 billion by 1995—a nearly eight-fold increase over the receipts documented in 1989.

The growth in books on expert systems has been almost as dramatic. Among these are a number of excellent manuscripts. However, the majority of these works have, in our opinion, failed to adequately address the needs of those who wish to understand the fundamentals of expert systems and to actually, and successfully, apply the methodology to real-life problems. Further, very few of the existing books have been specifically designed as textbooks, or, in particular, as textbooks for the large and growing audience that exists outside the computer science discipline. This text has thus been prepared in an attempt to fill this relative void. In order to accomplish this, the author has focused on one particular class of expert system, the type that uses the so-called *rule-based* format for knowledge representation. The rules within such a format are represented by a set of IF-THEN statements. Specifically, *if* a certain condition exists, *then* we take a certain action (or draw a particular conclusion). Through the development of such rules, we attempt to represent the decision-making process through which a human expert, or experts, solve certain types of problems. In this regard, a collection of rules is said to represent a knowledge base.

While alternative modes of knowledge representation exist, and have their particular uses, advantages, and disadvantages, the rule-based approach is the one that is, far and away, the most commonly encountered—as well as the most widely implemented in actual practice. Further, and from a more pragmatic perspective, the majority of supporting computer software for expert systems employs rule bases. This is particularly true in the case of software for use in the development of small-to-medium sized rule bases (i.e., from about 50 to 200 rules), the type most frequently encountered by the beginner—and the knowledge-base size for which the largest growth in *actual* implementations has most recently been noted. Happily, rule bases are also a particularly natural and flexible mode of knowledge representation. As a result, the time required to both learn and apply expert systems may be substantially decreased when rule bases are employed. Thus, our almost exclusive focus on rule-based expert systems should not detract, in the least, from the usefulness of the material presented.

This text also takes a somewhat different perspective of expert systems than that found in most other books on expert systems—or on artificial intelligence. In particular, we do not consider expert systems to be an *alternative to conventional computer programming*. That is, as we see it, expert systems is not an alternative to FORTRAN, BASIC, C, or any other computer programming language. Even further, *expert systems is not, in our opinion, computer programming*. Rather, expert systems is an approach to decision making or, at the least, a methodology for use in the support of decision making. Like most other systematic approaches to decision making, the expert systems process employs models. More specifically, the heart and soul of an expert system is its model of the knowledge base of a human expert. Consequently, the primary intent of this text is to demonstrate a philosophy and procedure through which one may learn how to properly model the knowledge-base component of a rule-based expert system. Those individuals who have the aptitude, training and experience necessary to develop such models are designated as *knowledge engineers*. And the intended audience for this text is those individuals who wish to become knowledge engineers, or to manage knowledge engineers, or to at least become more familiar with expert systems and the role played by the knowledge engineer.

However, even if one becomes an expert in rule-base development, it is still necessary to have an appreciation and understanding of the other elements of expert systems. Thus, this text also addresses the very essential topics of

- *Knowledge acquisition.* The approach used to elicit the rules from the expert—or from a set of representative examples of (good) decision making
- *Inference (or knowledge processing).* The strategies and procedures for deriving a conclusion, or conclusions, from a knowledge base
- *Validation.* The procedures and guidelines by which one may determine the integrity and evaluate the performance of an expert system
- *Implementation.* The guidelines for implementation, control, monitoring, and maintenance of a completed expert system
- *Staffing and training.* A set of guidelines and suggestions for the selection, education, and training of the knowledge engineer—and the role and placement of the knowledge engineer within the organization

In addition, the coverage of this text extends to such topics as uncertainty in rule bases, bridges to external programs, alternative modes of knowledge representation, and expert systems development, and the employment of hybrid expert systems (i.e., a combination of an expert system with a more conventional approach to decision analysis).

The approach taken in this text has been to address the development of *generic* expert systems, and to most definitely not tailor the text or examples to any particular software package for expert systems development. However, in order to provide the reader with a vehicle for actually building rule-based expert systems, and to thus evaluate one's abilities in this area, the text includes software (in the form of a floppy disk) and a user's manual (in the form of an appendix) for the exercise of a demonstration version of the EXSYS rule-based expert systems shell. This software may be used on an IBM personal computer, or compatible system.

EXSYS has been selected for a number of reasons, including the fact that it

- Requires absolutely no knowledge or experience in computer programming, or in the use of any computer language
- Employs a straightforward menu-based approach to rule-base development that serves, in general, to reduce errors in data entry as well as to expedite the construction of a knowledge base
- Represents one of the most popular, computationally efficient, and widely used of the existing expert systems development packages
- Includes features that serve to enhance the package's usefulness and versatility (e.g., multiple modes of inference, confidence factors, mathematical variables, rule checking, bridges to external programs, blackboarding, and report generation)

Exposure to the EXSYS package should provide the reader with the background and experience necessary to use and understand virtually any other rule-based expert systems development tool. However, should the reader (or instructor) so wish, any other rule-based expert systems development tool may be used in place of, or to complement, the EXSYS package. Again, it is to be stressed that while EXSYS has been provided with this book, the text is not dependent on the use of either EXSYS or of any other software package.

This book has been designed for use in either the classroom, as a textbook in support of an introductory-level course on expert systems, or as a text for self-study by those who wish to learn more about this topic—and who possess the motivation necessary to carry out this pursuit. Problem sets are provided at the end of most chapters, and these exercises should be attempted by the serious student. A separate solutions manual has been prepared and is available from the publisher.

The text has been designed for upper-level undergraduates or graduate students in the areas of business, engineering, and the social sciences. However, because of the author's own background and experience, emphasis has admittedly been placed on examples and practices from the first two disciplines. Still, each example stands alone in the sense that enough detail is presented so that virtually anyone, from most any field, should at least be able to follow the logic involved and, I believe, benefit from the discussion.

One particular point concerning reader prerequisites must be emphasized. This text has been designed to address the development of rule-based expert systems, as solved through commercially available (rule-based) expert systems software. As such, there are absolutely no prerequisites on the part of the reader, of any exposure or experience in computers, computer programming, or computer languages—including the so-called AI languages (e.g., LISP and PROLOG).

The author's personal recommendations on the design of a minimal course on expert systems is that it should cover, at the very least, Chaps. 1 to 8 and Chaps. 11 to 13. Particular emphasis should be placed upon the material in Chaps. 4 to 8 and 11 to 13. The extent to which the EXSYS software (or any other rule-based expert systems development software) is employed is a matter of personal preference. Typically, the software has been used in support of a term project, that is, the development of a small expert system for the solution of some problem of particular interest to the student. However, the author would advise caution concerning the manner in which supporting software is introduced. Specifically, it is intended that the software *not* be employed until after the first six chapters have been covered. This text has been written in an attempt to introduce the reader to rule-based expert systems, the development of knowledge bases, and an understanding of how conclusions are reached by means of this methodology. It is the author's conviction that this topic cannot be fully understood and appreciated if one devotes the majority of one's time to the "care and feeding" of expert systems software. Thus, the author most sincerely hopes that the reader will suppress the desire to tear open the software envelope and immediately begin building expert systems. With a bit of patience, you will discover that you are much better able to develop *successful* expert systems if you first understand and appreciate the difference between just any knowledge base and a well-designed knowledge base. Finally, without an understanding of just how the inference engine of an expert system actually works, you risk the very real likelihood of developing conclusions that you do not fully understand. Remember, your goal should be to become a knowledge engineer, not a software guru.

James P. Ignizio

ACKNOWLEDGMENTS

The author wishes to acknowledge the influence of the efforts of a host of other authors in the fields of expert systems and decision support systems. In particular, I wish to recognize the contributions of Guy Benchimol, Paul Harmon, Peter Jackson, David King, Thomas Laffey, Pierre Lévine, Henry Mishkoff, Harvey Newquist III, Tin Nguyen, David Peat, Deanne Pecora, Walton Perkins, Jean-Charles Pomerol, David Prerau, Elaine Rich, Herbert Simon, Beverly Thompson, William Thompson, and Donald Waterman. Special acknowledgment is due the works of Ken Pederson, that have served to influence and motivate portions of the material in this book relating to guidelines in rule-base design.

Thanks are also due to Dustin Huntington, President of EXSYS Inc., for his kind permission to include the EXSYS demonstration disk with each copy of this text. Mr. Huntington's advice and comments have also been invaluable in the development of the user's guide for EXSYS that appears as an appendix to the manuscript.

In addition, the author wishes to thank the many students who have used the draft copy of this manuscript in their classes on expert systems. Their input, in the way of comments, criticisms, and suggestions have been invaluable in strengthening and focusing the text. Particular thanks are due to those students who have contributed exercises for use with the text. These are Exercises 4.25, 7.14, 7.15, 8.10 (Rex Walheim); Exercise 5.12 (Susan Voss); Exercise 5.13 (Tony Yeung); Exercise 5.16 (Mark Valentine); Exercise 6.10 (Bert Bras); Exercise 7.16 (Jagadish Sorab); Exercises 8.11 and 8.12 (Jaymeen Shah); Exercise 8.14 (Mike Mergens); Exercise 8.15 (Firdhaus Alamsjah).

McGraw-Hill and the author would like to thank the following reviewers for their valuable comments and suggestions: Leonid Charny, Boston University; Steven J. Fenves, Carnegie Mellon University; Soundar R. T. Kumara, Pennsylvania State University; Yoh-Han Pao, Case Western Reserve University; and Efraim Turban, Eastern Illinois University.

Finally, I wish to thank Laura Ignizio Burke for her time and patience in reviewing the various drafts of the manuscript, and for her suggestions concerning the contents and organization of the final draft.

AN INTRODUCTION TO EXPERT SYSTEMS
The Development and Implementation of Rule-Based Expert Systems

CHAPTER

1

INTRODUCTION

PURPOSE

During the past decade there has been a virtual explosion of interest in the field known as *expert systems* (or, alternatively, as *knowledge-based systems*). Articles have appeared, and continue to appear, in trade publications, professional journals, and in the mass media which tout the implementation of this *new and novel* approach to a seemingly endless number of problems. In fact, some cynics might say that it is becoming harder and harder to find a problem for which expert systems has *not* been proposed. And it would seem that no computer software support system can now be considered complete without the inclusion of at least one expert systems development package, or "shell."[1] While such intense hype has certainly served to promote the expert systems methodology, it has also resulted in considerable misunderstanding, and even misuse, of the expert systems methodology. Instances of the implementation of expert systems by the wrong people, in the wrong manner, and on the wrong problem are, regretfully, more common than one might expect. This is unfortunate as it serves to obscure the irrefutable ability and potential of expert systems in decision making, including decision making in a host of important, real-world situations for which there are simply no appropriate alternatives.

In this text we shall attempt to separate fact from fiction and to introduce the reader to the field of expert systems, its history and foundations, and to its use in the analysis and solution of many important, difficult problems. We view

---

[1]An expert systems shell, as we shall discuss in detail later, is simply a tool used to facilitate the development of an expert system.

expert systems not as a panacea, but rather as simply an exceptionally interesting and potentially valuable tool for problem solving—and we strongly encourage its consideration—*but only for those problems for which it is truly appropriate and cost effective.* Fortunately, as we shall see, such problems are abundant.

We also hope to demystify this tool. As we shall see, the construction of an expert system is a relatively straightforward procedure, requiring very little in the way of background in sophisticated mathematical methods and absolutely none in computer programming. In fact, as shall be demonstrated, we can (and shall) build expert systems using statements in plain English that are implemented manually rather than on a computer.[2] This is not meant to imply, however, that the topic is either trivial or that the amount of effort involved in the construction of an expert system is insignificant. Rather, the development and implementation of expert systems for real-world problems require time, care, training, and experience, plus a certain natural affinity for dealing with *messy*, ill-structured problems. And *first and foremost*, one must select a problem for which the implementation of expert systems is appropriate. This implies, most strongly, that one must also be aware of potential alternatives to the expert systems approach. Those who identify appropriate applications of expert systems and who perform the process of development and implementation are called *knowledge engineers*. This text has been designed so that, upon completion of either a formal or self-study course based on the material covered, the serious student will have received the initial level of preparation necessary to be a successful and effective knowledge engineer.

PEORIA: A HYPOTHETICAL EXPERT SYSTEM

While we do not believe that a complete grasp of expert systems can be achieved without an understanding of the material to be covered in the text (or similar material from alternative sources), we shall attempt to at least briefly summarize the concept of the use of expert systems through a simplified, *hypothetical* example. This example reflects the features of an ideal expert system, and does not necessarily indicate the capabilities (as we shall see) of any single, existing expert system.

The example that has been selected is one concerned with the introduction of new products within an existing firm. The particular problem faced at the moment concerns what new product, or products, should be considered for introduction into the market at this time. It will be shown how an expert system might be used to assist us in making this decision.

Traditionally, new-product introduction is accomplished in a series of steps, utilizing input from numerous individuals having expertise in a variety of areas. In the terminology of expert systems, these human experts are called *domain*

---

[2]Real-world expert systems would, of course, require that one employ a computer for the computational efficiency that may be derived.

experts; that is, each possesses expertise in a very specific, and relatively narrow domain. In our example, the domain experts serve in the following manner:

- To select, from among a list of candidate new products, those that would seem to hold the most *promise*
- To develop, for each new product selected, a plan for the introduction of that product (typically to certain test market areas) on a trial basis
- To use the results of the test marketing efforts to decide which products will finally be incorporated into the firm's product line, how to initiate and control the manufacture of these products, how to distribute and advertise them, how to price the products, and so on

Certain firms are considered very successful, and consistently so, in the introduction of new products—and the reason for this success lies mainly in the skills of the human experts employed. Further, while a portion of these skills may have been learned through formal courses, books, and manuals, much of the expertise is gained through the experience, insight, and intuition of the expert.

However, human experts are subject to numerous drawbacks as follows:

- They may retire, join another firm, or start their own company.
- They can only be in one place at any one time.
- Their skills may diminish with time or illness (and many such individuals ultimately exhibit what is commonly known as *burnout*).
- Their performance may be severely impacted by emotions.
- They may become too easily distracted and, as a result, may miss the occurrence of a critical event.
- They require considerable time in training and at the conclusion of such training, they will not be equally proficient, although identically trained.
- They typically become overwhelmed by *too much* data.
- They find it particularly difficult to consider *interactions* in complex situations— and such interactions typically play a major role in real-world situations.

Thus, and in consideration of these drawbacks, let us examine a hypothetical expert system that has been developed for a firm for the specific problem of new-product selection, introduction, and marketing.

Since every expert system must have a name, let us call ours PEORIA (i.e., as influenced by the old saying, "But, will it play in Peoria?").[3] PEORIA contains, in its *knowledge base*, a representation of the rules used by the firm's experts

---

[3] Peoria, a city in Illinois, has become associated with this saying. Specifically, if something *plays* in PEORIA, it should *play* elsewhere. Thus, if a product is accepted in PEORIA, it should be accepted throughout the United States. This is a saying, however, that does not necessarily sit well with those who live in Peoria.

in making their respective decisions. In Chaps. 4 and 5, we shall examine just how such rules are represented and acquired. The *primary* role of the knowledge engineer is, in fact, that of acquisition and representation of such knowledge, or rules. Here, we need only accept that this has somehow been accomplished.

In addition to its knowledge base, PEORIA has direct access to all of the firm's databases[4] which include those concerned with the following:

- Customer surveys
- Products and strategies of competitors
- Status of all ongoing new-product efforts within the firm, as well as (legally obtained) intelligence about the research efforts of the firm's competitors
- Socioeconomic profiles of all primary population centers and very specific information about the firm's test-marketing locations
- Cost data (e.g., production costs, advertising costs, distribution costs)
- Characteristics of the existing production system (e.g., labor force, processors, storage space)

At the monthly meeting concerned specifically with new products, we may then access PEORIA either through the PEORIA computer terminal or directly through its associated voice recognition system. PEORIA, through its voice generation system, may then verbally respond to our questions while simultaneously displaying its comments on the computer screen. For example, we might start the session by stating: "Good morning PEORIA, we are ready to begin." PEORIA's response (both verbally and on the projection screen) might then be:

```
Good morning. Please indicate the subject
under consideration:

1. Selection of a candidate list of new
   products for trial marketing
2. Analysis of results of previous or ongoing
   trial marketing efforts
3. Comprehensive plan for the trial marketing
   of a new product
4. Comprehensive plan for the nationwide
   introduction of a new product
```

---

[4]Later, in Chap. 2, we shall discuss databases in more detail. At this time, it should be sufficient to think of a database as a set of numbers, or facts, while a knowledge base is a set of rules.

Let us assume that we respond verbally, that is, by simply stating the number corresponding to the appropriate selection from this screen menu. Let us further assume that the primary purpose of the meeting is to consider a new product that is intended for introduction into our population of retired citizens—a sector which we believe we have not successfully addressed in the past. Thus, we initially respond with the number 1.

PEORIA then flashes its next menu on the screen.

```
Select desired market sector: Age and sex

 1. Infants and toddlers
 2. Ages 2--10
 3. Ages 11--18
 4. Ages 19--25
 5. Ages 26--35
 6. Ages 36 -45
 7. Ages 46--55
 8. Ages 56--65
 9. Senior and/or retired citizens
10. Males
11. Females
12. Both males and females
```

Our response is to then select numbers 9 and 12.

Since PEORIA has direct access to our firm's databases, it can then scan each and every new product idea that has been generated (and documented) and is associated with senior/retired citizens. From these, it will select (using its programmed expertise) a ranked list of candidate new products, and flash these on the display screen. We can accept these recommendations or override them and select a candidate at something less than top ranking. For the purpose of discussion, let us assume that PEORIA displays the following summary on screen[5]:

---

[5]Note that PEORIA has listed a number of recommendations, rather than just a single one. The ability to develop *multiple* conclusions (as is typical of human experts) is one particularly important feature of an expert system.

```
I suggest that the following products, in order
of priority, be considered:

1. Senior citizen exercise videotape (Project
   2-1019)
2. Senior citizen travel guide (Project 2-0566)
3. Senior citizen dating service (Project
   2-3107)
```

As discussed, we may at this point either accept PEORIA's recommendation or override it and select a lower rated project, or even a project not on the list. However, an expert system will typically incorporate an explanation facility. To demonstrate, we may wonder why the senior citizen exercise videotape project received the highest ranking. If so, we can simply voice our question by stating: "How was number one selected?" In response, PEORIA will provide an explanation, again both verbally and via the screen display. For example, PEORIA might display the following message in response to our question:

```
   The senior citizen exercise videotape
(Project 2-1019) was among the top three rated
by a panel of senior and retired citizens.
   Project 2-1019 has less competition than the
other top-rated projects, and this competition
has performed poorly in such areas as
distribution, shelf space, and advertising.
   If a project receives both high test panel
ratings and has limited competition, it
receives a high ranking for consideration for
introduction. If it has both a high ranking and
high estimated profit potential, it will be
recommended for consideration.
```

Based on this explanation, let us assume that we decide to accept PEORIA's recommendation (i.e., the top-ranked alternative—the senior citizen exercise videotapes). We then continue our session with PEORIA by responding to its further queries. In doing so, we will likely observe two other interesting and useful features of an expert system. First, most expert systems have the ability to deal with uncertainty. For example, during a consultation session PEORIA may ask us a question such as:

> Do you expect labor costs to rise sharply over the next two quarters (−10/10)?

Our response to this question is expected to be a number between −10 and +10. PEORIA interprets a +10 to indicate that we are absolutely certain that labor costs will rise sharply during the next two quarters, while a response of −10 indicates our belief that there is absolutely no chance for such a rise. Numbers between −10 and +10 indicate various levels of uncertainty concerning a sharp rise in labor costs. For example, a response of 7, for instance, would indicate that we are relatively confident of a large rise in labor costs, while a response of 3, would indicate much less confidence in that reply.

The second feature is the ability of an expert system to deal with partial data. Consider, for example, the following question:

> List expected time to complete the product packaging effort:
>
> 1. Less than 1 month
> 2. Between 1 and 3 months
> 3. More than 3 months

Here, PEORIA expects a response of either 1, 2, or 3. However, what if we simply have no idea of how long it might take to complete the product packaging effort? Real-world decision making is, in fact, characterized by such gaps in our knowledge. In such an instance, we can simply respond by saying: "Unknown." PEORIA will accept this response and continue the consultation session. Continuing in this manner, we will ultimately develop a trial marketing plan for the senior citizen exercise videotape via PEORIA's assistance. At that time, we may decide to conclude this particular consultation session.

After actual implementation of the trial marketing effort, PEORIA will directly and automatically monitor the progress of the effort (again, through its access to the firm's databases) and keep us informed—and, if considered necessary by the expert system, possibly recommend an early termination of the trial program (e.g., some senior citizens may be injuring themselves and previously unforeseen legal ramifications may be developing).

Assuming that we do successfully complete the trial marketing effort, PEORIA will peruse the results and make its recommendation concerning whether this product should be considered for nationwide (and possibly, international) introduction, precisely how this should be done, the pricing and discount data, advertising modes and frequency, and so on—using the type of interactive sessions illustrated earlier. Should we agree with PEORIA's plan, then PEORIA will develop a printed report and transmit the plan to all associated personnel via electronic mail, of course, for actual implementation.

A Tentative Assessment of PEORIA

Our discussion of PEORIA implies the existence of a quite large and rather ambitious expert system with the following capabilities:

- Consultation interface via voice recognition and generation
- Extensive database interfaces
- The employment of multiple knowledge bases (e.g., possibly one for each area of expertise involved in the decision-making process)
- Explanation (i.e., the ability to explain why a question has been asked and how a conclusion has been reached)
- Development of multiple recommendations (i.e., the derivation of several alternative conclusions)
- Decisions under uncertainty (i.e., the ability to deal with responses for which varying levels of confidence exist)
- Decisions with unknowns (i.e., the ability to reach a conclusion with incomplete data)

Actually, each of these capabilities does exist at this time. However, the actual development of an expert system of the size and complexity implied by PEORIA, and the simultaneous incorporation of all of these features, would require a major effort. Even so, by the time you read this book, it is conceivable that one or more firms will have actually developed a system at least somewhat similar to PEORIA. In fact, with the exception of the voice recognition system, such an effort is presently underway in at least one company in the United States.

However, let us now consider a far more important question: *Should* such a system be built? Interestingly enough, this is a question that has been, and continues to be, overlooked in far too many (if not the majority) of the expert systems that have been developed, or that are proposed for development. We shall deal in detail with this fundamental question throughout this text. Here, however, we will simply address the appropriateness of the development of PEORIA (or any other expert system) by recognizing that such a system should not be considered unless there are definite advantages to such development—and unless such advantages significantly outweigh the cost and disadvantages of expert systems. Equally fundamental, an expert system is appropriate only if it is feasible to capture and represent actual expertise, either acquired directly from a domain expert or by means of access to documented examples of expert decision making. With all the hype about expert systems (and other areas of artificial intelligence) and the corresponding rush to *get in on the action*, these are two points that are all too often forgotten.

Considering either PEORIA or expert systems in general, let us examine some of the potential advantages over the use of human experts.

- The expert system is always and instantly available, and always performs at the same level of expertise.
- The expert system (i.e., as implied for PEORIA) has direct and instantaneous access to the necessary databases and is not bound to the limited, biased, and imperfect recollections of the human.
- The expert system is logical, objective, and consistent—and thus unswayed by emotional arguments that might influence a human.
- The expert system doesn't forget (e.g., any of the rules it employs to draw its conclusions) or make mathematical errors.
- The expert system is in (or may be placed into) a constant state of *awareness* (i.e., it will not overlook any critical events that it has been assigned, and has the means to monitor).
- The expert system makes its decisions with regard to the goals of the firm, rather than with regard to how such decisions might influence its personal promotions or pay raises.
- The expert system multiplies the expertise of the firm, that is, it is directly accessible by all other divisions of the firm whereas the human expert's access is limited by physical and geographical considerations.
- The expert system is, itself, a repository for the storage of the knowledge of those experts from whose input it was developed; it is a knowledge bank of considerable value and is thus a tangible, permanent asset of the firm.

The last advantage should absolutely not be underestimated. Traditionally, the expertise associated with the running of an enterprise is undocumented, residing in the (imperfect) memories of key personnel. An expert system provides a mechanism for the collection and storage of this expertise, *and this alone* may be reason enough to justify the consideration of expert systems. Further, while in some instances alternative means for problem solution may exist, no other approach features such a totally transparent knowledge base. That is, the knowledge within a rule-based expert system is in the form of statements in plain English (or whatever human language we wish to use).

While there are undoubtedly many advantages that we have failed to mention, the above list does point out those areas in which an expert system may be considered superior to the human expert. In fact, it is true that many of the expert systems that have been developed actually *outperform* those human experts (i.e., in the specific, narrow domain for which the expert system has been developed) from whom the expert system's knowledge base was constructed.

However, as with any methodology, there are certain limitations of expert systems as compared to human experts. In particular,

- A human expert is *streetwise*. That is, a human being understands and appreciates certain cultural factors about which an expert system may have absolutely no awareness, whereas even the most well-designed expert system can, at times, appear exceptionally naive.

- Human experts are generally aware of the scope and limits of their knowledge, whereas many (present) expert systems simply do not recognize their limitations. In other words, *they do not know what they do not know.*

- When faced with a new situation, human experts may sometimes develop entirely new, original, and even brilliant approaches to solving a problem. Specifically, the human expert exhibits creativity whereas a typical expert system is limited to that knowledge base as originally developed.

- In some cases, people simply wish to communicate with other people—and not with a computer program. Whether this is a result of distrust of computers, or simply the need to deal with other humans, it can restrict the acceptance of an expert system.

- Finally, the human expert is simply far more flexible than a computer program.

As a consequence, determination of the specific value (particularly in *measurable* terms) of the introduction of an expert system is not a clear-cut proposition. Further, one must realize that, when expert systems are implemented, they seldom replace the human expert. Rather, they can (and should) permit human experts to address their attention to other matters—matters that, in the past, had to compete with the decision-making processes now accomplished in a routine fashion by means of the expert system. Also, it would be rare indeed to have an expert system that could exist without continual upgrading and/or revision. As such, we often assign at least one human expert the responsibility of monitoring the performance of the expert system and detecting changes in the problem or problems which are addressed by the system. Whenever necessary, the human expert assists in the upgrading of the existing expert system.

Thus, unlike other areas of automation, the benefit of expert systems can seldom be measured in terms of the cost savings associated with the replacement of one or more human employees. Instead, the real benefit of an expert system is most likely to be in the area of improved decision making in conjunction with the prospect of permitting the human expert to address other, more important, more far-reaching problems. And it takes a progressive company to appreciate the importance of such relatively intangible benefits.[6]

---

[6]In this respect, there is an interesting analogy between quality control and expert systems. That is, the measure of the value of the production of *higher quality* products is not easy to determine—and is virtually impossible to state in terms of dollars alone. In the United States, many firms saw the inclusion of quality control as a concept that would impose additional manufacturing costs—and little else. However, as has been learned (the hard way) by the United States automotive industry, a firm's very existence may well depend on the incorporation of such practices.

EXPERT SYSTEMS: SOME DEFINITIONS

Using PEORIA as a basis, the reader should now have at least a somewhat better feel about just what might be meant by the term *expert system*. More formally, we might be tempted to define an expert system as *a computer program that exhibits, within a specific domain, a degree of expertise in problem solving that is comparable to that of a human expert*. And this definition is, in fact, quite typical of those provided in a fair number of texts on expert systems. When properly interpreted, this definition serves to capture, but only to a certain degree, the expert systems concept.

However, we believe that there is one major problem with the above definition. Specifically, it serves to rather strongly imply that an expert system is, quite literally, nothing more than a computer program—or, at best, some type of alternative to "conventional" computer programming. There are, in fact, a number of authors who use the term expert systems *programming* in their discussions of expert systems. This emphasis on computers and computer programming can, in turn, lead one to presume that the designation of knowledge engineer is just a fancy term for computer programmer—and that knowledge engineers are individuals who employ *expert systems programming* in place of more conventional programming languages, such as FORTRAN, BASIC, C, FORTH, or COBOL.

As a consequence, we feel that the definition stated above serves to capture just one perspective of expert systems—that one which emphasizes the computational support required in the implementation of this methodology. We, however, hold a quite different view of expert systems. Specifically, while we certainly agree that the computer and computer programming are an essential source of *support* for the implementation and solution of any expert system of meaningful size, they are not, we believe, the essence of expert systems. In fact, as will be illustrated in Chap. 6, one can develop expert systems and derive conclusions from them through a strictly manual process. In doing so, we should be able to note that the single major factor in the construction of any expert system is that of formation of the associated knowledge base (or, more accurately, the formation of a model of the knowledge base). Thus, as we see it, the primary role of the knowledge engineer is to acquire and represent, in the form of a knowledge base, the rules employed by a human expert in the solution of a particular problem. More precisely, *knowledge engineers develop models of the knowledge base*.

The role of the knowledge engineer, as we envision it, is analogous to many other disciplines. For example, engineers and scientists often focus their attention on the development of (mainly) quantitative models. Management scientists, operations researchers, and systems engineers, in particular, employ models to an extensive degree in their work. For example, a management scientist might construct an input-output model of an industrial sector.[7] From a somewhat simplified

---

[7]According to O'Connor and Henry [1975], input-output analysis is concerned with studying the interdependence of the producing and consuming units in a modern economy and with showing the interrelations among different sectors that purchase goods and services from other sectors and in turn produce goods and services which are sold to other sectors.

view, such a model serves to indicate the relationship among the various elements of the system, and the impact on the output of the system through changes to the input. The data associated with this model might then be fed to a computer program that serves to solve input-output models (i.e., a program that, in essence, finds the solution to a series of linear equations) [O'Connor and Henry, 1975]. Since such software exists and is readily available, the bulk of such an effort is (or certainly should be) concerned with the development of the input-output model. Further, the results produced by the computer (no matter how sophisticated the computer program or whatever underlying computer language is used) are totally dependent upon the accuracy and integrity of the *model*. While management scientists should certainly understand just how such a computer program produces its results, it is not necessary (and would, in most cases, be a waste of time) for management scientists to develop such software themselves—or to have any familiarity with the particular computer language used by the software.

In a similar manner, knowledge engineers should focus the bulk of their time and effort toward the development of a representative model of the knowledge base. Once this has been accomplished, they will most likely select an *existing* expert system software package for the performance of the inference process; that is, the derivation of the conclusions for the particular problem at hand.

Thus, let us amend the previous definition of an expert system to reflect the points made in the above discussion. Specifically, an expert system is defined herein as *a model and associated procedure that exhibits, within a specific domain, a degree of expertise in problem solving that is comparable to that of a human expert*. The model, in turn, is the representation of the knowledge base of the human expert. As noted, in this text our attention shall focus on models that employ the so-called rule-based mode of knowledge representation.

We should also note that not all expert systems are developed by way of an interaction (or, at least, a direct interaction) with the human expert (i.e., the so-called knowledge acquisition phase). Expert systems can be, and often are, developed in situations where there is either no human expert, or possibly no practical access to such an expert. For that reason, an alternative designation for such a system is that of a *knowledge-based system*. However, since expertise must ultimately be captured within such a system, the term *expert systems* is, we believe, an appropriate descriptor of the product to be developed.

THE IMPORTANCE AND FUTURE
OF EXPERT SYSTEMS

While, as we shall see, expert systems is sometimes misused, misapplied, and overly hyped, one cannot and should not dismiss it as either unimportant or a passing fad. The methodology of expert systems is definitely here to stay, and it will play an increasingly important role in future decision making. When the population of the earth had reached five billion (officially, as of July 11, 1987), it might seem strange to point out that one particularly scarce resource is that of human experts. However, this is indeed the case—and is most certainly a result of the increasing size and complexity of human endeavors.

For example, contrast the entrepreneurs of today with those of 50 years ago. Today, to start a company (even a *low tech* firm) with any reasonable chance of success, it takes much more than just a good idea and hard work. There are a multitude of local, state, and federal laws and regulations to consider—even in such a seemingly mundane matter as locating the site for the plant or office. A host of ever-changing rules concerning the labor force must also be addressed. Further, there is a bewildering array of potential approaches for financial support. And we have touched on but a few of the aspects that must be dealt with in the start-up process.

Now, one could attempt to learn (and stay current in) all of these aspects on one's own, which would be both foolish and futile. Alternatively, we turn to human experts: financial planners, lawyers, site-location analysts, computer systems designers, and so on. Anyone who has dealt with such individuals recognizes the following:

- Expertise is expensive and scarce.
- Levels of expertise can vary widely.
- Judgment, intuition, and experience play a key role in such expertise.

As such, another alternative would be to turn to expert systems for assistance. If such systems compare favorably in terms of their expertise (i.e., as contrasted with that available through the human expert) *and* are less expensive, they should certainly be attractive and in demand—and this is one reason for the intense interest in the expert systems concept in the commercial sector. Yet another reason for the consideration of expert systems is that of competition. In most endeavors, we are in competition with others (e.g., other firms, other countries). The employment of expert systems has already been shown to provide, for those using them, a competitive advantage. Thus, simply to remain competitive, the employment of expert systems may well become a necessity. In fact, we personally believe that this will ultimately become the major factor behind the growth and acceptance of expert systems.

While expert systems have considerable room for improvement, we should note that they are not just academic notions. Rather, expert systems have already been constructed and are being sold and/or implemented in such diverse areas as follows:

- Stock market advisors
- Commodity trading
- Financial planning
- Tax preparation and planning
- Granting of loans and determination of credit limits
- Diagnosis and treatment of various diseases
- Determination of the chemical properties of unknown compounds
- Scheduling and control of the automated factory

- Diagnosis and maintenance of complex machinery (e.g., locomotives, aircraft, spacecraft, and ships)
- Assignment of planes to airport gates and the scheduling of flights
- Loading of cargo on ships and aircraft
- Configuration and design of computers
- Layout and design of printed-circuit boards
- Selection and deployment of acoustic transducers within sonar arrays
- Facility location and layout
- Automation of the auditing procedure for foreign exchange transactions
- Scheduling and control of chemical processes
- Monitoring and analysis of dust within mines
- Monitoring and diagnosis of poultry
- Enhancement of fast food operations

While the above list is impressive, it is by no means complete nor does it indicate the vast potential for additional implementations of expert systems. Feigenbaum et al. [1988] have documented some 2000 expert systems used within the corporate world. And this number most likely represents only the tip of the iceberg. It should be obvious that, when dealing with the topic of expert systems, we are addressing a methodology with an established track record, as well as a definite and promising future. A recent prediction by Market Intelligence Research Company (MIRC) would tend to support this conclusion. MIRC notes that "... the U.S. expert systems market currently stands at $820 million per year, and will climb to at least $6 billion per year in total revenues by 1995" [*AI Week*, 1990].

PREREQUISITES OF THE READER

The only major prerequisites for an understanding and appreciation of the contents of this text are the following assumptions:

- An interest in the subject matter. Any topic is easier to learn if one is interested in the topic and in its potential implementation.
- A certain degree of intellectual maturity. That is, the reader should be capable of dealing with new ideas, terminology, and concepts and have an appreciation of the need to address any nontrivial problem in a systematic and logical manner.

Further, the mathematical requirements of the reader are minimal and should have been obtained through those courses typically dealt with in the first 2 or 3 years of undergraduate education (e.g., basic mathematics and elementary statistics). As such, the text has been designed for readers with a variety of backgrounds and interests—and has already been employed in courses attended by students in the fields of business, engineering, psychology, economics, management science, operations research, hotel/restaurant management, mathematics, statistics, law, computer science, and biology. Here, it has found use at both the graduate and undergraduate levels.

The one point concerning prerequisites that we particularly wish to stress is that *there is no presumption, on the part of this author, of any previous exposure to, or background in, computers or computer programming.* Specifically, this is most definitely not a text on computer programming (or particularly on the so-called languages of artificial intelligence such as LISP or PROLOG); rather, it is a text that addresses a methodology for the solution of certain real-world problems where typically this methodology happens to be implemented on a digital computer.

OVERVIEW OF THE MATERIAL
TO FOLLOW

In this chapter, we have provided a very brief introduction to the topic of expert systems and have attempted to indicate just why this methodology should be of interest to the reader. The next chapter will review some background material with respect to expert systems; in particular, the relationship of expert systems to data processing, management information systems, and decision support systems. An understanding of the similarities, differences, and relationships among these topics and that of expert systems should help the reader to more clearly visualize the intended role of expert systems. In addition, the topics of algorithms versus heuristic programming will be addressed—and will indicate to some degree the role that heuristics and heuristic programming play in expert systems.

Chapter 3 deals with a limited survey of past, present, and future expert systems—and even discusses the topic of "expert systems before expert systems." The readers will not only be introduced to some of the actual and potential applications of this methodology, but they will also be introduced to two conflicting perspectives on the history of the field. In particular, the skepticism in some quarters as to the origin of expert systems shall be noted.

The single most important part of any expert system is its knowledge base, that is, the set of heuristic rules (i.e., "rules of thumb") that serve to determine the conclusion at which the expert system ultimately arrives. In support of this, the heuristic rules must be acquired and then translated into a format compatible with that used by the computer. In Chap. 4, the topic of knowledge representation is addressed—with particular emphasis on the so-called rule-based (or production rule) format. Once again, it is to be emphasized that this text is focused specifically, and without apology, toward rule-based expert systems. In Chap. 5, the topic of how such knowledge might be extracted from either the human expert or from historical records is discussed. The topics within Chaps. 4 and 5 represent the major focus, or at least what should be the major focus, of the knowledge engineer.

Once the knowledge base has been established, the expert system utilizes an inference process through which it ultimately arrives at its conclusion. The portion of the expert system that actually accomplishes this is known as the *infer-*

ence engine, and is the topic of Chap. 6. Most typically, this inference process is carried out on a computer—by means of a commercially available expert systems software package. While it is seldom (if ever) necessary to construct such software, it is essential that one understand just what is going on inside such packages. In Chap. 6, various modes of inference (forward chaining, backward chaining, and combinations of the two) by means of a step-by-step procedure are presented; a procedure that may be implemented manually. At the conclusion of this chapter, the reader should have a very good understanding of just how expert systems actually work. As a result, the reader may approach the use of expert systems software in a reasoned and intelligent manner—attributes all too often lacking in those who learn expert systems through an immediate introduction to the software.

Through Chap. 6, only deterministic expert systems are addressed. That is, it shall be assumed that all rules in the knowledge base are certain and that any input from the expert system user is completely accurate. In actual practice, an expert system must often deal with uncertainty. In Chap. 7, the topic of uncertainty in expert systems is discussed and several approaches to the inclusion of uncertainty within such systems are presented. Next is a brief discussion of bridges within expert systems, that is, the provision for access to external programs. Without such access, the expert system is, generally, of only academic interest. The chapter then concludes with an illustration of explanation and justification capabilities within expert systems.

An often overlooked aspect of expert systems development is that of validation. In fact, expert systems have been, and are being implemented without ever having been thoroughly or adequately validated. In Chap. 8 three facets of validation are addressed: justification of the use of an expert system, validation of the rule base, and verification of the actual performance of the system.

One of the most promising extensions of expert systems is that of the development of *hybrid* expert systems. In this text, hybrid expert systems are those that combine two or more approaches to problem solving, where at least one of the approaches is that of an expert system. As just one example, one might combine an expert system with a simulation package. The expert system might then *take over* the running and interpretation of the simulation. One of the most valuable attributes of the hybrid expert systems approach is its ability, in many instances, to actually enhance the operating performance of expert systems, that is, to reach improved solutions. Hybrid expert systems are then the focus of Chap. 9 and are explained primarily through various examples.

Chapter 10 deals with alternative modes of expert systems development, including development languages, expert systems shells, and expert systems environments. In addition, a brief overview of supporting hardware is provided.

Chapters 11 and 12 deal with two closely related topics: implementation of the expert system and the requirements of a knowledge engineer. Other topics that are covered include the monitoring, control, and maintenance of expert systems. Chapter 13 concludes the body of the text and provides a summary of the material covered, thoughts on the future, and some final remarks on the relationship between expert systems and heuristic programming.

The final portion of the text is the appendix concerning the use and implementation of the EXSYS expert systems software package that accompanies the text. This short user's guide describes the fundamentals of this package and presents, through illustrative examples, how an expert system may be constructed and consulted. An additional, and considerably more detailed discussion of EXSYS may be found in the tutorial that is contained in the software that has been provided with the text.

Again, it is to be emphasized that this is not a text on EXSYS, or on any particular expert systems development package. The material within the text has been designed to encompass generic rule-based expert systems. No assumptions are made about which software package readers might be using now or in the future. The single purpose of EXSYS is to provide readers with ready access to a package that may be used to test the approaches discussed within the text and to augment their experience in the actual construction of expert systems knowledge bases, and of their employment in consultation. Further, any other rule-based expert systems software may be used in place of EXSYS, if one so wishes.

SUGGESTED PLAN OF STUDY

This text may be used in support of either a one- or two-term (i.e., semester or quarter length) course in rule-based expert systems. The primary philosophy underlying the text is that, *before* one begins to input numbers and symbols into a software package (e.g., an expert systems shell), one must first have a thorough understanding of just what an expert system is and just how it arrives at a conclusion. As such, even though the EXSYS software has been furnished with the text, it is hoped that the readers will resist the temptation to use this package until they have finished the first six chapters.

When using the text in support of a one-term course, there has been no difficulty in covering all 13 chapters, as well as introducing the EXSYS software. While an attempt is made to touch on most of the material in class, the lectures focus primarily on the material presented in Chaps. 4 through 8. We also cover the use of EXSYS in class, usually through demonstrations of the construction of small, illustrative knowledge bases. The material within the other chapters may, if one so wishes, be covered through reading assignments. In addition, students have been required to develop an original, prototype expert system—either by use of EXSYS or some alternative rule-based expert systems shell.

A two-term course should, in our opinion, cover (at minimum) all of the chapters plus the material on EXSYS. Our recommendation is to cover at least Chaps. 1 through 6 in the first term, and the remaining chapters in the second. In a two-term course, it should also be possible to place more emphasis on the problem sets provided in the text—as well as allow more time to be devoted to the discussion of these problems. Outside reading assignments (e.g., in the related journals and other books) should be assigned and at least one class project per term should be undertaken. For example, in the first term, students might be required to develop a relatively simple prototype knowledge base using EXSYS.

In the second term, more ambitious projects may be assigned (including, for example, the construction of hybrid expert systems)—and implementation of these knowledge bases on other expert systems software might be considered. Other projects that might be considered for assignment include the following:

- The development, by students with a background in computer programming, of their own expert systems shell (in whatever programming language the students are most familiar)
- An objective, systematic comparison of two or more expert systems software packages

In the event that readers intend to use the text as a self-study guide to expert systems, we recommend that they cover all of the text material and attempt the development, using supporting software, of one or more small expert systems. In addition, a serious attempt should be made to solve the problems provided at the end of most chapters.

In whatever manner the material may be covered, we still implore readers to always remember that the essence of any expert system, particularly a successful expert system, is the knowledge base.

EXERCISES

1.1. PEORIA, the hypothetical expert system used to illustrate the employment of expert systems, was supposedly derived by gathering expertise from individuals in a number of areas related to new product development. Can you think of precisely how you might go about determining just who these experts are—and how to differentiate between a *real expert* and a *nonexpert*?

1.2. PEORIA, as well as most real-world expert systems, includes both a knowledge base and a database (actually, a number of databases). Just precisely what are the differences between databases and knowledge bases?

1.3. Let us assume that you wish to sell a firm on the idea of developing an expert system to take over certain activities that are now being performed by one of their employees (who, incidentally, is about to retire). Prepare a short (e.g., one or two page) presentation in support of the funding of such an effort. What questions do you anticipate?

1.4. A certain U.S. firm concentrates solely on the manufacture of custom-designed pressure vessels (e.g., as used in chemical processing, food processing, and space vehicles). Over the past few years, they have witnessed a substantial reduction in the number of contracts they have received and have traced this situation to their inability to rapidly respond to requests for quotes. The most time-consuming portion of a response is that required in the development of estimates of cost and time (i.e., schedule) to build the given vessel. Quite simply, the four company experts who presently provide the cost and schedule estimates in support of such quotes are overworked and the firm has been unable to hire anyone (with sufficient experience and expertise) to assist them. The firm's management notes that, unless something is done to improve (e.g., automate) this process, their very survival is at stake. Consequently they are seeking an expert system to accomplish this task and have stated that they are willing to spend up to $5000 for such a tool. Comment on the aspirations of the company and the approach being considered.

1.5. In an article devoted to the determination of areas for expert systems application [Casey, 1989], the following statement appears:

> If you can solve the problem easier and faster in COBOL (i.e., rather than by means of an expert system), by all means go ahead. Acknowledging that conventional programming is perfectly adequate to do the job is not shameful.

How does this statement correlate with the view of expert systems as presented in this text? What do you think is the motivation for such statements?

1.6. Try to cite at least three *new* potential areas for the implementation of expert systems. Justify the use of expert systems (i.e., rather than the employment of some more conventional tool or of the hiring and/or training of additional human experts) for these areas.

CHAPTER
2

BACKGROUND

THE ART AND SCIENCE
OF DECISION MAKING

Decision making ranges from the routine and swift to the complex and time consuming. For example, before going to bed at night we may have to make a decision as to what time to set our alarm clock in order to make an early morning meeting across town. Rather obviously, one does not need to perform any lengthy, detailed analysis of such a situation. We can quite easily derive a solution by simply estimating the time required to shower and dress, eat breakfast, and drive across town. On the other hand, consider the decision concerning whether a firm should construct a new plant or, alternatively, modernize its existing facilities. This is a considerably more involved matter and one that would definitely benefit from a systematic approach to analysis and solution. Our focus, in this text, is on these latter situations. In such circumstances, decision making implies the existence of a minimum of the following four factors:

1. There must be a problem.
2. There must be a decision maker.
3. There must be a need to solve the problem.
4. There must be alternative solutions to the problem.

Given that these four elements do exist, there are a variety of methods through which we may derive candidate solutions to the problem under consideration—for presentation to the decision maker. The discipline devoted to the development and implementation of such tools may be called, for our purposes, *decision analysis*. Those who work within this discipline and who ultimately present the alternative solutions to the decision maker, are called *decision analysts*.[1]

As mentioned in Chap. 1, expert systems are perceived as simply another tool (albeit certainly an important one) for decision analysis. Consequently, to better understand and appreciate expert systems, it is vital to understand and appreciate decision analysis, its supporting elements, its evolution, and its role in the decision-making process. Further, with such a background, the relationships, differences, and similarities among such topics as artificial intelligence, expert systems, heuristic programming, and decision support systems should be made considerably more distinct. As a result, a clearer picture of the role and purpose of expert systems should, hopefully, emerge. In particular, it is anticipated that through such discussion, one may more fully appreciate just when and where to employ expert systems.

Obviously, decision making is hardly a new concept. Human beings have been making decisions ever since human life first appeared on this planet. Cave dwellers had to decide where to live, what to hunt, when to hunt, and when to move on. In making these decisions, it is extremely doubtful that they employed any rigorous approach to assist them in either substantiating or improving those decisions made. That is, decisions were reached strictly by intuition, experience, and judgment.

In more recent times, and in particular in the past few centuries, humans have developed, and have begun to rely on, more formal and rigorous means for assistance in their decision making. Such means have been achieved through an increased dependence on the use of decision models, and particularly on quantitative models[2] and analytical methods. Today, for example, we note the reliance of corporations and institutions on such techniques as follows:

- Spreadsheets and databases
- Statistical analysis
- Simulation
- Methods of mathematical optimization (e.g., calculus, linear programming)

While this more formal approach to decision making has certainly not supplanted the use of intuition, experience, and judgment, it has found wide acceptance and use as an adjunct to the decision-making process. Typically, when

---

[1] The focus of some fields, such as management science, operations research, and systems engineering is, in fact, specifically directed toward decision analysis. However, it should be obvious that decision analysis is a matter routinely faced by virtually any discipline—whether or not those in that discipline have been exposed to any formal methods in support of decision making.

[2] We shall use the terms *quantitative models* and *mathematical models* interchangeably.

we utilize this more explicit, analytically based approach to decision support, we call it decision analysis to distinguish it from the qualitative aspects involved in making decisions. However, ultimately both qualitative and quantitative factors must (or at least should) be taken into account in the decision-making process.

We wish to stress one additional point. We normally do not make decisions through decision analysis. Rather, we use the information provided by decision analysis to assist us in, hopefully, making *better* decisions. Thus, the purpose of decision analysis is to provide the decision maker with information for use in the support of the decision-making process, where such information has been derived through a logical, scientific, and systematic process. It is then up to the decision maker to decide how to interpret and use this supporting information in the determination of the ultimate decision.

DP, MIS, AND DSS

One way in which decision analysis might reasonably be viewed is as *a process which involves the transformation of data into* (useful) *information in support of the decision-making process*. As our civilizations have evolved, we have become great collectors of data. Unfortunately, data alone are of little benefit. To have value, data must be transformed into a format from which we can perceive such useful information as trends (e.g., whether the demand for a product is rising or falling), measures of central tendency (e.g., what average price is paid for a new home), and measures of dispersion or variability (e.g., over what range of temperatures would we expect a piece of machinery to operate). *One fundamental rule concerning data is that, to be of value, data must be in the right form, in the right place, at the right time.*

Data Processing (DP)

The simplest method for the transformation of data into information is that of *data processing*, or DP. Typically, the DP approach is used to transform a set of *raw* data into the following information:

- Statistics (e.g., numbers representing the mean, median, mode, and variance of the data)
- Pictorial representations (e.g., histograms, pie charts, bar charts)

For example, consider a problem in which data have been collected on engine failures for a specific type of military aircraft at several different bases. To simplify our discussion, assume that each base has the same number of total aircraft and each flies the same number of missions each month. Twelve months of data are given in Table 2.1.

The data in Table 2.1 are termed *raw data* as they are presented simply in the form in which they were originally collected. We may also consider these data to be our engine failure *database*. Simply looking at the data, there is not much that

TABLE 2.1
Engine failure data

Month	Base A failures	Base B failures	Base C failures
January	5	3	6
February	1	2	7
March	4	2	5
April	3	1	2
May	7	0	2
June	4	2	3
July	1	2	2
August	7	2	2
September	4	3	3
October	6	1	5
November	6	1	7
December	5	2	7

we can readily perceive—at least on any scientific basis. However, if we process these data using elementary statistical analysis, we may note the following:

- The average number of engine failures each month, per base, is 3.5.
- The average number of engine failures for each individual base is
     Base A: 4.42 per month.
     Base B: 1.83 per month.
     Base C: 4.25 per month.

Figure 2.1 depicts that portion of the data pertaining to base C only, in the form of a frequency polygon. Similar figures could (and should) be constructed for the other two bases.

Now, even though our data processing has been elementary and incomplete, we should still find it easier to make the following observations:

- The average monthly number of engine failures at bases A and C are more than twice those of base B.
- From Fig. 2.1, a trend in engine failures at base C seems possible. That is, failures appear to increase in the winter months and decrease in the summer.

The first observation may cause us to wonder if there is some problem (or problems) at bases A and C. For example, could their maintenance plans and/or personnel be less effective than those employed at base B? Could their pilots be less skilled or could they be placing greater demands on the engines? Or, alternatively, have the personnel at base B discovered some useful new policies, procedures, and guidelines for use in reducing engine failures?

The second observation would most likely bring into question the aspect of weather and temperature effects on engine failure—and this would be particularly

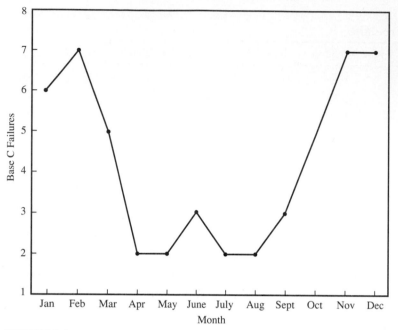

FIGURE 2.1
Engine failures at base C.

true if base C suffers from more severe winters than the other bases. Using this information, the decision maker might then conclude that the following studies need to be undertaken:

- The effects of weather and temperature on engines, causing failure for the specific type of aircraft under consideration (and procedures for mitigating such effects)
- An evaluation of the maintenance procedures and personnel at the bases
- An evaluation of any significant differences in pilot skills and/or types of missions flown between bases

Thus, from this simple illustration, we can see the usefulness of even a very rudimentary level of data processing.

Management Information Systems (MIS)

The next level of sophistication in the processing of data into information is called *management information systems*, or MIS. While there is no uniform agreement about the precise definition of MIS, the general intent of the earliest such systems was to provide information directly, and in real time,[3] to the decision makers—and in a format compatible with their style and needs for decision making. Sanders [1970] defines management information as "relevant knowledge, produced by the output of data processing operations, and acquired to achieve specific purposes." Such information is thus the basis upon which managers may pursue their duties, specifically the duties of *planning, organizing, staffing,* and *control.* Although management information systems certainly existed before the advent of the computer, it is now customary to think of an MIS as a system that furnishes management information by means of a digital computer and connecting information network.

The typical MIS concept involves a computer console display at the decision-maker's desk. Using this console, the decision maker may ask for information concerning some specific topic and this information would then be immediately displayed—typically in the form of tables or graphs—for consideration by the decision maker.

As an example, the manager in question may be concerned with production scheduling. He or she would then access the computer and request information pertaining to what is believed necessary to produce a schedule for the next production period (e.g., a 30-day period). Such information might include the following:

- Raw material inventory levels expected over the next 30 days
- Orders and contracted delivery dates
- Production-line status (e.g., work-force levels, machinery that is down for maintenance, and so on)
- Production-rate limits

The data used to support these topics are addressed and processed through the data processing operation. The resulting information is then further processed to establish it in a format that is deemed most useful for a specific decision. This final level of information, typically in the form of specialized tables, figures, and graphs, is then made available to the manager at the computer console. Based upon his or her interpretation of such input, the manager establishes the production schedule and transmits this schedule, by way of the computer link, to the production supervisor for implementation.

Decision Support Systems (DSS)

As may be noted from the above discussion, management information systems are relatively passive entities. While they remove much of the drudgery of data processing and development of visual aids, and substantially decrease the time required to obtain such information, they still play a limited role in decision making.

---

[3]The term *real time* is used to indicate that any information sent is *current.*

However, at about the same time that MIS was becoming popular, developments were taking place in other fields (e.g., management science, operations research, operations analysis, systems analysis, and systems engineering) that addressed the implementation of certain analytical methods for decision analysis. In particular, representative mathematical models of certain classes of problems were being formulated and various methods for providing solutions to the models were constructed. Included among such methods are the following:

- Mathematical programming (i.e., finding the optimum solution to an *objective function* subject to the satisfaction of a set of *constraints*), including linear, nonlinear, integer, and dynamic programming
- Marginal analysis
- Input-output analysis
- Queuing theory (the modeling and solution of problems involving waiting lines)
- Inventory theory
- Project scheduling (e.g., PERT, CPM)
- Simulation
- Reliability and quality control
- Forecasting
- Group technology (i.e., the classification and grouping of parts)
- Material requirements planning (MRP)

Assuming that we can represent our specific problem using one or more of such models, the associated methodology may then be used to develop a proposed solution. However, as the critics of such approaches have noted, to accomplish this, one is required to transform a real-world problem into a mathematical model—a process that will invariably require the use of simplifying assumptions, along with the potential disregard of factors that cannot be easily quantified. Even so, such methods have found wide acceptance and are regularly used to solve, or to assist in solving, many real problems.

While some advocates of decision support systems might disagree, one may think of a DSS as a combination of an MIS and the analytical tools as listed above. Thus one conception of a decision support system is that of a computerized system for accessing and processing data, developing managerial displays, *and* providing recommended courses of action as developed through the use of modern analytical methods. Using this definition, a block diagram for a general decision support system is depicted in Fig. 2.2. Everything above the dashed line in this figure is assumed to be contained within the DSS, in particular within the computers and computer networks employed by the DSS.

Recalling our hypothetical example as presented earlier to illustrate the use of an MIS, let us now consider how a DSS might be used to approach the same problem of providing assistance in production scheduling. Just as in the case of the MIS, the DSS would access the database and develop displays in the appropriate format. However, assuming that our DSS includes a supporting

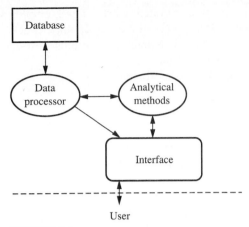

FIGURE 2.2
A generic decision support system.

tool for the solution of scheduling problems, the manager will also be provided with a recommended schedule (or perhaps, several alternative schedules from which to choose) for production as generated by the scheduling methodology. The manager may then either accept the DSS recommendation or develop his or her own schedule—which may be compared with the one developed by the DSS through a simulation of the proposed schedule, for example. Thus, a decision support system is certainly a far more active participant in the decision-making procedure than either DP or MIS.

Throughout our discussion of DSS, we have referred rather casually to *analytical methods*. Such methods normally invoke the use of algorithms for the derivation of solutions for the particular class of mathematical model under consideration. To more fully appreciate the DSS concept, as well as the difference between DSS and expert systems, we need to understand algorithms—the topic of the section that follows.

ALGORITHMS AND RELATIONSHIPS

One formal definition [Gear, 1973] of an algorithm is "a method for solving a problem using operations from a given set of basic operations, which produces the answer in a finite number of such operations." Typically, these *basic operations* are simply elementary mathematical procedures such as addition, subtraction, multiplication, and division. Note most carefully that this definition implies that an algorithm *converges,* that is, it will always reach *the* answer (i.e., the *optimal* solution) in a finite number of steps.

Algorithms may be applied to either a single mathematical *relationship,* or (and more likely) to a set of such relationships, for the purpose of deriving a solution. A mathematical relationship is simply a mathematical statement that relates the various components of a system. In other words, a relationship is a representation of our knowledge of how a particular system works. For example, one well-known mathematical relationship is $F = m \cdot a$ (force equals mass times acceleration). Another would be $P = s - c$ (profit equals sales minus costs). Such relationships are also called *mathematical models*.

Consider the two mathematical relationships listed below:

$$5 \cdot x + 2 \cdot y = 9$$
$$3 \cdot x - y = 1$$

These relationships represent a set of simultaneous linear equations. They involve two *variables*, x and y. The first equation indicates one relationship that exists between these variables (i.e., the sum of 5 times x and 2 times y is equal to 9) while the second indicates another relationship (i.e., 3 times x minus y is equal to 1). A number of algorithms exist in the form of step-by-step procedures, for deriving the values of x and y (i.e., the solution to the problem). When applied to this problem, we would determine that the answer is

$$x = 1 \quad \text{and} \quad y = 2$$

This is, in fact, the only correct solution to this problem and any algorithm designed for such a problem class will guarantee the development of this solution.

Conventional decision analysis involves the application of an algorithm to the mathematical model of a problem to derive a solution to that problem. Further, the approach traditionally used *mixes the model (i.e., the relationships) with the steps of the algorithm.* Consider, for example, the process used to find the average sales price for automobiles sold by a distributorship during a particular period. Obviously, the single relationship involved is simply

$$AP = \frac{SP}{N}$$

where AP = the average price of all cars sold
SP = the sum of the prices of all cars sold
N = the total number of cars sold

A step-by-step procedure (i.e., an algorithm) for determining the average price is listed below:

1. Remark: Let N represent the number of cars sold during the period
2. Let total price = 0.00
3. Input N
4. Let $i = 1$
5. Input price of car i
6. Total price = total price + price of car i
7. Let $i = i + 1$
8. If $i \leq N$ then return to step 5

9. Average price = (total price)/N

10. Stop

Notice carefully that the average price relationship is a part of the process (i.e., step 9) and that the order of the steps of the process are critical. To find the correct value, we must proceed from one step to the next, repeating a step (or series of steps) only when actually specified in the procedure (i.e., note step 8). The reader is also advised to take special note of the "if-then" statement of step 8. Unlike the "if-then" statements (i.e., production rules) of an expert system, this particular statement is used simply to control the order of the solution process. We shall elaborate further on the distinction between these two different types of "if-then" statements in the chapters to follow, particularly in Chap. 4.

Such comingling of relationships and procedures, and the need to maintain a strict order of the steps of a procedure, are fundamental to conventional, algorithmic approaches to decision analysis. However, as we shall see later, expert systems are designed and operate in quite a different manner.

HEURISTICS AND
HEURISTIC PROGRAMMING

Heuristic rules, or heuristics for short, are rules that are developed through intuition, experience, and judgment. Typically, they do not (as did the relationships described previously) represent our knowledge of the design of, or interrelationships within, a system; rather, they represent guidelines through which a system may be operated. Heuristics do not necessarily result in the best, or optimal, result. Further, unlike the relationships discussed in the previous section, such rules are not usually in the *public domain*. That is, while anyone can go to a text of elementary statistics to determine how to compute an average, heuristic rules are usually far more private—and personal. Typically, they evolve through years of experience and reside in the memory of an individual. Heuristics are often called *rules of thumb*.[4] For example, consider the following heuristics:

- Don't ask the boss for a raise if he is in a bad mood.
- Avoid Houston's Southwest freeway during the rush hour.
- Sell a stock if the dividends are to be cut.
- Buy gold as an inflation hedge.
- Schedule any dental patient requiring a root canal at some time prior to 2 P.M. (i.e., to reduce the likelihood of the dentist having to work beyond quitting time).

---

[4]The term *rule of thumb* comes about from the rather appalling laws, that existed into the nineteenth century, that permitted a husband to beat his wife as long as the diameter of the stick used did not exceed that of his thumb. In the Netherlands, we are told, this is called *vuistregal*, or the *rule of fist*, which sounds even more ominous.

- If you are a criminal lawyer, try to reject any jury candidates that have personally been the victim of a crime, or whose immediate family members have been the victims of a crime.
- Don't pursue a new project unless the estimated rate of return exceeds 30 percent.

In virtually any situation, expertise involves the development and use of heuristics. Consider, as just one example, the profession of police detective. For the most part, each individual who reaches the rank of police detective within a given police force has gone through essentially the same training and has faced similar experiences. However, some detectives will ultimately exhibit far more proficiency in their work than will the bulk of the detective force. These are the experts. Such experts exist in virtually all professions, and they often stand head and shoulder, in terms of their performance, above their colleagues. In all likelihood, one reason for their success is that they have been able to develop a set of heuristics which permits them to solve problems faster, and generally better, than their fellow workers.

One of the general characteristics of many heuristics is their focus on *screening*, *filtering*, or *pruning*. Each of these terms represents just another way to state that heuristics may be used to reduce the number of alternatives (or alternative paths in the search procedure) that are considered. Typically, an expert learns through time and experience that certain approaches tend to work well, while others do not. Consequently, even though it is conceivable that a better solution might be missed, the apparently less-attractive approaches are rejected for consideration early in the search process.

When one or more heuristics are combined with a procedure for deriving a solution from these rules, we have a *heuristic program*.[5] As in the case of algorithms, heuristic programming involves finding a solution to a problem using operations from a given set of basic operations, where such a solution is produced in a finite number of such operations. However, and this is the main difference between algorithmic procedures and heuristic programming, the solution found may or *may not* be the theoretically best possible answer (i.e., the optimal solution).

Note that when one uses heuristics, and heuristic programming, one is implicitly accepting the notion of *satisficing*. Satisficing, in turn, is a concept proposed by Simon [1957] for use in the explanation of how individuals and organizations actually arrive at decisions. Specifically, we typically do not seek the *optimal* solution; rather we seek an *acceptable* solution—one that satisfies our predetermined aspirations. We do this because there usually is not the time, resources, nor need to obtain the elusive best of all solutions. Heuristics (and heuristic programs) are then intended for use in obtaining acceptable solutions. *However, we can only justify the use of heuristics in those cases for which more*

---

[5]The term *heuristic program* (or *heuristic programming*) should not be confused with computer programming. A program is a solution. Programming is a procedure for obtaining a solution. Thus, heuristic programming is a procedure for finding the solution to a model consisting of heuristic rules.

formal analytical methods (in particular, methods that develop optimal solutions) would prove less effective. Let us thus consider some examples wherein we might, or might not, wish to employ heuristic programming.

For our first example, consider the heuristic rule given earlier for scheduling patients requiring a root-canal. In developing such a rule, the dentist's receptionist may have observed that, even with complications, most root canals are completed in 2 hours or less. Since the dentist wants to leave work at 4 P.M., it seems reasonable that we do not schedule patients after 2 P.M. However, in place of this heuristic we could use a more rigorous, analytical approach to determine root-canal patient scheduling. For example, we could hire a management consulting firm. This firm might then examine collected data (or establish a system for the collection of such data) on root-canal patients and develop more precise figures on the expected time for completion of such work and the probability of the process exceeding x number of minutes. As a result, it may be found that patients can be scheduled up to 2:21 P.M. with the probability of going past 4 P.M. being no more than 5 percent. While this may be the case, it is highly unlikely that such precision is necessary—or worth the cost required for the analysis. Thus, the employment of heuristics seems reasonable in this instance.

Next, consider a case in which the use of heuristics can actually result in substantial degradation in the decision-making process. Let us assume that a small firm assembles and sells special-purpose computers from kits (i.e., all of the parts for each computer form a single kit). The firm projects a demand of 1200 of these computers per year (at a virtually constant rate) and, because of the delicate nature of customer relations, does not wish to run short. Each time the firm places an order, a cost of approximately $100 is incurred (i.e., the costs associated with simply placing the order—not the cost of the parts). Further, it is estimated that the inventory holding cost[6] per computer kit, per year, is about $24.

Joe Neanderthal, the firm's production manager, prides himself in his "seat of the pants" decision-making ability, and views analytical methods with considerable disdain. Based on his intuition, he *feels* that orders for the kits need not be placed more often than twice a year—and this is the heuristic rule he follows. When questioned about whether this policy might not be improved, Joe's reply is "It works; and if something ain't broke, don't fix it!"; a haunting refrain that is replayed throughout virtually all institutions and organizations. In fact, such an argument is itself a heuristic rule—a rule used mainly to defend the *status quo*.

Consider then the cost of the production manager's policy. Specifically, these costs are due to the placement of the orders and the inventory charges, and may be calculated as follows:

$$\text{TC} = \frac{1}{2} \cdot Q \cdot C_h + D \cdot C_r/Q$$

---

[6]Inventory holding cost is that cost associated with keeping a unit of inventory in stock. This might include such charges as storage fees, insurance fees, and security.

where TC = total cost of inventory policy per year
Q = number of kits in each order (the unknown)
C_h = inventory holding cost per kit, per year (i.e., $24)
C_r = cost of placing an order (i.e., $100)
D = total demand per year

Here, we have constructed a mathematical model, or relationship, that serves to relate total inventory costs (TC) to the inventory policy (i.e., the number of kits per order and the time between orders). Notice that the first term of TC is that associated with inventory holding costs (i.e., the average inventory per period is simply one half of Q, the quantity ordered at the start of a period) while the second is that portion of the cost incurred for the orders placed.

Notice carefully that the mathematical model indicates rather precisely just how two elements of the inventory system (i.e., number of orders and amount of inventory ordered) are combined to determine the total cost of inventory. To construct a mathematical model, one must consider how the components of the system interact. On the other hand, consider the heuristic rule used by Joe. Like virtually all heuristic rules, it does not reflect (at least directly) any precise knowledge of components of the system, or how the system actually works. Instead, the heuristic simply reflects a policy through which the (alleged) domain expert has been able to reach satisfactory conclusions. In this particular case, the policy is quite naive.

Under Joe's heuristic policy, the total yearly inventory cost is $7400. However, anyone with a quantitative background would (or should) recognize that calculus could be used to solve for the *optimal* value of Q. In particular, this problem is known as the *EOQ* inventory model and a simple formula exists [Ignizio and Gupta, 1975, p. 221] that may be used to find the optimal solution. Using that formula, we would find that the optimal policy would be to place an order for 100 kits each month, resulting in a total yearly cost of just $2400; or a savings of $5000 per year over the heuristic policy of the production manager. And this solution may be derived, by hand calculations, in a matter of minutes. Consequently, in this case it would be foolish and costly to not consider the use of existing analytical methods. Also, note that it would be just as foolish to attempt to capture the *expertise* of the production manager; that is, through an expert system.

As one more example of the implementation of heuristic programming, consider the problem faced in a specific type of facility location problem [Ignizio, 1971]. That is, we wish to locate warehouses throughout the United States to minimize the sum of the distances between any customer and the warehouse closest to that customer. For example, we might wish to locate exactly 30 warehouses from among a list of 100 possible location sites. One approach to solving such a problem would be to enumerate all of the possibilities, that is, list and evaluate all possible combinations of 30 sites from among the 100 possible sites. However, this would require one to examine approximately 2.95×10^{25} combinations—a task that would be foolhardy even using a computer.

Alternatively, we could formulate this problem as an *integer programming model* [Kaufmann and Henry-Labordère, 1977; Taha, 1975] and solve it by the solution methods (i.e., algorithms) associated with integer programming. However, at some point, even such mathematically sophisticated approaches will no longer work. This is because such a problem exhibits what has been called *combinatorial explosiveness*. That is, the time required to solve such problems increases *exponentially* with problem size; and at some point it is simply unrealistic to consider solving such problems by exact, analytical approaches—even on the most powerful existing computer. Thus, in such an instance, we might be well advised to resort to heuristics and heuristic programming simply because of the computational complexity of the problem.

Now, let us compare the heuristic programming approach with the algorithmic process as previously described. In particular, recall that we noted, in the application of an algorithm to a mathematical model, that we form a solution process that mixes the model (the relationships) with the steps of the algorithm. Further, the results achieved depended upon our following these steps in precise order.

In heuristic programming, we have, in essence, the same situation. That is, the heuristic rules comingle with the steps of the solution procedure. However, in this instance, our solution procedure is not an algorithm, as it does not guarantee an *optimal* solution. One designation for the solution procedure is that of an inference process—the procedure which serves to infer conclusions from the set of heuristics employed. Others sometimes call this the *logic* of the heuristic programming procedure. To illustrate, consider a heuristic program for the scheduling of parts for processing on a machine. The heuristics used to provide a solution (schedule) for this problem involve the following:

- Schedule jobs with shorter processing times before those with longer ones.
- If two (or more) jobs are tied for processing times, give priority to the job that is most tardy—or most likely to become tardy (a tardy job is one that will not be finished until after its due date).

Application of these rules, through a heuristic program, will certainly result in a schedule. Hopefully, such a schedule might even be a good one—but there are no guarantees as to how close, or far, we might be from the optimal schedule. In order to keep the discussion simple, let us restrict our heuristics to just the first (i.e., process jobs with shortest times first). The steps of a heuristic program for solution follow:

1. Remarks: Let the number of jobs to be scheduled equal M. Let the processing time of job i on the machine be denoted as t_i. Let $J = p_1, p_2, \ldots, p_M$ represent a given schedule of jobs, partial or complete, where p_k is the position in the schedule of the kth job. Initially, J is the null set.
2. Let $k = 0$. Let $J = \phi$ (where ϕ represents the null, or empty set)
3. Let $T_{\min} = \infty$ (i.e., let $T_{\min}$ equal some number larger than any job processing time)

4. Let $i = 1$

5. If i is already a member of the set J then go to step 7

6. If $t_i \leq T_{min}$ then let $T_{min} = t_i$ and also let $r = i$

7. Let $i = i + 1$

8. If $i \leq M$ then go to step 5

9. $k = k + 1$

10. Let the next element in J be job r

11. If $k \leq M$ then go to step 4

12. Stop

Note that the heuristic rule is implemented in the conduct of steps 3 through 10, which serve to order unassigned jobs according to the shortest processing time. And again note the necessity to follow, precisely, the steps in order and the comingling of procedures and heuristic rules. We should also observe that the steps of the above procedure include a number of "if-then" statements (i.e., steps 5, 6, and 8). Again, these differ considerably from the "if-then" rules of expert systems in that these statements are employed simply to control the solution process.

We may thus conclude that heuristics and heuristic programming, *when and where appropriate,* may enhance one's decision-making procedure. As such, these methods often form a portion of the tools incorporated into decision support systems (e.g., to augment the set of analytical tools), and serve to alleviate the limitations of the more rigorous analytical techniques. In Chap. 9, we will return to the topic of heuristic programming through the introduction of hybrid expert systems—decision analysis methodologies that often involve the combination of expert systems and heuristic programming.

ARTIFICIAL INTELLIGENCE

Let us now proceed to what may at first appear to be an entirely different, unrelated topic: the field known as artificial intelligence, or AI. However, AI is concerned with precisely the same problem that DSS and heuristic programming are concerned with, that is, decision making. One fundamental difference is that the objective of those in the AI community is considerably more ambitious than that of the DSS sector. The purpose of AI is not simply to support decision making, or to just enhance decision making; rather, the ultimate goal of AI is to develop an *intelligent machine* that will itself make decisions. In particular, this intelligent machine should exhibit intelligence on the same order as that of a human. Thus, such a machine will be able to learn through experience, to recognize the limitations of its knowledge, and to exhibit true creativity. One difficulty with this goal, as stated, is that there is no agreement as to precisely how one defines, much less measures, *intelligence.*

Elaine Rich [1983] in her text on AI, does not attempt to define intelligence but she does offer an intriguing definition of artificial intelligence. She states that "artificial intelligence is the study of how to make computers do things at which, at the moment, people are better." Using this definition, we may avoid the problems of either the definition or the determination of the existence of intelligence and, instead, simply compare the computer's performance (in some area) with that of humans. For example, at the moment, humans are able to visually recognize other humans much better than the computer. We can pick out faces of individuals we know even when they are in a crowd, and even when their faces are partially obscured. We can often even recognize people from the backs of their heads, or from just the way they walk. While programs do exist that permit computers to recognize a limited number of faces, under specific conditions, their performance still pales in comparison with that of the human. As such, it is generally accepted that the study of the recognition of human faces by way of the computer is a *legitimate* area of AI. An interesting, and somewhat controversial aspect of Rich's definition of AI is that, once we do achieve face recognition on the computer at a level equal to or better than that exhibited by humans, such study will no longer be AI.

A Brief History of AI

While the human race has long conjectured about the possibility of building a synthetic being (e.g., an intelligent machine), it has not been until this century that such dreams finally seemed realizable to a relatively large number of people. This perceived feasibility is due, of course, to the development of the digital computer.

The digital computer has been in existence since World War II wherein they were used by the British and Americans in such tasks as numerical computations and code breaking. However, while the computer possesses an astonishing capability to store and manipulate data, few of us would consider such activities to constitute any form of true intelligence (and this is so even without a formal definition of intelligence). We probably feel this way because such tasks, even though once accomplished solely by humans, are considered to be *mechanical* in nature. We may thus look on the computer as an exceptionally powerful *calculating machine*; but not as a machine that actually has intelligence. More specifically, the computer (at least in its traditional role of storing and manipulating data) may be looked upon as a rather *dumb* device that does precisely what it is told—and only what it is told—and has no awareness whatsoever of what it is doing or why it is performing a particular task. As such, it is somewhat akin to a dog trained to retrieve the morning newspaper.

However, among the early users and designers of digital computers were a small group of *radicals*. These individuals wondered if one might not be able to use the computer as a means for simulating various aspects of human intelligence. As a result of this shared interest, a conference was held in the summer of 1956 in Hanover, New Hampshire, the home of Dartmouth College. Now known as the Dartmouth Conference, this meeting was attended by only a handful of scientists in such diverse fields as mathematics, electrical engineering, neurology, and psychology.

The proposed intent [Rose, 1984] of the Dartmouth Conference was to explore the conjecture "that every aspect of learning or any other feature of intelligence can in principle be so precisely described that a machine can be made to simulate it." While there were certainly no *breakthroughs* announced at the conference, it did provide a sharper focus for the field now designated as artificial intelligence. It was, in fact, one of the conference participants (John McCarthy) who, for better or worse, suggested that the term *artificial intelligence* be used to designate this new area of science.

Another outcome of the Dartmouth Conference was the establishment of certain forecasts for artificial intelligence [Peat, 1985; Simon and Newell, 1958]. Specifically, it was predicted that, *by 1970,* a computer would do the following:

- Be a grandmaster at chess.
- Discover significant, new mathematical theorems.
- Understand spoken language—and provide language translations.
- Compose music of classical quality.

Rather obviously, and with our advantage of hindsight, these forecasts were overly ambitious. However, these forecasts did serve to encourage wider interest in artificial intelligence and establish certain goals for those within the field. But they also served to have a very negative impact as, with the passage of time, it became clearer and clearer that these forecasts were not going to be met and had been, at the very least, unrealistic. The early enthusiasm in artificial intelligence was, by the mid-1960s, considerably subdued. In fact, AI seemed to virtually "vanish into the woodwork" at about that time—not to be resurrected until the 1980s.

During this period of disillusionment, sometimes referred to as the *AI winter*, some of those in the AI community felt that it was time to reconsider the goals of AI. In particular, they felt that it was time to regroup and to reconsider the fundamental premise of a general purpose intelligent machine. For the most part, AI scientists were agreed that knowledge was the essential ingredient in intelligence. The computer, despite its enormous capacity for data storage, simply could still not store all of the knowledge necessary to simulate all facets of human intelligence. However, these individuals wondered if it might not be possible to at least store that knowledge associated with a very narrow domain—and then exhibit abilities on the computer similar to that of a human for this restricted area. They then decided that, at least for the time being, developments and research in AI should be as follows:

- More modest
- More focused
- Directed toward a narrow sector of expertise—rather than general, overall intelligence

And the name given to this subfield of AI was *expert systems,* or *knowledge-based* systems.

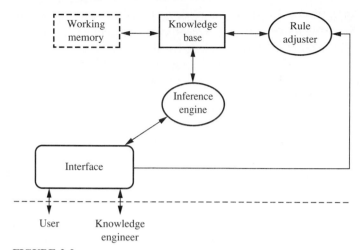

FIGURE 2.3
A generic expert system.

The Expert Systems Concept

In Chap. 1, a hypothetical and rather ambitious expert system, the system designated as PEORIA, was described. Here, let us consider the concepts and components of an expert system in general. In the chapters to follow, these concepts and components will be addressed in detail.

Figure 2.3 depicts one possible representation of an expert system. The components above the dashed line are those within the computer. Below this line, access capabilities for two types of human users are noted. The first is that individual designated as the *knowledge engineer*. As discussed, the knowledge engineer is the person responsible for placing the knowledge into the expert system's knowledge base; the portion of the expert system shown at the top of Fig. 2.3. He or she accomplishes this through the *interface* (shown directly above the dashed line) and the *rule adjuster*.

The knowledge engineer is also the interface between the human expert (if there is one) and the expert system. That is, the knowledge engineer somehow must capture the expertise of the human expert and then express this expertise in a format that may be stored in the knowledge base—and will be used by the expert system. In the ideal expert system, there would be no need for a knowledge engineer. The domain expert would interact directly with the expert system and would replace the knowledge engineer in the figure.

The second type of individual with access to the expert system is designated, in Fig. 2.3, as simply the *user*. This designation refers to anyone who will be using the expert system as a decision-making aid (i.e., as a consultant). And the successful knowledge engineer must always keep in mind that the expert system is ultimately intended for the benefit of the user, *not* for that of the knowledge engineer or the domain expert.

The interface handles all input to the computer and controls and formats all output. In PEORIA, we were presented with voice-generated replies and simultaneous computer screen menus, and we responded with either verbal replies or typed input. The interface would handle such chores. A well-designed interface would be one that exhibits ease of use (i.e., so-called user friendliness), even for the novice user. The interface also handles all communication with the knowledge engineer (or domain expert) during the development of the expert system's knowledge base. Another property that is sometimes exhibited in expert systems is that of *explanation*. That is, some expert systems (recall the discussion of PEORIA) have a limited ability to explain the reasons for any questions asked of the user, as well as the rationale for the conclusion reached. Again, this function would be the responsibility of the interface.

The *inference engine* is employed during a consultation session. During consultation, it performs two primary tasks. First, it examines the status of the knowledge base and working memory so as to determine what facts are known at any given time, and to add any new facts that become available. Second, it provides for the control of the session by determining the order in which inferences are made. An alternative designation for the inference engine, and perhaps a more appropriate one, is that of *knowledge processor*. As the knowledge processing element of an expert system, the inference engine serves to merge facts with rules to develop, or infer, new facts.

The *knowledge base* is, as we have emphasized repeatedly, the very heart of any expert system. A knowledge base will typically contain two types of knowledge, that is, facts and rules. The facts within a knowledge base represent various aspects of a specific domain that are known *prior* to the exercise (i.e., the consultation session) of the expert system. The rules within the knowledge base are simply heuristics of the type discussed earlier. If the knowledge base has been constructed through interaction with a human expert, these rules represent the knowledge engineer's perception of the heuristics that are employed by the expert in decision making. The primary emphasis of this text is the understanding, development, and use of the knowledge base—in particular, a knowledge base using a rule-based format.

The *working memory* of an expert system changes according to the specific problem at hand. The contents of the working memory consist of facts. However, unlike the facts within the knowledge base, these facts are those that have been determined for the specific problem under consideration *during (and at the conclusion of)* the consultation session. More specifically, the results of the inference process are new facts and these facts are stored in the working memory.

The final module to be discussed is the *rule adjuster*. In most existing expert systems, the rule adjuster serves merely as a rule editor. That is, it enters the rules specified by the knowledge engineer into the knowledge base during the development phase of the expert system. It may also allow for various checks on these rules (e.g., consistency, completeness), and these checks will be discussed in detail in Chap. 8. In more ambitious expert systems, the rule adjuster may be used in an attempt to incorporate *learning* into the process. In such instances, we would *teach* the expert system by providing it with a set of examples and then critique its performance. If its performance is unsatisfactory, the rule adjuster automatically revises the knowledge base. If satisfactory, the rule adjuster may simply reinforce the existing knowledge base.

An expert systems "shell" includes all of the components listed in Fig. 2.3 *minus* the knowledge base. Using a shell, it is up to the knowledge engineer to develop the knowledge base and to then insert this knowledge base into the architecture to form a complete expert system, as intended for a specific domain. The use of a shell thus frees the knowledge engineer from the need to repeatedly develop all the supporting elements of an expert system, and thus to focus his or her attention on the development of the knowledge base.

The architecture of the generic expert system, as depicted in Fig. 2.3, should serve to indicate at least some of the differences between this approach and that of algorithmic procedures and heuristic programming. In particular, note that the knowledge base is separated from the inference engine. In other words, and unlike algorithms and heuristic programming, an expert system separates the heuristic rules from the solution procedure. The knowledge base contains a description, or model, of *what we know* (i.e., about deriving a solution to a given problem). The inference engine contains a description of *what we do* to actually develop the solution. While the knowledge base changes from domain to domain, the inference engine remains the same. As we proceed through Chaps. 4 through 6, in particular, we should be able to more fully appreciate the implications of this architecture and the associated procedures for solution.

The Expert System as a Component of DSS

Recall the architecture of a generic DSS, as previously depicted in Fig. 2.2. While that figure implies that a DSS employs strictly analytical methods (algorithms), virtually all existing decision support systems (of which we are aware) include the use of heuristics and, in many cases, heuristic programming. These heuristics exist, usually, within the user interface and/or within the portion of the system denoted as the analytical methods module. The inclusion of these elements serves to enhance the capabilities (as well as, quite often, the acceptance) of decision support systems and they serve as a valuable—albeit sometimes hidden—adjunct to the DSS concept.

More recently, DSS designers have begun to include heuristcs—and expert systems—elsewhere within the DSS architecture. When expert systems are employed in such a manner, they are termed *embedded* expert systems. Figure 2.4, which is simply a combination of the elements of Figs. 2.2 and 2.3, depicts the general architecture of such a *hybrid* DSS.

The combined system of Fig. 2.4 would seem to provide a logical extension, as well as an enhancement, to either the DSS concept or the expert systems approach. That is, it allows for the choice between analytical tools (mathematical models and algorithms) and expert heuristics (knowledge bases and inference strategies), depending on the specific characteristics of the problem at hand. In

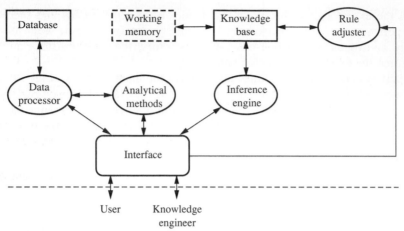

FIGURE 2.4
The combined DSS and expert system.

fact, it permits the use of both approaches to the same problem—with the analytical tools handling that portion of the problem amenable to such an approach and the expert system dealing with the remainder. It also permits the expert system to have direct access (through the interface) to the database. Recalling the design of PEORIA, such access is often of considerable value. However, for the sake of clarity, we shall generally restrict our discussion of expert systems to those of the type depicted in Fig. 2.3 (i.e., stand-alone expert systems). In Chap. 7, ways in which the concept may be broadened will be considered briefly.

On the Origin of Expert Systems

When one examines the bulk of the artificial intelligence or expert systems literature, one is given the definite impression that expert systems is a relatively new concept (with the earliest such system developed in the late 1960s) and is a concept that owes its existence primarily (if not solely) to the so-called AI community (which, most recently, consists of—or would appear to consist of—a large proportion of individuals from the field of computer science). There are others (e.g., individuals in the fields of management science, operations research, and engineering in particular) who take strong exception to this perspective. Rather than viewing expert systems as an entirely new idea, they would consider expert systems simply *another format* for the implementation of heuristic programming. Heuristics, in turn, can be traced back to at least Socrates (469 B.C.) and Pappus (300 A.D.) [Koen, 1985]. They were also considered and used by such scientists as Descartes, Leibnitz, Bolzano, and Mach and, more recently by Polya [1957]. Under this particular view, expert systems is then neither new nor a product of either AI or computer science.

However, while heuristics and heuristic programming have been in use for centuries, these approaches have rarely been considered *scientifically acceptable*. A paper describing a heuristic programming approach to some problem would, until fairly recently, have had little chance for publication in most professional journals (outside, perhaps, of computer science). In such fields as operations research, heuristic programming was felt to lack the rigor or credibility of algorithmic approaches—*even, oddly enough, when the approach solved problems that were far beyond the capabilities of the analytical techniques*. Expert systems, however, have received rather obvious, and quite prominent acceptance. This acceptance is due, in large measure, to various contributions by the AI community coupled with its highly successful promotion of the expert systems approach. Heuristic programming [Fuller, 1978; Ignizio et al., 1972; Ignizio, 1980; Müller-Merbach, 1981; Polya, 1957; Simon and Newell, 1958; Tonge, 1961; Wiest, 1966], has generally been an unstructured field of endeavor, with little in the way of a general set of guidelines, rules, and procedures. Further, those who practiced within this field were all too often looked upon with some degree of contempt by their more *mathematically gifted* contemporaries. This is in considerable contrast to the treatment of heuristics (as expert systems) by the AI community.

The AI community, while using heuristics as the essential ingredient of the expert systems knowledge base, has augmented its use with a very formal set of policies and procedures. That is, some much needed discipline and rigor to the heuristic programming approach have been added. Further, the importance and credibility of the heuristic (or expert systems) approach have been recognized and efforts in the area have been encouraged. Thus, while heuristic programs and expert systems share much in common, including the fact that both suffer from the same limitations (i.e., an inability to guarantee a mathematically optimal solution and lack of any formal proof of convergence to optimality—or even near optimality), it is expert systems that has proven to be both commercially and scientifically acceptable.

The primary enhancements brought to heuristic programming from the AI community are not trivial. For example, the separation of the inference process (i.e., the solution procedure) from the knowledge base (i.e., the heuristics) is a concept which at first glance might seem minor. However, do not be misled. This separation has resulted in the ability to focus our efforts much more intensely toward the development of the knowledge base model—rather than have to spend an inordinate (and generally unnecessary) amount of time and resources in the development and implementation of the entire solution process (i.e., in particular, the inference process). Since it is the model that determines the outcome, whether one uses algorithms, heuristic programming, or expert systems, it should be apparent that this is the area to which we should devote the bulk of our efforts.

Expert Systems in the 1990s—and Beyond

While some may trace expert systems back to antiquity (i.e., in the form of heuristic programming) and others to the 1960s, the bulk of humanity first became acquainted with the topic in the 1980s. Articles during this period, with but a few exceptions, painted an excessively optimistic picture. Funding and interest in expert systems soared. In a matter of years, and sometimes only months, professional societies related to the field grew from a handful of members to thousands—with the preponderance of the membership really not quite sure just what expert systems were, but feeling that they simply could not afford to "miss the boat" on such a major development.

One of the major players behind the interest in AI and expert systems in the early 1980s was the Department of Defense (DoD) and, in particular, DARPA (Defense Advanced Research Projects Agency). In 1983, DARPA announced a 10-year plan for its *strategic computing initiative,* or SCI. The purpose of SCI is to develop machine-intelligence technology—with expert systems at the heart of virtually all of its proposed projects. Projections and pronouncements made by DARPA, and by those researchers funded by DARPA, were rather reminiscent of those made at the Dartmouth Conference—and have recently generated a similar backlash. Specifically, it soon became clear that the projections for SCI were overly ambitious and DARPA seemed to virtually reverse field—moving in only a few years from avid fan to (in some cases) outspoken critic of AI and expert systems.

In 1985, Stefik [1985] made the following, prophetic observation:

> Throughout the history of AI research, grand predictions have at times been made about the emerging power of computers. These predictions have often been picked up and translated prematurely into concrete projections that have led to disappointment as the real difficulties in implementation became known. Some critics [of SCI] suggest that this time the overselling is being done by DARPA, with two possible dire consequences: a setback (perhaps even a "dark age") for computer science if the field loses credibility; and, far more serious, the possibility that defense-policy planners will take plan predictions at face value and make decisions on unrealistic expectations of machine-intelligence technology.

Vacillation in support has hardly been unique to expert systems. Consider, if you will, the case of neural networks—another subfield of artificial intelligence. In the 1960s, interest in the field of neural networks was surging—with considerable support from DoD. However, by 1970 a single negative assessment of the field, by Minsky and Papert [1969], served to virtually kill all federal funding for neural networks (even the Soviet Union followed suit). Almost two decades passed before neural networks reemerged as a *legitimate* field of research. In 1988 [Schwartz, 1988] the deputy director of DARPA's Tactical Technology Office, stated: "neural networks is a more important technology than the atom bomb." The reversal of opinion could hardly have been more dramatic.

For the majority of the 1980s, the media painted an overly optimistic picture of expert systems. More recently, the assessment has more often been overly pessimistic. The truth lies somewhere in between. Expert systems is a powerful, practical, and eminently useful tool—but only when employed on the right problems, in the right manner, by the right people. And in these respects, expert systems is no different from any other methodology. Criticism of expert systems is, for the most part, misplaced. Those failures and disappointments that have occurred have rarely (if ever) been failures of expert systems; rather they were failures on the part of those who misused the technology—either through ignorance or avarice. Placing the blame on expert systems for such failures is just as wrong as placing the blame on automobiles when assessing the devastation caused by drunken, irresponsible drivers. Hopefully, this text will do something to dispel the unfounded criticism of expert systems—as well as to temper unwarranted expectations.

Certain Differences of Opinion

Much of the criticism now being directed toward expert systems is, we believe, due to a rush to become involved with the methodology coupled with a failure to take the time and effort to truly understand and appreciate the concept, its history, scope, and limitations. In addition, we must admit to disagreement with a number of commonly held perceptions of, and practices within, expert systems. These include, in particular, the following:

- The implication, in the literature, through omission of statements to the contrary, that expert systems can and should be used on virtually any problem. And a failure to emphasize the importance of having a reasonable familiarity with alternative solution procedures.
- The implication that, to understand and use expert systems, you must be familiar with certain *AI languages* (particularly LISP and PROLOG).
- The emphasis, in too much of the expert systems literature, on those factors that really only *support* expert systems (e.g., the computer and computer programming).
- Statements that imply that expert systems is just an "alternative to conventional computer programming."
- The widely held belief that a knowledge engineer is synonymous with a computer programmer—or computer scientist.
- The implication that one may learn how to use expert systems by simply learning how to run a commercial expert systems software package—and the resulting development of expert systems *software technicians,* rather than competent knowledge engineers.
- The belief that, just because a person is doing a job, he or she is an expert in that job—and the concomitant cloning of mediocrity.
- The widespread belief that the best, if not only way to validate the performance of an expert system is to compare its performance (i.e., conclusions) with those arrived at by the human expert from whom the knowledge base was cloned.
- Exaggerations and oversimplifications of the alleged *modularity* of the expert systems knowledge base (i.e., that one may simply add or delete rules to a knowledge base, without any impact on the solution).
- The belief that potential expert systems developers should look at (only evidently) applications that have the potential of either saving the company or

earning for the company several million dollars a year. This further implies that the only expert systems worth building are those involving many hundreds or thousands of rules.

We shall, in this text, ultimately address each of these issues. The reader may then draw his or her own conclusions.

It is our firm belief that one's introduction to any field, or any solution methodology, is best obtained through a balanced presentation of the scope and limitations, and advantages and disadvantages of the discipline. And this is the philosophy that has been followed in the preparation of this book.

SUMMARY

A rather broad range of topics has been covered in this relatively short chapter. The purpose, however, was primarily to explain the relationships among such

TABLE 2.2
A comparison among approaches

	Algorithmic procedure	Heuristic program	Expert system
Model	Quantitative	Quantitative or qualitative	Quantitative or qualitative
	Precise relationships	Imprecise relationships	Imprecise relationships
	Opaque	Opaque	Transparent
	Structured development process	Unstructured development process	Relatively structured development process
	Revision relatively difficult	Revision relatively difficult	Revision relatively easy
Variable (attribute) values	Numeric	Numeric or symbolic	Numeric or symbolic
	All values must be known to solve	All values usually must be known	Values may be known or unknown
Solution process	Convergent	Satisfices	Satisfices
	Dependent on order	Dependent on order	Relatively independent of order
	Comingles *what we know* with *what we do*	Comingles *what we know* with *what we do*	Separation of *what we know* from *what we do*
Conclusions	Usually single (without explanation)	Usually single (without explanation)	Often multiple (with explanation)

topics as algorithmic methods, heuristic programming, and expert systems. In support of this, we have attempted to summarize some of these notions and results in Table 2.2. In addition, in Table 2.3, we have listed the terminology typically employed within each sector. As may be noted in both of these tables, the differences between approaches are quite often more of semantics than of real substance.

The discussion in this chapter has skirted to some degree the issue of where and when expert systems should be applied (i.e., just how does one identify a problem for which an expert system might be a rational solution procedure). However, Table 2.4 serves to summarize the points either explicitly made or implied thus far. In particular, Table 2.4 lists the attributes of a problem that *might* be a candidate for solution by expert systems. However, in the final decision about whether or not to employ expert systems, one must (or should) consider the following factors:

- Economic benefits of the expert system, and a comparison of such benefits with those offered by appropriate, alternative approaches
- Feasibility (i.e., are there the time, funds, personnel, and equipment necessary to develop a useful system?)
- Management and labor's support and acceptance

Rather obviously, each of these factors is identical to those that should be considered in the development and implementation of virtually any complex system.

TABLE 2.3
Terminology employed within each sector

Algorithmic procedures	Heuristic programming	Expert systems
Entity	Entity	Object
Variables and values	Variables and values *or* attributes and values	Attributes and values
Relationships	Heuristics	Heuristics (or rules)
Algorithms	Inference—or logic	Inference
Analysts (or modelers)	Analysts (or modelers)	Knowledge engineers
Solution derivation	Solution derivation	Consultation session
Search	Search	Chaining
Model development	Model development	Knowledge acquisition
Model format	Model format	Knowledge representation
Implementation	Implementation	Implementation
Models, or sets of relationships	Models, or sets of relationships	Knowledge base

TABLE 2.4
Attributes of problems addressable by expert systems

A measurable and significant difference between conclusions reached by the expert versus a nonexpert—and an unwillingness to accept decisions reached by nonexperts
A shortage of the expertise necessary to deal with the problem—coupled with the high cost of recruiting and/or training a new expert
The existence of expertise (i.e., human expert or documented examples of good decision making)
A relatively narrow domain of expertise
A relatively stable domain
A domain with a preponderance of heuristic procedures (i.e., as opposed to algorithmic procedures)
A preponderance of attribute values that are symbols (as opposed to numbers)
The need for a transparent knowledge base (i.e., a listing in English of the rules employed to solve problems within the domain)
The need to reach decisions in the absence of complete information (i.e., attribute values that may be unknown)
A combinatorially exposive problem (i.e., one that would require an excessive time for solution by algorithmic procedures)

EXERCISES

2.1. List at least ten heuristic rules that you, yourself, employ in everyday life. How did you arrive at these rules?

2.2. Develop a mathematical model for the following problem and discuss the physical meaning of each component and relationship used in the model. A firm is considering the funding of six different research efforts. Each effort must be either fully funded or else rejected for funding, and each requires exactly a year to complete. The following table lists other data pertinent to the problem. Given that the firm wishes to select a mix of projects that will maximize expected revenue, subject to a total research budget of 12 units, develop the mathematical model that best describes the situation.

Project	Budget requirements	Expected revenue
1	3	9
2	2	10
3	4	14
4	1	4
5	5	10
6	3	12

2.3. Develop a *heuristic program* that will provide the solution to the problem presented in Exercise 2.2 (and one that would be applicable to much larger problems). Describe the motivation for the use of the heuristic program selected.

2.4. One of the early forecasts for AI was that a computer would, by 1970, be able to compose music of *classical quality*. Do you believe that such a forecast will ever come true? Why, or why not?

2.5. We have noted that there are some people who believe that expert systems and heuristic programming are, for all practical purposes, one and the same thing. What are your own feelings about this matter? Should one really care?

2.6. Examining both Table 2.4 and Exercise 2.2, what are your conclusions with regard to the suitability of expert systems for solution of the type of problem described in the exercise?

CHAPTER
3

PAST, PRESENT, AND FUTURE EXPERT SYSTEMS

ON EXPERT SYSTEMS
BEFORE EXPERT SYSTEMS

At the close of Chap. 2, some differences of opinion as to the origin of expert systems were briefly alluded to. That is, not an insignificant number of individuals feel that expert systems is nothing more than heuristic programming under a different name, and subject to a different set of jargon. Reinforcing this perception is the fact that, today, we may note rather compelling evidence of a practice that this author terms *rebranding*; that is, we see software packages that were developed in the past as *heuristic programs* now being sold as *expert systems*. In many instances, the *only* thing changed has been the product name. In others, a heuristic programming package may have been simply recoded; say from FORTRAN into LISP (a highly questionable practice for a variety of reasons, both ethical and practical), and then designated (and sold) as an expert system.

Another factor that disturbs some individuals is their perception that some portion (if not a majority) of the AI and computer science communities has failed to recognize that, not only was there *life before the computer*, there were also expert systems before the computer. Consider Weiss and Kulikowski's definition of an expert system [1984]. They state that an expert system is a system that

- Handles real-world, complex problems requiring an expert's interpretation
- Solves these problems using a computer model of human reasoning, reaching the same conclusions that the human expert would reach if faced with a comparable problem

Our dissidents feel that the phrase *computer model* in the above definition could be replaced simply by the word *model*, and the definition would still hold true. (For example, addition is addition, whether done by hand or on a computer.) Under this slightly modified definition of an expert system, one may note numerous examples of expert systems (or at least what appear to be expert systems) prior to either the existence of the digital computer or the field of artificial intelligence. As just one example, consider the detailed operation procedures, or DOPs, used for decades in engineering. These DOPs are typically brief manuals that represent a compilation of rules (many of which are heuristic) and procedures, as developed by an expert engineer, for use by the technician in the field when dealing with a specific problem. Such manuals have found particularly widespread use in the aircraft and aerospace industry. DOPs have long been routinely used for preflight and inflight assistance in both the identification and correction of aircraft malfunctions.

Consider, for example, the following excerpt from a DOP as written for the Saturn/Apollo moon landing program in 1962 [Ignizio, 1962]:

> Step 7: **If** the antenna to transmitter coaxial cable has been disconnected from the antenna port, **then** examine the coaxial cable connecter for any signs of breakage, discoloration, or fraying. **If** such signs are detected, **then** remove and replace the connecter via the procedure outlined in Section 3.2. **Otherwise**, connect the RF generator to the antenna port **and** initiate the test sequence as outlined in step 8.
>
> Step 8. **If** the RF generator is connected to the antenna port, **then** turn it on and set the frequency to 220.5 Megacycles (+/− 0.1 Megacycle). **If** the generator has been on for at least 2 minutes, **and** the frequency is stable, **then** initiate the VSWR test as outlined in Section 2.2.

Examining these few steps carefully, we may note that the contents of each step represent extracts from the knowledge base of an electrical engineer with specific expertise in the testing of RF (radio frequency) systems; specifically, telemetry antennas. Note carefully that each step consists of one or more "if-then" (or "if-then-otherwise") type of statements. That is, *if* a particular situation exists, *then* take a certain action. Such statements are designated by those in AI as *production rules*, or simply rules, and form the basis of so-called *rule-based systems* (which, in turn, are used as the foundation for the vast majority of existing expert systems). Further, the specific rules listed above, as the author of the referenced document will attest, were developed through experience and judgment. Thus, with the single exception of the use of a computer to implement these instructions, such DOPs would certainly appear to satisfy most of the criteria typically associated with an expert system.

A much earlier example of the employment of heuristic programming/expert systems is given by the knowledge base (i.e., set of heuristic rules) used by Chinese civil engineers, several thousand years ago. These rules were followed in the construction of dams and waterways, and allowed for the widespread dissemination of the expertise of the best and brightest engineers in the country.

Even with the rather dubious requirement that an expert system *must* be implemented on a computer, there are numerous examples of techniques, developed under the original designation of a heuristic program, that would also appear to satisfy most, if not all, of the criteria normally used in the AI community's definition of an expert system. Consider, as just one example, a heuristic program that is concerned with facility location, a problem briefly mentioned in Chap. 2. Specifically, in the early 1960s, the author developed a heuristic program for the location of such entities as warehouses, missile sites, fire stations, and so forth [Ignizio, 1971]. A heuristic program was used because of the size and computational complexity of the problems that were faced. While the solutions obtained by this method are not always optimal, the problem sizes for which it is intended simply rule out the use of exact methods. The primary objective of the heuristic program is to select a set of facility sites from among a large number of candidates to provide satisfactory *coverage* to a set of *customers*. The heuristic program was written in FORTRAN and has been implemented, with success, in several real-world situations.

In the years prior to the development of this book, an attempt was made to precisely define the differences between expert systems and heuristic programming. This led us to wonder, in fact, if it were not possible that some individuals, particularly those within the AI community, relied *solely* upon the mode of representation (of the model and solution approach) on the computer to distinguish between the two approaches. This led to the decision, on our part, to perform a modest, informal survey to test this hypothesis. While we cannot and do not claim any statistical significance for the results of this survey, we do believe that they are both interesting and enlightening. The original heuristic program as developed for facility location was quite easily restated as a set of production rules. Following this rather modest reformulation, we used two different methods to develop an equivalent *expert system*. In one instance, we developed the expert system using the LISP language. In the other, we employed a commercial expert system shell: the EXSYS package.

Over the years since then, we have shown the FORTRAN versions to some individuals and the LISP and EXSYS versions to others. All of those who have been involved in this survey are either members of the AI community or have specific interest and expertise in the area of expert systems. In nearly every instance, those shown the FORTRAN program did not consider it to be an expert system. However, the vast majority (thus far, more than 90 percent) of those shown either the LISP or EXSYS programs readily accepted these particular depictions as *legitimate* expert systems, and realize that these conclusions were reached despite the fact that all three versions solve the same class of problem, using the same set of heuristics, in virtually the same manner, and reach precisely the same conclusion.

The little experiment described above was rather enlightening. Even more educational, however, have been the results of a study that we have been conducting since 1980. Here, we have been examining documentation associated with expert systems efforts within the military sector. And here we have noticed that a surprising number of *expert systems* that have been developed appear, to me, to be remarkably similar to earlier, *heuristic programs*. We have the impression, in fact, that the military (i.e., the taxpayer) is sometimes paying twice for certain software efforts; first as a heuristic programming effort and later as an expert systems effort. Moreover, we are not the only ones harboring this suspicion. In fact, one of the statements made by an early critic of expert systems is that the difference between an expert system and a heuristic program is akin to the difference between a black-and-white movie and its colorized version. Nothing has really changed except for the visual impact. Practices such as the ones previously described do little to alleviate such suspicions.

Obviously, there is considerable confusion concerning the differences between expert systems and heuristic programming—as well as questions as to whether or not they are really just one and the same thing. We shall return to this issue in Chap. 13, the concluding chapter of this text. In the meantime, the reader should simply remain aware of this situation—and of the difficulty in drawing truly meaningful distinctions between these two concepts. You may also wish to review the tables presented at the conclusion of Chap. 2, in particular Tables 2.2 and 2.3.

As an exercise, we *strongly* suggest that the reader review two relatively recent papers. The first is presented as a *heuristic program*, and deals with the loading of military cargo planes [Cochard and Yost, 1985]. The second is presented as an *expert system*, and deals with exactly the same problem [Anderson and Ortiz, 1987]. These two papers have created quite a furor, particularly among those who insist that there are no meaningful differences between heuristic programs and expert systems. Both papers will be addressed in more detail in Chap. 8. Here, we will simply note that we believe that those who would seek to determine the differences between heuristic programming and expert systems must look beyond the surface, that is, beyond the terminology used in discussions and the software employed in implementation.

EARLY EXPERT SYSTEMS

Leaving aside, for the moment, the more controversial aspects of the origin of expert systems, let us now focus on those expert systems that are typically regarded by the AI community as being the first such systems developed. These are the expert systems of the late 1960s and the 1970s.

DENDRAL: An Expert
in Chemical Identification

Work on DENDRAL, generally considered to be the very first expert system, began in the mid-1960s at Stanford University under the direction of Joshua

Lederberg, Edward Feigenbaum, and Bruce Buchanan [Lindsay et al., 1980]. Lederberg was the recipient of the Nobel Prize in Chemistry and Feigenbaum and Buchanan are recognized as pioneers in expert systems research. The purpose of DENDRAL, which did not actually become operational until the early 1970s, is the identification of the molecular structure of unknown compounds, a problem of considerable computational complexity. DENDRAL, unlike many of the early expert systems, found acceptance and is still in use by chemists all over the world. In fact, for some tasks, DENDRAL is generally acknowledged as performing *better* than any human expert.

Lederberg had first developed a *conventional* approach to this problem in 1964, and this was designated as the DENDRAL *algorithm* (recall our discussion of algorithms and heuristics in Chap. 2). The purpose of the collaboration with Feigenbaum and Buchanan was to determine if heuristics could be used to develop results comparable to the algorithm, but in less time. The heuristic method uses, as does the algorithm, the set of spectroscopic data from the compound to be identified. The heuristic program then reduces the large number (typically millions) of potential structures down to some reasonable number. It accomplishes this task by filtering out (or *pruning*) all structures that could not be associated with the sample. These filtering rules are those heuristics typically used by expert chemists. The remaining possibilities are then compared to the unknown compound on the basis of matching spectrogram patterns. DENDRAL utilizes production rules and was implemented in the LISP programming language. DENDRAL does not have an explanation facility. That is, it simply reaches a conclusion and this conclusion is presented to the user.

HEARSAY I and II: Speech Recognition

HEARSAY I (1969) and HEARSAY II (1971) were developed at Carnegie-Mellon University in an attempt to demonstrate the possibility of a speech recognition system [Reddy et al., 1973]. Specifically, the goal of the system was to have a computer understand spoken input. The input to the HEARSAY system is a speech waveform. From this waveform, a set of hypotheses about what may have been said is developed. A best guess from this set is then presented as the output.

One of the more innovative concepts developed by the HEARSAY project was that of the use of *multiple* knowledge bases. These knowledge bases communicate by use of a *blackboard*, which serves as the working memory that is shared by all knowledge bases. Each knowledge base, in turn, concentrates on a different aspect of the speech recognition problem. The blackboard is controlled by an agenda scheduler who serves to determine which knowledge base is to use the blackboard next. Several present-day expert system shells, in fact, incorporate the blackboarding concept.

At the completion of the HEARSAY project in 1975, the system had a vocabulary of about 1000 words and was able to correctly interpret spoken input roughly 75 percent of the time. This was accomplished in a time span of only a few times longer than was taken by a human listener. A fast mainframe computer was used to achieve these results.

One important result of the HEARSAY project was the demonstration that an expert systems approach (i.e., the use of heuristics) was superior to what had been the conventional approach to speech recognition. The conventional methodology relied upon statistical tools, that is, the analytical approach. Another result of HEARSAY was that it spawned several follow-on efforts dealing with the interpretation of several types of signals, in particular, acoustic signals such as those obtained through sonar contacts. Included among these are the HASP/SIAP systems [Nii et al., 1982].

INTERNIST/CADUCEUS:
An Expert in Internal Medicine

The INTERNIST project was started in the early 1970s, and continues today under the name CADUCEUS [Pople, 1984]. The codevelopers are Harry E. Pople, Jr., and Jack D. Myers of the University of Pittsburgh. Pople is a computer scientist and Myers is a physician specializing in internal medicine. One of the truly striking things about INTERNIST/CADUCEUS has been its ability to remain a viable project over such an extensive period of time.

The goal of INTERNIST is to perform a diagnosis of the majority of diseases associated with the field of internal medicine. This, in itself, is an ambitious endeavor as there are hundreds of such diseases. However, not only is INTERNIST/CADUCEUS intended to diagnose each disease, it is supposed to consider all the possible combination of diseases that might be present in the patient. It is estimated that the number of such combinations is on the order of 10 to the 40th power. Consequently, as in the case of DENDRAL, we are faced with a problem that exhibits combinatorial explosiveness; a problem for which the heuristic approach is most appropriate.

MYCIN: An Expert in Blood Infections

MYCIN is, at this time, probably the most widely known of all expert systems thus far developed [Shortliffe, 1976]. And this is despite the fact that it has never been put into actual practice. Many early texts on expert systems have, in fact, focused primarily on the MYCIN system—and the project has served to substantially influence much of the subsequent work in the construction and implementation of expert systems. MYCIN was developed at Stanford University in the mid-1970s.

MYCIN was designed solely as a research effort to demonstrate how expert systems might actually be constructed for reasonably large and complex real-world problems. The particular role proposed for MYCIN was that of providing assistance to physicians in the diagnosis and treatment of meningitis and bacteremia infections. MYCIN is thus somewhat akin to INTERNIST/CADUCEUS in its purpose, except that it focuses on a far smaller number of diseases and thus requires a considerably smaller knowledge base.

The knowledge base of MYCIN contains the heuristic rules used by physicians in the identification of certain infections. EMYCIN (for *Empty* MYCIN) is the name given to MYCIN when this specific knowledge base is removed. In many cases, one may collect a knowledge base associated with a different domain and insert this into EMYCIN, where the result is a new, working expert system. This *plug-in* concept is termed, as has been discussed, an expert system shell. The result of incorporating a knowledge base associated with pulmonary disorders into EMYCIN resulted in a new expert system known as PUFF.

PUFF: An Expert in Pulmonary Disorders

PUFF was developed in 1979, using the EMYCIN shell [Aikens et al., 1984; Harmon and King, 1985]. The purpose of PUFF is to interpret measurements related to respiratory tests and to identify pulmonary disorders. PUFF interfaces directly with the pulmonary test instruments used in such measurements. At the conclusion of the testing, PUFF presents the physician with its interpretation of the measurements, a diagnosis of the illness, and a proposed treatment scheme. The first version of PUFF had 64 production rules. A more recent version (coded in BASIC) had about 400 rules.

The validation procedure for PUFF was to compare its diagnosis with that of two expert pulmonary physiologists. The results of PUFF and that of the physiologists agreed in more than 90 percent of the test cases. PUFF is said to now be used on a routine basis.

One particularly interesting feature of PUFF is that, to all outside appearances, it is indistinguishable from more conventional laboratory *tools*. The acceptance of PUFF, by the medical profession, may well be based, in large measure, on this perception. That is, PUFF is viewed as an ordinary piece of laboratory equipment, rather than as an intelligent (and possibly superior) competitor.

XCON (R1): An Expert in Computer Configuration

XCON (originally titled R1) was developed for the configuration of VAX computers at the Digital Equipment Corporation (DEC) [McDermott, 1982]. The initial effort was a collaboration between investigators at Carnegie-Mellon University and those at Digital Equipment Corporation. R1 utilized the OPS4 expert system building tool (since replaced by OPS5).

A VAX computer may be configured in an enormous number of ways, and DEC attempts to configure each computer according to the specific requirements of each customer. Such a problem might be thought of as a type of loading problem wherein a box (the computer frame) is to be loaded with equipment (the parts of the computer) to achieve a specific purpose (satisfy the computing requirements peculiar to each customer). Of course, the combinatorial complexity of this problem is enormous and, as such, certain heuristic rules must be employed to reach an acceptable configuration within a reasonable time frame.

The R1/XCON effort began in 1978 and a prototype expert system, based mainly upon the contents of a VAX configuration manual, was demonstrated in April 1979. By September of that year, the system was able to correctly configure more than 75 percent of all customer orders that it was given. By 1981, the system had been implemented and was being employed on a routine basis. However, updating and refinement of the system continue as the system attempts to track changes in design as well as strive for improved productivity. At the time of this writing, XCON consisted of more than 8000 production rules running under the OPS5 environment (an LISP-based system that typically operates in a forward chaining mode).

XCON has received enormous coverage in the literature and is usually presented by the AI community as the first truly successful, commercial implementation of the expert systems concept. Today, however, some of this initial exuberance has been diminished. Newquist [1988*b*] states, in fact, that the joke going around inside of DEC is that "XCON replaced 75 people who were expert configurers, but it now takes an additional 150 people to maintain XCON." While this may be an exaggeration, rumors of problems do continue to surface and claims are made that XCON has either gotten *too large* or that OPS5 simply does not provide the flexibility required to support such a massive, dynamic expert system. Newquist, however, notes that this may simply be a case of poor planning—or no planning at all. That is, when XCON was being developed, little thought was given to the rather essential problem of the future maintenance of such a system.

DELTA/CATS: An Expert in the Maintenance of Diesel-Electric Locomotives

DELTA/CATS-1 is (or perhaps was) an expert system developed by the General Electric Company in the early 1980s [Bonissone and Johnson, 1983]. DELTA stands for *diesel-electric locomotive troubleshooting aid* while CATS represents *computer-aided troubleshooting system*. The purpose of DELTA/CATS-1 is to assist railroad personnel in the maintenance of General Electric's diesel-electric locomotives.

DELTA/CATS-1 consists of the knowledge base (i.e., set of heuristic rules) that was acquired through interviews with a General Electric employee, David Smith. Smith had been with GE for more than 40 years and was recognized as the firm's expert in diesel-electric locomotive maintenance. The development effort began in 1981 and the first field prototype was completed in 1983. Over this period, the number of production rules increased from 45 to 1200. The system was originally developed in LISP and then converted to FORTH for increased transportability and speed of execution. Both forward and backward chaining are utilized.

A particularly interesting feature of DELTA/CATS-1 is its interface with visual support systems. For example, schematics of the locomotive may be printed out and a videodisk player can be used to show just where particular components are located within the locomotive, as well as how to access and maintain these parts.

More recently [Newquist, 1988*b*], rumors have circulated that DELTA/CATS-1 is having problems similar to those cited for XCON, and that the system may, in fact, have been shelved. Newquist even reports that GE had approached at least one expert systems vendor to take over the entire system, as long as the vendor would provide maintenance support. The vendor is said to have refused. As in the case of XCON, a system that has been used as a model of successful expert system implementation has, evidently, been undone for lack of a plan for its maintenance. Considering that DELTA/CATS-1 was itself designed for maintenance (i.e., of locomotives), this seems to be a particularly ironic situation.

SOME MORE RECENT EXPERT SYSTEMS

The expert systems described in the previous section have become almost legendary in the AI sector. However, as noted in the description of both XCON and DELTA/CATS-1, even the best of ideas can be undone for the lack of planning; particularly planning for the maintenance of any real-world implementation of expert systems. The issue of the maintenance of expert systems will be addressed in Chap. 11. However, let us now consider some of the more recent expert systems.

GATES: An Airline Gate Assignment and Tracking Expert System

GATES [Brazile and Swigger, 1988] is in use (evidently in prototype form) at New York's JFK International Airport. The system is being used by TWA to assist ground controllers in the assignment of gates to arriving and departing flights. The knowledge base was acquired from an experienced ground controller who solved such problems on a daily basis.

The gate assignment problem can become quite complex, and requires rapid solution during intervals of flight delays, bad weather, mechanical failures, and so forth. Optimization methods (e.g., linear integer programming) had been attempted but were simply unable to cope with the real-time demands of the problem. Thus GATES was developed, using PROLOG, and implemented on a personal computer. The system handles about 100 or more flights a day, has direct access to TWA's database, and can create gate assignments in about 30 seconds. Previously, the human experts had needed between 10 and 15 hours to prepare an assignment, and as much as an hour to modify the assignment each morning.

QMR: Medical Diagnostic Expert System

Using the massive knowledge base first developed for INTERNIST, QMR (quick medical record) assists physicians in the diagnosis of an illness based upon the patient's symptoms, examination findings, and laboratory tests [Kane and Rucker, 1988]. QMR, which is resident at the University of Pittsburgh, incorporates over 4000 possible manifestations of diseases and is said to perform at a level comparable to practicing physicians.

FXAA: Foreign Exchange Auditing Assistant

Chemical Bank does $750 billion a year in foreign exchange trading. This involves thousands of transactions a day with the paperwork resulting from such transactions weighing in at about 10 pounds per month. As such, the audit volume is well beyond the capabilities of unassisted human auditors. One particularly important type of audit is that of the recognition of irregular transactions. FXAA [*AI WEEK*, 1988*b*] has been developed to provide the necessary auditing assistance.

Tests of the system demonstrated a productivity increase, through FXAA, of 30-fold and provision for increasing the frequency of auditing from every 18 months to a monthly audit. Using the ART expert system environment, FXAA is a rule-based expert system that has evidently made a major impact within Chemical Bank.

Jonathan's Wave:
An Expert in Commodities Trading

A number of firms and individuals have developed expert systems for stock and commodity trading. While it is still too early to assess the success or failure of these programs, they have attracted considerable attention, and customers. Jonathan's Wave is just one of these, developed specifically for commodities futures and commodities futures options trading [Newquist, 1988*a*].

Developed by Michael Archer, Jonathan's Wave runs on two 286-based personal computers. The knowledge base is written in C while the inference engine is written in PROLOG. Incorporated in the program are the knowledge bases of several approaches to commodities trading. Based upon their suggestions, and their past performance, the system determines the trades to be made. In this manner, the system acts somewhat as though it were using *multiple* experts to reach its conclusion.

Insurance ExperTax:
An Expert in Tax Planning

Coopers and Lybrand have created Insurance ExperTax [*AI WEEK*, 1988*a*] to assist in the identification of tax planning and accrual issues. John Bailey, chairman of Coopers and Lybrand's insurance industry practice, states: "Overall, insurance companies have not yet fully analyzed the impact of the recent changes in the tax laws and could accelerate their efforts in this area. . . . Through Insurance ExperTax, companies now have access to the knowledge and tax planning skills of Coopers and Lybrand's top insurance tax professionals from across the country. Our system can help these companies focus on the actual effects of tax reform."

Insurance ExperTax took more than a year to develop and consists of more than 3000 rules. Created in LISP and running on the IBM PC, the tool is an industry-specific enhancement to ExperTax, Coopers and Lybrand's expert system first developed in 1986.

HESS: An Expert Scheduler
for the Petrochemical Industry

HESS was developed at the University of Houston in support of product scheduling at a major petrochemical firm's refinery [Deal, forthcoming]. The knowledge base in HESS was developed via the acquisition of heuristic rules from two refinery product schedulers. Their function was to determine what product, or products, to produce, at what time, and through which processors. Their performance was measured against the costs of production, production ruins (i.e., products that do not meet specifications and must be either recycled or downgraded), and lost customer sales.

HESS was developed using the EXSYS expert system shell, and through a 12-month effort. HESS, which stands for *hybrid expert system* (see Chap. 9) *scheduler*, consists of approximately 400 production rules and runs on an IBM PC or compatible. Not only does HESS accomplish the scheduling task previously done by the human expert, it does so on a much more consistent basis. Savings through the implementation of such a system, at a typical U.S. refinery, have been estimated (by refinery personnel) to be on the order of several million dollars each year. Despite this, the package has yet to be fully implemented as a result of the lack of access to the firm's databases. The importance of such an access, or *bridge*, will be discussed in Chap. 7.

An Expert in Poultry Farming

The Georgia Tech Research Institute is developing an expert system for the poultry farmer. The system utilizes the Nexpert Object expert system shell [Baker, 1988]. The system analyzes data from the poultry farm's environmental control system. Using information on feed and water consumption, temperature, humidity, and ammonia levels, the system may be used to alert the farmer to any diseases the chickens have, or may get.

DustPro: An Expert in Mine Safety

Using the Level 5 expert system shell as a development vehicle, the U.S. Bureau of Mines has developed an expert system named DustPro [Newquist, 1988c]. DustPro replaces the limited number of human experts that assess the air quality of mining operations. Based on the amount of coal and silica dust in the air, mining operations must be adjusted to ensure that safety requirements are satisfied.

There are about 2000 working mines in the United States. This means that 2000 human experts would be needed to provide constant surveillance at each mine. DustPro alleviates this requirement. DustPro interfaces with monitoring systems (that monitor methane gas emissions and dust). It runs on a PC and takes about 15 minutes to reach a conclusion. The knowledge base consists of roughly 200 rules. By late 1988, the Bureau had delivered 200 copies to mining operations worldwide.

TOP SECRET: An Expert
in Security Classifications

Within the Department of Energy (DOE), there are more than 100 classification guides to nuclear weapon security data [Newquist, 1989]. One of the more onerous tasks within DOE is to attempt to correctly classify a given document through the use of these guides. Document classification determines who is permitted to view a document, and who is not—a potentially critical factor in national security.

The Albuquerque Operations Office of DOE selected EXSYS to perform this classification task. The knowledge base of this shell contains the rules from the classification guides that determine just how to classify a document (e.g., as confidential, secret, or top secret) and the system is being used to relieve the work load of the people previously assigned to this effort.

Codecheck: An Expert
in Computer Program Assessment

Abraxas Software [Sherer, 1989] has introduced an expert system for evaluation of C source codes. Termed Codecheck, the package is a rule-based expert system that checks C source code for such things as complexity, formatting, and adherence to standards. A code that satisfies such checks is more likely to be maintainable and portable.

Abraxas personnel note that the most common cause of hard to maintain software is the programmers' tendency to write overly complex code. Codecheck identifies those portions of the code that may be simplified. In addition, Codecheck evaluates the portability of the source code by comparing it with the numerous standards now existing for C programs and identifies any code that will not port between DOS, OS/2, UNIX, VMS and Macintosh. The introductory price for Codecheck was $295.

Expert Systems for Faster, Fast Food Operations

Finally, we may note that expert systems have even permeated the fast food market. A recent article [Eskow, 1990] describes the introduction of expert systems into such companies as Wendy's, McDonalds, Pizza Hut, Burger King, and Kentucky Fried Chicken. Here, such systems serve to reduce inventory, speed up service, and even act as training assistants. Wendy's, for example, uses expert systems to plan faster and more efficient delivery of inventory—and plans to expand the package into a more *far-reaching* decision support system. McDonalds incorporates expert systems in its European operations which, incidently, are entirely PC-driven. Such packages provide valuable, timely assistance to managers who are neither familiar, nor entirely comfortable with the pace of activities in such operations. In a sector in which there exists such fierce competition, any improvement in cost reduction and enhanced operations can simply not afford to be overlooked.

AN EVALUATION OF PROBLEM TYPES

Although we have presented just a few examples of expert systems, we might note that they are fairly representative of the bulk of applications thus far developed. That is, the majority of applications involve *classification* (or diagnosis). For example, in the medical expert systems, we are given certain data (symptoms) with regard to a patient and attempt to diagnose (i.e., classify) the associated cause, or disease. In maintenance applications, precisely the same type of problem is faced. Here, the symptoms are the data on machinery performance while the diagnosis involves the identification of a defective or failed component. Further, once a classification has been made, the specific class is matched to an associated treatment.

The remaining set of applications involves what is defined in this text as *construction* problems. XCON and HESS are representative of this type of application. Note that XCON attempts to construct a VAX computer, while HESS attempts to construct a schedule.

Classification and Construction Problems: Definitions

We may view classification as an attempt to draw boundaries about existing elements. For example, a certain set of existing symptoms point to a particular disease. Construction, on the other hand, seeks to determine the arrangement (and, quite often, the order) of elements. Thus, the development of each new element of a schedule (i.e., the order and assignment of a task) would represent a problem of construction. In most instances, the problem type also determines the mode of inference most applicable. That is, classification problems usually require backward search (backward chaining) while construction problems typically require forward search (forward chaining).

Another, more visual, means for discriminating between these two fundamental types of problems is available by means of noting just how each type of problem is *mapped*. To illustrate, consider Fig. 3.1. On the left of this figure, we have five objects. Associated with each of these objects are certain attributes and values. On the right side of the figure, we have mapped these five objects into two groups. Now, if the boundaries of these two groups were known *before* the mapping, we would have a classification problem. However, if the process of mapping also includes the development of the group boundaries, then the problem is one of construction.

To further clarify this concept, consider a problem in which the five objects of Fig. 3.1 are five different automotive engine parts. Associated with each object is a set of data pertaining to various quality tests. Further, we simply wish to distinguish between parts that are acceptable and those that are not. Thus, a priori, we have but two classes. Using the data set, a quality control engineer (or an expert system) may then assign each object to one of the two classes. And this is a typical classification problem.

Next, let us assume that the problem involves the loading of five items (the five objects on the left side of the figure) onto a fleet of trucks. Initially, we are not sure how many trucks (i.e., groupings) are necessary. Associated with each item are such attributes as weight, volume, cost, and priority. Using the values of these attributes, the cargo loader (or expert system) will then determine

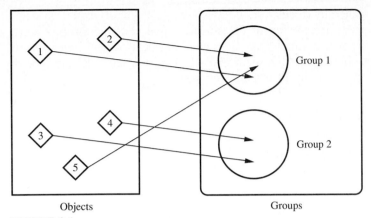

FIGURE 3-1
Mapping of objects to groupings.

the loading scheme. Thus, Fig. 3.1 depicts a scheme in which items 1, 2, and 5 are loaded on one truck while items 3 and 4 are loaded on another truck. Again, note that the determination of the number of trucks used (e.g., number of groups) is an integral part of this problem, which is representative of a typical problem of construction.

Other authors sometimes further partition problem types (e.g., into those of planning, prediction, design, monitoring, repair, control, and interpretation). However, we believe that at a somewhat higher level, the vast majority of expert systems applications involve either classification or construction, or both—and that these two types serve sufficiently well to indicate the problem faced and the possible approach to be used.

A more recent, and more precisely defined, attempt toward problem classification has been accomplished by Ten Dyke [1990]. He lists four types of applications for AI (or expert systems):

- *Class I*. Characterized by a need to select a solution from a fairly well-defined set of possible alternatives—such as the medical diagnosis problem. This class coincides to what I have termed as *classification* problems.
- *Class II*. Characterized by a need to create a plan or configuration, such as computer system configuration and scheduling. This class coincides to what I have termed as *construction* problems.
- *Class III*. Characterized by the need for true creativity. Such problems include those of design, including those where the very nature of the problem itself might have to be redefined.
- *Class IV*. Characterized as applications that humans can handle and computers can't. Included among this class are such problems as face recognition, reasoning by analogy, and learning how to talk.

Herein, we shall generally focus on the first two classes of expert systems applications—under the designation of classification and construction problems. Problems of class III will normally require something beyond rule-based expert systems, while those of class IV are most likely beyond the scope of any type of expert system and are primarily of research interest.

FUTURE EXPERT SYSTEMS

Most of the expert systems that have thus far been discussed in the literature are essentially *stand-alone* systems. However, in the very near future it is likely that a large portion (if not the majority) of the expert systems that are developed will be *embedded* systems, that is, systems that form only a part of the overall software package. Hybrid expert systems, to be discussed in Chap. 9, are just one example of such an approach.

Another form of the embedded expert system is that of the so-called *intelligent interface*. Intelligent interfaces shall rely, more and more, on expert systems to better achieve *user friendliness* in software. Such a system will immediately determine whether or not the user is a novice or expert, and tailor its actions accordingly. That is, the novice user will require more help, support, and guidance, while the more experienced user will need but minimal assistance. Ideally, such systems should permit one to employ a software package without first having to attempt to read, and make sense of, the jargon-laden instructions of the accompanying user manuals.

Another trend that we expect to continue is the increased development of smaller expert systems—expert systems having 200 or fewer rules. This particular prediction is, however, somewhat at odds with a commonly held belief of the AI community. Specifically, Feigenbaum (who, as you may recall from our discussion of DENDRAL, is one of the pioneers of expert systems) has stated that expert system developers should look at applications that have the potential of either saving the company or earning for the company several million dollars per year [Feigenbaum, 1986]. This has generally been interpreted to mean that only large-scale expert systems (many hundreds or thousands of rules) are worth considering. Such a view may have made sense some years ago when expert system development was somewhat of a trial and error process, and when much of the software for support had to be developed by oneself—and in a difficult language such as LISP, and with the support of expensive LISP machines. However, with the advent of powerful, inexpensive expert system shells—and with implementation on the personal computer—the development of small expert systems are highly cost effective.

Consider the example set by Du Pont [Press, 1988]:

> The feasibility of low-cost delivery of expert systems is dramatically illustrated by the experience of the AI division at Du Pont. Division director Ed Mahler believes small is beautiful—or at least not too risky. Mahler has deployed well over 200 expert systems, and they are not Feigenbaum's million-dollar projects. Each is quite small (averaging 80 production rules), but Mahler estimates that aggregate savings to Du Pont at tens of millions annually.

Such results are not unique to Du Pont. More and more organizations are addressing these *small* problems in order to obtain a quick payoff for their investment in expert systems. And such problems are generally ideal candidates for the rule-based expert systems that are the focus of this text.

SUMMARY

The applications of expert systems appear to be increasing at an almost exponential rate. Walker and Gerkey [1988] state that in 1968 there were approximately 1 to 10 significant applications of artificial intelligence. In 1983 there were about 200 to 300 significant applications. And, in 1988, there were 1500 to 2000 significant applications. Since the majority of the applications of AI are expert systems, the increase in the growth of expert systems is on the same order. This coincides closely with the results presented by Feigenbaum et al. [1988], who claim to have documented some 2000 expert systems used within the corporate world. Further, a large number of expert systems applications most likely remain uncounted, being either proprietary or classified in nature.

However, among the expert systems that have been implemented, there are questions concerning the actual success of at least some of these implementations. As may be noted from the discussions of XCON and DELTA/CATS-1, some of the most highly touted expert systems have evidently run into considerable difficulty. Some, like HESS, have failed to receive user acceptance. Others have not progressed past the prototype stage. The mere existence of one or more publications concerning an expert system does not guarantee that the system was a success in practice. In some cases failures have definitely occurred, and many of these failures have been due to a neglect of the critical factor of expert system maintenance. In others, failure may be traced to the choice of the wrong development software, and/or the wrong hardware. However, one should not be overly discouraged by the existence of such problems. As discussed before, in most, if not all cases, the problems encountered in the implementation of expert systems have not been a fault of the methodology. Rather, the fault most usually lies with certain errors on the part of those who have attempted to implement the methodology.

Despite the problems and questions concerning expert systems, the methodology can, if properly understood and implemented, provide one with an exceptionally powerful and useful tool for dealing with a variety of important, real-world problems. Hopefully, upon completion of this text, the reader will have an appreciation of both the scope and limitations of this tool.

EXERCISES

3.1. If you have read both the paper on DMES [Cochard and Yost, 1985] and the one on AALPS [Anderson and Ortiz, 1987], answer the following questions:
   (*a*) Is the problem being addressed appropriate for expert systems? Why or why not?
   (*b*) Would you classify DMES as a heuristic program or an expert system? Why or why not?
   (*c*) Would you classify AALPS as a heuristic program or an expert system? Why or why not?

(*d*) What type of problem does DMES and AALPS address, one of classification or one of construction?

3.2. Given the following list of problems, state whether they are of the classification or construction category:

(*a*) A loan officer at a bank evaluates loan application forms and determines whether or not such loans should be granted, and the associated rates.

(*b*) A radar operator determines, from images presented on the radar screen, the type of aircraft under surveillance.

(*c*) A ship loadmaster (i.e., for ocean vessels) determines the loading scheme for each ship in port. This includes consideration of where to place each object in the ship, interactions with other objects, and the maintenance of an even keel.

(*d*) A university official must determine where and when to schedule classes to avoid conflicts (e.g., only one class to a room), satisfy the constraints on faculty (e.g., work-load limitations and course preferences), and attempt to satisfy student requirements (e.g., a student might need a certain class in order to graduate).

(*e*) A technician oversees a bank of sensor readouts in a power plant. Based upon what he sees, he must determine whether or not some action needs to be taken.

(*f*) A police detective is responsible for missing person investigations. In some cases the person has run away; in others, the person may have been kidnapped or met with some other form of foul play. The detective must, using the information available, prioritize the case before him and decide on the action to be taken.

(*g*) A distributed computer system is to be built. Each *node* of the system contains a computer, and every node is to be linked to one or more other nodes in the system. The entire network is supposed to serve as a processor of certain data by means of various algorithms. Each processing task requires a different amount of computer support. We need to determine the number and type (e.g., memory availability and processing speed) of the nodes in the system.

(*h*) The owner of a vineyard produces several types of wine from a variety of grapes that he grows. At this time of year, he must determine which grapes need to be picked and the processing schedule for the harvested grapes.

3.3. Examine articles that have appeared within the past 12 to 24 months in such journals as *AI Expert*, *IEEE Expert*, and *AI Magazine* that pertain to the application of expert systems. Try to categorize each of these applications (i.e., as a problem of classification or construction).

3.4. A number of early expert systems experienced initial success but, more recently, have been targets of criticism. Specifically, they have evidently grown too big, too complex, and too inefficient. What might be some of the reasons for this situation? Who do you believe should bear the blame? And what might be done to alleviate such problems in future expert systems?

KNOWLEDGE REPRESENTATION

COMPONENTS OF KNOWLEDGE IN EXPERT SYSTEMS

We are all familiar with the saying "knowledge is power." This is never more true than in expert systems. The knowledge that is contained within an expert system consists of

- *A priori knowledge:* the facts and rules that are known about a specific domain prior to any consultation session with the expert system
- *Inferred knowledge:* the facts and rules[1] concerning a specific case that are derived during, and at the conclusion of, a consultation with the expert system

In this chapter, our concern will be with the manner in which these types of knowledge, and in particular knowledge of the first type, may be represented within the digital computer. In particular, our attention will be focused on the use of rule bases for the representation of expert knowledge.

In the brief introduction to expert systems in Chap. 2, it was noted that knowledge is contained within both the expert system's knowledge base and its

---

[1] Actually, rules are generated during a consultation session only if our expert system is capable of *learning*. In general, inferred knowledge consists simply of new facts, or conclusions.

working memory. The knowledge within the knowledge base is that of the first type, that is, a priori facts and rules about the specific domain. The knowledge within the working memory is dynamic as it changes for each problem addressed— and is of the second type, that is, inferred knowledge about the particular problem under consideration. To clarify these concepts, consider the following example in which we are concerned with the knowledge base of a loan officer at a bank.

THE KNOWLEDGE BASE
OF AN EXPERT LOAN OFFICER

Let us introduce Dan Smith. Dan has been one of the senior loan officers for a bank in Houston for nearly 20 years, and he is generally considered to be the most capable of the loan officers at the bank. Thus, in the specific domain of bank loans, Dan is an expert.

As a loan officer, it is Dan's job to decide the disposition of loan applications received by the bank. Specifically, he must evaluate each application to determine whether or not a loan should be granted and, if granted, the rate and duration (and any other pertinent terms) of the loan.

There are two basic types of mistakes that a loan officer can (and will, regardless of his or her expertise) make. First, a loan may be denied to a person who might actually have made all the payments. Denial of such a loan represents, on the part of the bank, a lost opportunity. Second, a loan may be granted to someone who later defaults on the payments. In this instance, the granting of this type of loan results in reduced revenue. The loan officer must thus establish some sort of policy to balance off these two events in order to attempt to maximize the bank's profits, minimize its risks, and still maintain good customer relations.

The policy ultimately arrived at will be one that combines two sources of knowledge. That is, the loan officer will use what he or she has been taught through formal course work and training. The knowledge obtained through formal study, or available in the public domain (e.g., in books, manuals) is termed *deep knowledge* and forms just one part of the expert's knowledge base. However, in addition to this deep knowledge, an expert develops, through experience, his or her own set of heuristic rules. These heuristic rules are termed *shallow knowledge*; and they are the rules that we particularly wish to include in the knowledge base. Through the use of the expert's heuristic rule set, and in conjunction with the facts that are accessible, he or she is able to obtain better results, and in less time, than that of a novice.

Returning to the specific case of Dan Smith, let us consider the portion of the knowledge base which is stored within his memory. Irrespective of the applicant under consideration, Dan has somehow accumulated a set of heuristic rules and facts concerning the granting of loans. For example, some of the facts that may be stored might include the following:

• Loan applicants fill out formal application forms.
• At this time, the bank is particularly concerned with avoiding bad loans.
• The XYZ Corporation is considering personnel layoffs.

Also stored in Dan's memory are the heuristic rules he uses in the disposition of any loan application. For example, one such rule may be "whenever the bank expresses particular concern about bad loans, place additional emphasis on the applicant's long-term employment prospects." This might be represented by the following statement, or rule:

> *If* the bank is worried about bad loans
> *Then* emphasize the applicant's employment prospects

Remember that the above facts and rules are those appropriate for any case. Let us now turn to the actual decision-making process for a specific loan applicant. Our applicant, Pete Jones, first fills out an application form on which he is asked to list the values for such attributes as his age, annual income, employer, length of time with employer, education, address, marital status, number of dependents, previous credit history, home ownership status (i.e., does the applicant own or rent his or her residence and what are the monthly payments), number of automobiles (and monthly payments), and so on.

The information on this form represents facts about a particular individual (i.e., Pete Jones in our example). While Dan Smith will certainly consider these facts during his decision-making process, they most likely will be forgotten once the decision has been reached. And there is, in fact, no point in remembering all the details of every loan application. Thus, we can consider the facts derived from Pete Jones' application form to be in Dan's *working memory*, where this working memory is physically represented by the application form itself. Some of the facts that may become available through Pete Jones' application are that he works for the XYZ Corporation and makes $50,000 per year, or

- Applicant's name = Pete Jones
- Applicant's salary = $50,000
- Applicant's employer = XYZ Corporation

Now, based on the data (i.e., facts) on the application form, there are typically certain formal policies that the bank follows with regard to the evaluation process. They may, for example, have a policy that a loan should not be granted if the monthly payments associated with that loan exceed some percentage of the applicant's take-home income. Such rules are those associated with deep knowledge in that they are in the *public domain* (i.e., available to all key bank personnel). Consequently, if Dan examines the form and finds that Pete is in clear violation of several such rules, he may immediately conclude that no loan should be made. Such clear-cut cases are not, however, situations in which expertise is most valuable. Rather, *it is quite often in the marginal cases where an expert can make a significant difference*.

For the purpose of discussion, let us assume that the facts available through Pete's application form are such that, while marginal, they do not serve to rule out the granting of a loan. However, since the application is marginal, this is a signal to Dan to give it even more attention than usual.

For example, Dan might first look at Pete's address, which happens to be in a section of town that is in somewhat of a state of decline. Dan recalls that it is taking longer and longer to sell houses in that area, and that house prices there have declined relative to the rest of the housing market. Further, Pete has been with his present employer for only two years, and that particular employer (i.e., the XYZ Corporation), is not noted for employment stability. Coupled with this is the fact that Pete's present mortgage payments are relatively high and the amount paid for his house now seems somewhat on the high side. Dan concludes that there is good reason to expect that the applicant might soon be placed in a position that would result in a default on the loan. And he has made this decision based on the rules and facts in his knowledge base coupled with the facts in working memory (i.e., the application form). At this point, a loan investigator might well conclude that the loan should be denied. However, Dan has learned that a personal interview is often of prime importance to his decision (and this is yet another heuristic rule in his memory, or knowledge base). Thus, he decides that even though the situation appears bleak, it may be worthwhile to talk directly with Pete.

During the interview, Dan asks a number of carefully calculated questions; questions that are intended to draw out additional information from the applicant. However, Dan has also learned, through his experience, to first relax the applicant by making *small talk*. In doing so in the past, he has found that quite often some unexpected, but pertinent information may result. In fact, when chatting with Pete, Dan discovers that he is married to the daughter of the vice president of the XYZ Corporation. This vice president, in turn, is a major shareholder and very securely positioned in the company. Dan concludes that Pete's loan should be granted. He also adds a new heuristic to his knowledge base; that is, if the applicant is married to the child of one of the executives of the firm at which the applicant is employed, then the loan application is to be given additional weight.

CONSIDERATIONS IN KNOWLEDGE REPRESENTATION

In the above example, a large part of Dan's knowledge base (i.e., including some portion of the deep knowledge available in the public domain) is retained within his memory. Somehow, some way, the facts and heuristic rules that he employs have been transformed into a format that can be both stored and retrieved from his brain. Investigators in artificial intelligence and psychology (among others) are keenly interested in just how this is achieved, and several theories have been proposed.

Our concern, however, shall be more immediate and far more pragmatic. We wish to store the knowledge base of an expert system in the memory of a digital computer, using means presently at hand. We are not, in fact, really concerned about whether or not the method of storage we employ has some analogy to the manner in which knowledge is stored in the human brain. We simply seek a fast, effective procedure for use with the computer. This is not at all meant to imply that we should consider the study of the brain and human memory to be unimportant. It most definitely is; but until that study provides results that may be used to refine or replace existing methods, we shall simply remain interested observers.

Thus, our major concerns deal with how to represent the facts and rules within the knowledge base to

- Provide a format compatible with the computer.
- Maintain as close as possible a correspondence between this format and the actual facts and rules (i.e., the rules as they are perceived by the domain expert).
- Establish a representation that can be easily addressed, retrieved, modified, and updated.

Elaborating further on the last two points, it would be highly desirable to use a format that is *transparent*, that is, a representation scheme that may be easily read and understood by humans.

Several modes of knowledge representation have been proposed. As we have emphasized, the primary focus will be on rule-based systems for knowledge representation, since it is through this process that the knowledge bases of the expert systems to be described in this text will be developed. However, before we cover rule-based systems in detail, let us first discuss certain alternative modes of knowledge representation.

ALTERNATIVE MODES
OF REPRESENTATION

In this section, briefly described will be the pertinent features of such modes of knowledge representation as OAV (or object-attribute-value) triplets, semantic networks, frames, logic programming, and neural networks. To begin, OAV triplets will be addressed. Not only are they a mode of knowledge representation in themselves, they also form the building blocks of virtually any other approach to knowledge representation.

OAV Triplets

Object-attribute-value triplets provide a particularly convenient way in which to represent certain facts within a knowledge base and may be extended (as we shall see) to provide the basis for the representation of heuristic rules. Each OAV triplet is concerned with some specific entity, or object. For example, our object of interest might be an airplane. Associated with every object is a set of attributes that serve to characterize that object. Using the airplane as an example (i.e., as the object), some of its attributes include the following:

- Number of engines
- Type of engine (e.g., jet or prop)
- Type of wing design (e.g., conventional or swept back)

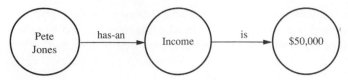

FIGURE 4.1
OAV network.

For each attribute, there is an associated value, or set of values. For instance, in the case of the C130 military cargo aircraft (known as the Hercules), the number of engines is four, the type of engine is prop, and the wing design is conventional. Notice in particular that values in OAV triplets may be numeric or symbolic. We may list these facts as shown below:

- Number of engines = 4
- Engine type = prop
- Wing design = conventional

Observe that, in this list, the object itself (i.e., the C130 aircraft) is never explicitly stated. Actually, the above statements represent AV (attribute-value) pairs. However, associated with any AV pair is some object. Thus, any AV pair implies an OAV triplet.

Yet another way to represent an OAV triplet would be through the use of a network representation as indicated in Fig. 4.1. The basic building blocks of a network are its nodes (i.e., the circles) and branches, or edges (i.e., the lines connecting two nodes). In Fig. 4.1, the object is Pete Jones, the attribute is his income, and the specific value of his income is $50,000.

Semantic Networks

A semantic network may be thought of as a network that is composed of multiple OAV triplets in network form as illustrated in Fig. 4.1. However, rather than pertaining to just one attribute for a single object, semantic networks may be used to represent *several* objects, and *several* attributes per object. Returning to our aircraft illustration of the previous section, we might develop a partial semantic network as illustrated in Fig. 4.2. Here, we note that the C5A is a special type of aircraft (i.e., a large military cargo plane). Further, since the C5A is an aircraft, it *inherits* the properties associated with aircraft in general (e.g., it flies, has wings, carries people). Such an inheritance property can prove to be of considerable value in the reduction of memory storage requirements. That is, since a C5A is an airplane, there is no need to store, at the C5A node, the fact that it can fly, has wings, and can carry people. Thus, the semantic network scheme provides for a convenient approach for the representation of *associations* between entities.

We might also note that the OAV triplet is actually just a restricted subset of semantic networks wherein the only relationships that may be used are those of "is-a" and "has-a." OAV nodes, in turn, may be any of three types: objects, attributes, or values.

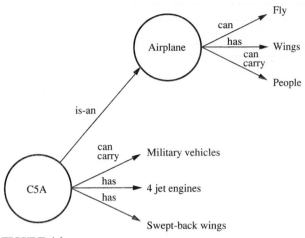

FIGURE 4.2
Semantic network.

Frames

While semantic networks provide a relatively versatile means for knowledge representation, the use of frames represents an alternative approach that serves to capture most of the features of the semantic network while providing certain additional aspects. In fact, we may think of a semantic network as being a subset of the concept of frames.

The employment of frames represents a particularly robust way in which to present knowledge. A frame contains an object plus *slots* for any and all information related to the object. The contents of such slots are typically the attributes, and attribute values, of the particular object. However, in addition to storing values for each attribute, slots may contain default values, pointers to other frames, and sets of rules or procedures that may be implemented.

Figure 4.3 illustrates a frame-based representation for the object *dog*. Note that the slots within this frame include values (e.g., Beagle), defaults (e.g., four legs), and procedures (e.g., for a medical examination). The procedures, in turn, could well point to other frames. The versatility of the frame-based mode of knowledge representation should be obvious.

The primary drawback to the use of frames is, ironically, caused by the very robustness of such a mode of representation. Frames have so many capabilities as to make their use a rather complex matter. Jackson [1986] states that "many people are unhappy with frame- and object-based systems because they seem to depart from logic and because their flexibility in matters of context and control can make their behavior both hard to predict and difficult to understand." As a result, to obtain any reasonable proficiency in the use of frame-based tools in expert systems, a lengthy training period is required. Despite such drawbacks, frames can prove quite useful, if not essential, in the design of large-scale, complex expert systems—particularly those involving a large amount of a priori facts (i.e., data)

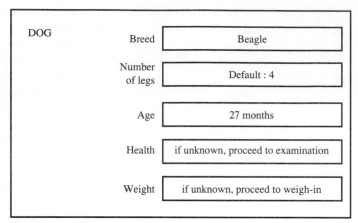

FIGURE 4.3
A frame-based representation.

and multiple objects. While frames are not focused on in this text, it is strongly encouraged that the serious student investigate this topic—*after* he or she has attained a reasonable level of competence in the use of rule bases.

Representation via Logic Statements

The most common form of logic is that known as *propositional logic*. A proposition, in turn, is a statement that may be either true or false. Propositions may be linked together with various operators (termed logical connectives) such as AND, OR, NOT, and EQUIVALENT. Linked propositions are termed compound statements. To demonstrate, consider the statements X, Y, and Z where the first two statements (X and Y) are true while Z is false. Thus we may conclude that

- X AND Y is true and X AND Z is false. If two statements are connected by AND, both must be true for the compound statement to be true.
- X OR Y is true and X OR Z is also true. If two statements are connected by OR, that statement is true as long as either one or both statements are true.
- NOT Z is true. The NOT connective simply negates the statement. Since Z was false, NOT Z must be true.

Predicate calculus represents an extension of propositional logic. The fundamental elements of predicate calculus are the object and the predicate. A predicate is simply a statement about the object, or a relationship that the object possesses. Predicates may address more than one object and may be combined by use of logical connectives. Some examples of the use of predicate calculus follow:

- Mammal(dog), which is read as "a dog is a mammal"
- Four_legs(dog), which is read as "a dog has four legs"
- Mammal(chicken), which is read as "a chicken is a mammal"
- Sister(joan, jack), which is read as "Joan is Jack's sister"

The first of these statements is true, the second is in general true, and the third is definitely false. Unless we know Joan and Jack personally, the validity of the fourth statement is unknown. Using predicate calculus, we can then represent such compound statements as "Joan is Jack's sister and Fred's cousin" as

Sister(joan, jack) AND cousin(joan, fred)

We can also represent various relationships, or rules, by means of predicate calculus. For example, consider the heuristic that states that, if interest rates are rising, bond prices will fall. Using predicate calculus, and the logical IF statement, we can represent this heuristic as

fall(bond_prices) if
    rise(interest_rates)

where we have used a format similar to that employed in PROLOG, a computer programming language popular in artificial intelligence.

One major advantage of the use of logic for knowledge representation is that logic-based languages, such as PROLOG, do exist. However, such languages have been criticized for a certain lack of flexibility; a criticism that is becoming less valid with recent enhancements to their procedures. A more immediate and pragmatic drawback of the use of logic for knowledge representation is the fact that one must learn some logic programming language (e.g., PROLOG) in order to develop expert systems.

Neural Networks

Obviously, somehow, some way, the human brain stores knowledge. What is not so obvious is the precise manner in which this is accomplished. Neural networks represent mankind's attempt to replicate, in hardware, theories pertaining to the brain. Specifically, it is thought that knowledge is stored in neurons (or, actually, in the connections between neurons). Figure 4.4 depicts a simplified representation of only two neurons within the neural network of the human brain.

In the human brain there are more than 10 billion neurons, and each neuron is connected to one or more other neurons, resulting in a massively interconnected network. At each neuron, impulses are received by the dendrites and transmitted by the axons. If the output of the axon is at a *high-enough* level, the signal will *jump* the synaptic junction and trigger the connected neuron (or neurons). It is believed that knowledge might then be represented by the weightings on each neuron to neuron interconnection, which in turn influence the level of strength of the interconnecting impulses.

The attempts to duplicate the neural network structure of the brain have been, at best, extremely modest. Typically, electronic amplifiers are used to represent the neurons and resistors to correspond to the interconnecting weights. And existing systems have but a few layers of relatively few neurons. Despite this, neural networks can be used to accomplish some intriguing tasks, including some

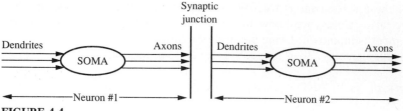

FIGURE 4.4
A portion of a neural network.

success in speech recognition. In particular, they provide a robust approach to the general problem of pattern recognition. Probably the biggest single disadvantage of the neural network approach to knowledge representation is the fact that any knowledge that exists is almost totally opaque.[2]

Since neural networks are often excellent choices for problems of classification, they may be combined with an expert system to perform certain tasks. That is, the neural network may be used to classify and, based upon this class, the expert system may then be used to determine the specific course or courses of action to take. Such combinations are termed, in this text, *hybrid expert systems* and will be addressed in detail in Chap. 9.

REPRESENTATION VIA RULE-BASED SYSTEMS

Undoubtedly, the most popular mode of knowledge representation within expert systems, at least at this time, is the mode obtained through the use of rules, or rule-based systems. Alternatively, such rules are referred to as IF-THEN, or production rules. We have selected rule-based expert systems as our approach to knowledge representation for a number of reasons, including their popularity and widespread use. However, it should be stressed that this decision does not imply that rule-based systems are necessarily the *best* approach—or, in particular, the best approach for every situation. There are those who present quite persuasive arguments for other approaches; in particular for the employment of frame-based representation. Our choice of rule-based knowledge representation has been made for the following reasons:

- The majority of existing expert systems development packages (and this is especially true for expert systems shells) employ rule bases.
- Rule-based expert systems development packages are normally much less expensive (in terms of both the initial cost of the package as well as the overall cost of using the package) than those employing alternative modes of representation.

---

[2]One of the major areas now being addressed and funded in neural network research is, in fact, the development of rules for representation of the process employed by the neural network.

Specifically, they cost less to purchase, normally do not require any expensive hardware (most run on inexpensive, general purpose personal computers), and require minimal expenditures toward training.[3]

- The widespread availability of rule-based expert systems shells permits the knowledge engineer to focus his or her attention on the most critical phase of the development of an expert system, that is, on the knowledge base.
- Rules represent a particularly natural mode of knowledge representation. Consequently, the time required to learn how to develop rule bases is minimized.
- The learning curve for rule-based expert systems is much steeper than for any alternative mode of representation (i.e., it takes less time to learn how to use and implement rule-based expert systems).
- Rules are transparent, and are certainly far more transparent than the modes of knowledge representation employed by rule-based systems' two major competitors: frames and neural networks. Further, such transparency often leads to an increased willingness, on the part of management, to accept the solutions obtained. And the importance of this last factor should not be underestimated.
- Rule bases can be relatively easily modified. In particular, additions, deletions, and revisions to rule bases are relatively straightforward processes. And this is particularly so in the case of well-designed rule bases.
- Rule-based expert systems can be employed to mimic most features of frame-based representation schemes (the amount of work that may be needed to accomplish this, however, is not necessarily trivial).
- Validation of the content of rule-based systems (i.e., the determination of the completeness and consistency of the representation) is a relatively simple process. Similar validation of frames or neural networks, on the other hand, is normally difficult to impossible.

Finally, it is our opinion that the rule-based approach provides the most appropriate introduction to the topic of expert systems. With the rationale for our selection of rule-based knowledge representation behind us, let us now address the topic of the development of rule bases.

An Assumption: Deterministic Rule Bases

For the discussion to follow, and for all material through and including Chap. 6, we shall assume that our rule bases are deterministic. That is, we assume that there is no uncertainty with regard to either the facts used (e.g., those input from

---

[3]The costs of training are all too often grossly underestimated by firms considering the in-house development and maintenance of expert systems. In particular, the costs associated with the employment of frame-based systems can be on the order of ten to a hundred times that of the initial investment in the system. Such expenditures may be minimized through the employment of rule-based expert systems.

the user) or the conclusions reached. We shall then deal with uncertainty in rule bases in Chap. 7.

Production Rules: An Overview

Rule-based modes of knowledge representation employ what are termed *production rules* or, for short, simply *rules* [Fikes and Kehler, 1985; Hayes-Roth, 1985]. Such rules are typically of the IF-THEN variety. However, in some instances this is extended to include IF-THEN-ELSE rules. For example, we might have the IF-THEN-ELSE rule as shown below:

Rule 1: *If* the student's GRE score is 1350 or more
       Then admit the student to the graduate program
       Else, do not admit the student

Which is equivalent to two IF-THEN rules or,

Rule 1*a*: *If* the student's GRE score is 1350 or more
       Then admit the student to the graduate program

Rule 1*b*: *If* the student's GRE score is less than 1350
       Then do not admit the student

For clarity of presentation, we shall focus primarily on just IF-THEN rules. In fact, it is generally advisable to avoid the use of ELSE statements in rule-based expert systems. This is true for three reasons. First, a number of commercial expert systems development packages simply do not permit the use of IF-THEN-ELSE rules. Second, validation of such rules is considerably more difficult than for their IF-THEN equivalents. Third, when encountered in the inference process, such rules will tend to always reach a conclusion (i.e., either the one in the THEN portion of the rule or the one in the ELSE portion). This can result in some unanticipated results. Thus, whenever one comes upon such a rule, we strongly advise the formation of the corresponding two equivalent rules.

    An alternate designation for IF-THEN rules is that of *condition-action* or *premise-conclusion* statements. The reason for this terminology should be obvious from the above example. That is, given the condition (or premise) that a student has a GRE of 1350 or better, we take the action (or arrive at the conclusion) of admitting the student to our graduate program. In this text, we shall refer to the IF statement as the premise and to the THEN statement as the conclusion.

    We may also note that production rules contain (or imply) OAV triplets. Returning again to our previous example and, specifically, to rule 1*a*, we may note that there is one OAV triplet implied in the IF, or premise, portion of the rule:

    Object = student
    Attribute = GRE score
    Value = 1350 or above (i.e., the value to be tested against)

and another OAV triplet is implied in the THEN, or conclusion, portion:

Object = student
Attribute = admission status
Value = yes (i.e., the value to be assigned)

It is particularly important to notice the distinction between the values listed in a rule premise and those listed in a conclusion. In a rule premise, we are *testing* (or comparing) the value in the statement with any value provided. For example, in the premise to the above rule, we would test the student's actual GRE score value against the value of 1350 (or larger). However, in the conclusion, we are *assigning* a value to an attribute. Referring again to the above rule, if the student's GRE score is 1350 or above, we may then assign the value of *yes* to the attribute *admission status*.

We should also realize that there may be several premise and conclusion statements within a single rule. Each of these are termed *clauses* (i.e., premise clauses and conclusion clauses). For example, the rule below has two premise clauses and two conclusion clauses:

Rule A: *If* the condition of the sky is clear
         and the outside temperature is low
         Then the chance of frost is increased
         and outdoor plants should be covered

Another rule with multiple clauses in the IF and THEN portions is demonstrated by rule B.

Rule B: *If* the age of the car is new
         or the condition of the car is good
         Then the car should start
         and the trip should be safe

Further, note that while premise clauses may be connected by AND as well as OR operators, the conclusion clauses may *only* be connected by AND statements. That is, *all* of the conclusion clauses in a production rule must be true.

Clauses connected by AND operators are denoted as *conjunctive* clauses. Those connected by OR operators are termed *disjunctive* clauses. When we refer, in this text, to the premise of a rule, we are referring to the complete set of premise clauses—in whatever manner they may be connected. The same is true of a reference to the conclusion of a rule, except that in this case the only acceptable statements are either a single conclusion clause or a set of conjunctive clauses.

Attribute-Value Pair Properties

As noted, each premise and conclusion clause contains attributes and values. Further, there must be an associated object, either implied or explicit. Consider, for example, the rule shown on the next page:

Rule 1:   *If* grade point average (GPA) equals or exceeds 3.5
          Then accept into honor society

In this rule, the attribute for the premise clause is *grade point average*, and the value to be tested against is *3.5 (or greater)*. The object has not been specified, but is implied by the rule and/or the particular situation under consideration. That is, the implied object for the premise clause is a *student*. The same object just happens to be implied in the conclusion to this rule. As we discussed earlier, when clauses contain only attributes and values, as in the case of the rule under discussion, they are sometimes called attribute-value, or AV, pairs.

In the conclusion clause, the attribute-value pair is *accept* and *into honor society*. Actually, this is a poor choice of wording for this conclusion. A better conclusion clause might be: "Then student's acceptance status is yes." While the English may not be quite as natural as before, this restatement permits us to more clearly isolate the attribute and the value. In the restatement, the attribute is clearly *student's acceptance status* while the value to be assigned is *yes*. In general, rules should be written so that identification of the attribute and value is straightforward—while the rules remain intelligible.

The AV pair is the fundamental building block of a premise or conclusion, and thus the fundamental building block of a production rule. Associated with each AV pair is a set of properties. The most typical of these are summarized below:

- Name (the name of the attribute)
- Type (the class of values associated with the attribute, that is, symbolic or numeric)
- Prompt (the query presented to the user, when necessary)
- Legal values (the set of acceptable values for the attribute)
- Specified values (one or multiple)
- Confidence factors (i.e., as associated with the attribute value, or values)

Each of these properties should be defined before we proceed further.

NAME. The name of the attribute is simply the wording selected to identify the attribute of the (explicit or implicit) object associated with the clause under consideration. For example, some of the attributes typical of an automobile are color, number of doors, model, horsepower, number of cylinders, and mileage. Be careful that it is an attribute that has been identified, not the object. That is, "the car = red" should be replaced by a statement such as "the color (of the car) = red."

TYPE. Attribute values may be either numeric or symbolic. For example, the temperature of a patient may be given in degrees Fahrenheit—a numeric value. Alternatively, we might specify the temperature values to be symbolic, such as *high* or *normal*. Yet another set of symbolic values would be *yes* and *no*, for example, such as with respect to the presence or absence of some feature.

PROMPT. Associated with *certain* attributes are user prompts, or queries. When necessary, the user replies to this prompt with a value for the attribute under consideration. For example, the user (a medical doctor) may be asked: "What is the temperature of the patient?" Notice carefully that only certain attributes should be assigned prompts. Specifically, the only attributes that should normally be provided with prompts are

- Attributes that appear in a premise statement and never appear in any conclusion statement of the rule set
- Attributes for which the user can conceivably provide a response

The rationale for the second guideline should be obvious. However, for the first, note that there is no need to prompt the user for the value of an attribute if that attribute ever appears in a conclusion clause. This is because the inference strategy may be employed to deduce the value for that particular attribute—and we would prefer to let the expert system do the work rather than the user.

LEGAL VALUES. Associated with every attribute is a set of legal, or acceptable, values. For example, the legal values for a person's weight would simply be the set of nonnegative real numbers (and we might further limit these to some realistic subset). The legal set of values for the color of a particular type of flower might be red, white, yellow, and pink. Some expert systems shells include a check of these values. Thus, if the user replies with a nonlegal value, this is detected and the user may be asked to reply again (with an acceptable value). In the case of expert systems that provide menu-driven prompts, the set of legal values is simply presented to the user and he or she can only select from that list.

SPECIFIED VALUES. Legal values simply represent the complete set of acceptable values for a given attribute. Specified values, on the other hand, indicate the actual set of values that are either to be tested against (i.e., in a premise clause) or that will be, or have been, assigned (i.e., in a conclusion clause). More specifically, we are concerned with whether or not *multiple* specifications are permitted. Consider first the use of multiple values for an attribute in a premise clause. As an example, in the case of a flower, the legal values may be red, white, yellow, and pink. However, some flowers (or arrangements of flowers) may be multicolored. Thus, if multiple values are permitted (and this is a function of the expert systems shell used), the user could reply with more than one value. To illustrate further, consider the user prompt shown below:

```
Query:  Indicate the color of the flower:
        1. Red
        2. White
```

```
3. Yellow
4. Pink
Input Choice(s): _____
```

This particular prompt has been presented in menu format. If the associated expert systems package permits, the user may respond with, say, 1 *and* 2, indicating that the color of the flower is red *and* white. Thus, color has *two* values attached to it.

Multiple values may also be allowed (where, again, this is dependent upon the particular software package employed) for attributes that appear in conclusion clauses. In other words, it may be permitted to assign (i.e., conclude) multiple values to the attribute in a conclusion clause. For example, an expert system may have been designed to provide advice on vacation planning. Thus, it might make sense to provide more than one suggested vacation location to the user. Consequently, for the single conclusion attribute *suggested location*, we might have England, Germany, and France.

CONFIDENCE FACTORS. If the expert systems development package permits, we may deal with uncertainty in either conclusions (i.e., the conclusion attribute values assigned) or premises (i.e., the premise attribute values used). Since we shall not deal with uncertainty and confidence factors until Chap. 7, we shall merely note that this too is an AV pair property.

In order to clarify the above discussion, let us consider some further examples. To begin, let us examine rule 1, as presented earlier and restated below:

Rule 1: *If* student's GPA exceeds 3.5
  Then student's acceptance status is yes

Referring to the AV pair of the premise of this rule, the associated properties are

- Name: student's GPA
- Type: numeric
- Prompt: "What is the GPA for this student?"
- Legal values: all numbers from 0 to 4 (i.e., where A = 4.0)
- Specified values: single (i.e., a given student has only a single GPA)
- Confidence factors: none

If we examine the conclusion clause of this rule, the properties of the AV pair are

- Name: student's acceptance status
- Type: symbolic
- Prompt: none (recall that no prompts are provided for attributes that appear in conclusion clauses)

- Legal values: yes or no
- Specified values: single (i.e., the student is either accepted or rejected)
- Confidence factors: none

As one more example, let us consider the properties of the rule listed below:

Rule A: *If* client is risk adverse
        Then invest in blue chip stocks

Once again, we suggest that the wording of this rule would benefit from change. Specifically, in the premise clause we might think that the attribute is *client*. But what if there exists some other rule within this knowledge base that checks the client's age, marital status, and so on. Actually, the client is the object, not the attribute. The attribute, in this case, might be designated as *client's risk profile*. Turning to the rule conclusion, the attribute is *invest* but would be better described as *client's investment strategy*. Let us then rewrite rule A as:

Rule A': *If* client's risk profile is risk adverse
        Then client's investment strategy is blue chip stocks

Alternatively, we may drop the word *client's* from the premise and conclusion to shorten the wording. Using the latter form, the property list for the AV pair of the premise is

- Name: risk profile (or, client's risk profile)
- Type: symbolic
- Prompt: "What is the client's risk profile?"
- Legal values: risk taker, risk adverse, neutral, unknown
- Specified values: single
- Confidence factors: none

And, for the AV pair of the conclusion clause, we have

- Name: investment strategy (or, client's investment strategy)
- Type: symbolic
- Prompt: none
- Legal values: junk bonds, growth stocks, certificates of deposit, blue chip stocks
- Specified values: multiple
- Confidence factors: none

Note the ability to use *unknown* as a legal value for the AV pair of the premise clause. Further, for the AV pair of the conclusion, multiple values are allowed—and thus multiple conclusions may be reached.

There is one further point to be addressed with regard to attributes and their values. Consider the rule below:

Rule A: *If* voltage = 12
            and terminals = clean
            Then battery status = alright

Note that we have employed the use of the equals sign. Now, depending upon the expert systems package used, equals signs can mean different things. For example, in the first premise clause of the above rule, we may either be testing the value of voltage against the *number* 12—or the *string* 12. That is, 12 may be considered either a number or a symbol, depending on the software package. In some instances, we distinguish between numbers and symbols by placing the numbers in parentheses (or brackets), as follows:

$$If \text{ voltage } = (12)$$

In other cases, the attribute is placed within parentheses, or brackets, if it is one that has a mathematical (rather than symbolic) value. This would be written as

$$If \text{ [voltage] } = 12$$

In any case, check with the software package being used whenever equals signs are used.

Clause Properties

As we have discussed, there are two types of clauses: *premise clauses* and *conclusion clauses*. Other properties associated with clauses are summarized in the list below:

- Single versus multiple (or compound) clauses
- Conjunctive versus disjunctive (multiple) clauses
- Free (premise) clauses
- Specified (premise) clauses (i.e., specified true or specified false)

Let us examine each of these properties in turn. First, a premise or conclusion may consist of a single clause or a set of clauses. In the latter instance, we are said to have multiple clauses.

Multiple clauses, in turn, may be either conjunctive clauses (each clause connected by the AND operator) or disjunctive (each clause connected by the OR operator). However, recall that disjunctive clauses are not permitted in the conclusion of a rule. Also, note that the premise of a rule may be quite complex. For example, consider the rule shown below:

Rule interview: *If* the applicant's education level is college graduate
                    or applicant's experience level is more than 10 years

    and applicant's references are excellent
     or applicant's references are good
   Then applicant's interview status is yes

In this rule, we are simply evaluating the applicant to determine whether or not he or she should be invited for an interview. To be invited, the applicant must either be a college graduate or have at least 10 years of experience. Further, the applicant must have references that are considered either excellent or good.

  Another property of a clause is that associated with premise clauses only. This is the property of being either free or specified, and if specified, of being either true or false. If the value of a premise clause attribute is not yet known, that clause is designated as a *free clause*.[4] Note most carefully that we have drawn a distinction between *not yet known* and *unknown*. We shall describe the implications of this distinction in the material to follow. If a clause is not free (i.e., the value of its attribute is known, or specified), then such a clause is either *true* or *false*. To illustrate these notions, consider the following simple premise clause shown below:

$$If\ A\ =\ X$$

Now A must be some attribute for the object about which the clause is concerned. X is then one possible (legal) value for this attribute, and we must test this clause to see if A does indeed equal X. If we do not know the value for A, and have yet to seek this value, the clause is free. However, if we do know the value for A, and this value is indeed X, then the clause is true. Otherwise (i.e., if the value of A is known but is something other than X), the clause is false. Summarizing as follows:

- If the value of a premise clause attribute is not yet known, and has not been sought, then that clause is termed a *free clause*.
- If the value of a premise clause attribute is known, and the clause is satisfied, then that clause is termed a *true clause*.
- If the value of a premise clause attribute is known, and the clause is not satisfied, then that clause is termed a *false clause*.

  To further illustrate these particular properties, let us use the simple knowledge base listed below:

Rule 1: *If* maximum hotel daily rates $\leq$ \$75
    and desired location is downtown
    Then hotel selection = Fred's Beds

---

[4]More precisely, a premise clause is free if the value for the attribute of that clause is not known *and* that value has not yet been sought.

Rule 2: *If* maximum hotel daily rates > $75
or bed types = king-size
Then hotel selection = King's Castle

Initially, that is, before we consult this rule base, all attribute values are not yet known and thus all premise clauses are free. (Further, all conclusion clauses are unassigned.) Now, if we consult the knowledge base and, in response to user prompts, reply that maximum daily charges are to be less than $75, the first premise clause of the first rule is true while the first premise clause of the second is false. Next, we might reply that we desire king-size beds. In this instance, the only clause condition that changes is that of the second premise of the second rule. Specifically, that clause is now true.

The properties of free, true, or false would seem to be straightforward. And indeed they are; however a certain degree of confusion may occur when one employs *unknown* as an attribute value. Note carefully that we must differentiate between *unknown* and *not yet known*. *Not yet known* means that the value for a respective attribute has not yet been determined (i.e., we have yet to ask the user for the value—or to attempt to infer a value). Thus, the associated clause is free.

Unknown, however, can be employed in one of two ways. First, it may simply be a legal value for a given attribute. Consider, for example, the rule stated below:

Rule unknown: *If* the patient's temperature is unknown
Then next activity is to take patient's temperature

Let us then assume that, in response to a user query concerning the value of the patient's temperature, we respond with the value *unknown*. All this means is that the premise clause is true, and the rule is triggered.

The second manner in which *unknown* may be employed is slightly more complex, and a function of the specific mode of inference used by the software package. In this case, a value of *unknown* is assigned to an attribute whenever its value cannot be determined from the inference procedure. To illustrate, consider the following rule from a rule set used for the maintenance of certain equipment. Further, let us assume that the complete set of legal values for pressure reading are *low*, *nominal*, and *high*.

Rule pressure: *If* pressure reading = low
Then examine casing for cracks

Let us assume that, during the inference process, no value could be assigned to the attribute *pressure reading*. Some inference procedures will then automatically assign a value of *unknown* to pressure reading—and thus the premise clause of rule pressure is then *false*. Other inference procedures will simply leave the premise clause free.

Rule Properties

As with AV pairs and clauses, there are certain important rule properties. Some of the more typical rule properties include

- Name (the name of the rule)
- Premise (the IF portion of the rule)
- Intermediate conclusion (the THEN portion of the rule)
- Conclusion (the THEN portion of the rule)
- Notes (notes associated with the rule)
- References (formal references associated with the rule)
- Confidence factors (a measure of confidence in the rule's conclusion)
- Priority (a property used by the inference process)
- Cost (a property used by the inference process)
- Chaining preference (the *normal* or default mode of search used by the rule)
- Status: active, inactive/discarded, triggered, fired

Let us address each of these properties in turn.

NAME. Each rule should have a distinct, as well as descriptive name. Actually, in this text, we have not always followed this guideline. However, it is a good idea to do this when building any actual knowledge base. Specifically, rather than just labeling a rule by a number or letter (e.g., rule 1, rule A, and so on), it is best to label the rule with a name that serves to concisely describe the contents and/or purpose of that rule.

PREMISE. Every rule consists of one or more premise clauses. The complete set of premise clauses is termed the rule premise. A rule premise may consist of conjunctive or disjunctive clauses. When we refer to the status of a rule premise, realize that this is a function of the status of the *collection* of individual premise clauses.

INTERMEDIATE CONCLUSIONS AND CONCLUSIONS. Every rule consists of one or more conclusion clauses. In the case of multiple conclusion clauses, the clauses must be conjunctive. There are, in turn, two types of rule conclusions: intermediate conclusions and (final) conclusions. An intermediate conclusion is one that is the conclusion clause of one rule while also serving as a premise clause for another. A (final) conclusion clause is one that does not appear as a premise clause for any other rule. Consider, for example, a rule base used to classify animals. Some rules may have final conclusions such as "Then the animal species is lion." Others may have intermediate conclusions, used to filter out various possibilities. For example, the (intermediate) conclusion of another rule may be

"Then the animal genus is a mammal." Two representative rules, for this situation, are listed below:

Rule mammal: *If* the animal's birth process = gives live birth
                Then the animal genus = mammal

Rule lion: *If* the animal genus = mammal
                and animal color = tawny
                and animal's eating habit = carnivorous
                Then animal species = lion

Note that the conclusion to *rule mammal* is an intermediate conclusion since it serves as a premise clause for *rule lion*. The conclusion to *rule lion*, on the other hand, is a final conclusion for this particular rule set.

NOTES AND REFERENCES. It is essential that a rule base be documented. While you, the developer, may know the reason and source of the rules, others will not. Further, with the passage of time, even the developer will find it difficult to recall the origin and specifics of each rule. Many development packages permit the inclusion of notes and references, and this is a feature that should most definitely be employed in any actual knowledge-base development.

(RULE) CONFIDENCE FACTORS. When uncertainty is employed, we may associate confidence factors with each rule. As mentioned, this feature will be described in Chap. 7. Here, we may simply note that the confidence factor of a rule's conclusion is a function of the confidence factors of the rule and the rule premise.

PRIORITY AND COST. In some development packages, we are permitted to assign a priority and/or cost to each rule. Such properties are normally employed as a means to decide, during the inference procedure, the specific rule to be dealt with at a particular instance. Typically, the procedure will select the rule with the highest priority or the lowest cost.

CHAINING PREFERENCES. As we shall see, the inference process involves a search procedure. In some cases, the search moves in a *forward* direction—from premises (or facts) to conclusions. In others, the search moves *backward*—from a hypothesized conclusion to the premises necessary to infer that conclusion. However, in addition to such *normal* modes of search, or chaining, some development packages permit the employment of a mixture of search methods. In such instances, we might label rules according to their preferred or default method of chaining, either backward or forward. This feature will be illustrated in detail in Chap. 6 when we address mixed modes of chaining.

RULE STATUS. During consultation, the status of each clause and rule is subject to change. Keeping track of such changes is an essential part of the inference process. We shall describe that process later; first, we need to become acquainted with the terminology used. A summary of this terminology and associated definitions is provided below:

- The premise of a rule is true whenever a test has been made and it has been determined that the premise has been satisfied.
- The premise of a rule is false whenever a test has been made and it has been determined that the premise has not been satisfied.
- If the premise of a rule is true then that rule is said to be *triggered*.
- If the premise of a rule is false then that rule may be *discarded* or, in some cases, made *inactive*.
- If a rule is *fired* then this implies that the action implied by the conclusion clause(s) is taken. The values associated with each attribute of the conclusion clauses for this rule are said to be assigned.
- A rule that has been fired is no longer active. It is either discarded or, in some cases, made inactive.
- If a rule is to be fired, that rule must first have been triggered.
- If a rule has been neither fired nor discarded, that rule is designated as being *active*.

Example 4.1. In an attempt to clarify the above discussion, let us turn to an example of aircraft identification. A small portion of the rules that might be employed in the identification of four specific aircraft are listed below. Let us, for the moment, ignore just how such rules were developed (and whether they are valid, and if valid, whether they are the most efficient representation available). These are matters that will be dealt with later.

Rule C130: *If* engine type is prop
 Then plane is a C130

Rule B747: *If* engine type is jet
 and wing position is low
 Then plane is a B747

Rule C5A: *If* engine type is jet
 and wing position is high
 and bulges on fuselage are absent
 Then plane is a C5A

Rule C141: *If* engine type is jet
 and wing position is high
 and bulges on the fuselage are present aft of the wing
 Then plane is a C141

Now, let us assume that a plane has just been observed and we have been told (only) the following facts:

- The engines on the plane are jet engines.
- The wing position is high.

In this case, the premise of rule C130 is false and thus rule C130 is discarded. The premise of rule B747 is also false (notice that the premise clauses are connected by the AND operator and thus if one premise clause is false the entire premise of the rule is false) and it too is discarded. Further, we do not have enough information to determine if the premise of rules C5A and C141 are true or false (i.e., the status of the third premise clause for both rules is still free). However, should we also be told that there are no bulges on the fuselage of the aircraft, we may discard rule C141 (i.e., the premise is now false) and trigger rule C5A. Finally, since rule C5A is triggered, it may be fired. Firing rule C5A simply means that we draw the conclusion that the observed aircraft is indeed a C5A.

Example 4.2. As another illustration, consider the rules listed below:

Rule 1: *If* A = X
  or B = Y
  Then D = XX

Rule 2: *If* A = X
  and C = Z
  and B = NOT Y
  Then D = YY

Let us assume, for sake of illustration, that A = X, B = NOT Y, and C = NOT Z. From these facts, we may note that the premise of rule 1 is true. That is, since the premise clauses are connected by OR, only one premise clause need be true for the entire premise to be true. Rule 1 is thus triggered. In the case of rule 2, the premise is false (because C = NOT Z) and thus that rule may be discarded. Finally, since rule 1 is triggered, it may be fired which leads to the conclusion that D = XX.

Rule Conversion: Disjunctive Clauses

In Example 4.2, we should have observed just how important it is to be careful when evaluating the conditions of a set of premise clauses that are connected by the OR operator (i.e., rule 1 in Example 4.2). We could avoid this by *simplifying* rule 1 of that example. That is, rule 1 may be represented by two rules which, together, are equivalent to rule 1. These rules are shown below.

Rule 1*a*: *If* A = X   Rule 1*b*: *If* B = Y
  Then D = XX   *Then* D = XX

While such conversions are not necessary in the general methodology of expert systems, they often make it easier for the beginner to follow the inference process of an expert system when manual demonstrations are employed.[5] Further, some expert systems development packages do not permit the use of disjunctive premise clauses. However, such a conversion does result in an enlargement of the number of rules necessary to represent the knowledge base of an expert system.

---

[5] Also, as we shall see later, the use of only conjunctive clauses can serve to make rule checking considerably simpler.

Despite this, the beginner may be well advised to consider such a conversion—as well as determine the restrictions of the software that is to be used.

Multiple Conclusions

Now, if you have really been paying close attention, you should have noticed a potential problem with the rules in Example 4.2. That is, *both* rules will be true whenever

$$A = X$$
$$C = Z$$
$$\text{and } B = \text{NOT Y}$$

and this means that both would be triggered in this instance. If we were to fire both rules, we would conclude that

$$D = XX \quad \text{and} \quad D = YY$$

which can create some definite problems. For example, if D is replaced by *aircraft type*, XX is replaced by C141 and YY is replaced by C5A, we have drawn the conclusion that the plane is simultaneously a C141 and a C5A, and this of course is impossible.

We must stress, however, that it may be quite reasonable for an expert system to draw multiple conclusions—and this is particularly so if we are dealing with uncertainty. That is, if the rules in Example 4.2 were modified to read as follows:

Rule 1: *If* A = X
     or B = Y
     Then D = XX (cf = 0.3)

Rule 2: *If* A = X
     and C = Z
     and B = NOT Y
     Then D = YY (cf = 0.6)

And if cf is used to represent the confidence factor for a rule (i.e., some rough measure as to the plausibility of the conclusion drawn by the rule), then we very well might find that both rules fire, concluding that

$$D = XX \quad \text{with} \quad \text{cf of } 0.3$$
$$\text{and } D = YY \quad \text{with} \quad \text{cf of } 0.6$$

This is like stating that the plane is either a C5A or a C141, but I feel somewhat more confident that it is a C5A.

There are also instances in which multiple conclusions may make sense even though uncertainty is not being employed. To illustrate, consider the three (deterministic) rules listed below:

Rule A: *If* client's risk profile is risk adverse
     Then client's investment strategy is blue chip stocks

Rule B: *If* client's investment portfolio is less than $50,000
          and client's age is more than 60
          Then client's investment strategy is high-grade bonds

Rule C: *If* client's risk profile is risk taker
          and client's age is less than 45
          Then client's investment strategy is growth stocks

Further, let us assume that the client under consideration is 63 years old, risk adverse, and has an investment portfolio of $35,000. In this instance, rule C may be discarded and both rules A and B are triggered. Notice that the attribute for the conclusion of both triggered rules (as well as for all the rules) is *client's investment strategy*. Now, if multiple values are permitted for this attribute, we may fire both rules A and B to conclude that

> Client's investment strategy is blue chip stocks
> Client's investment strategy is high-grade bonds

In essence, we have concluded that either strategy (or, a combination of both investment opportunities) is advisable. Thus, in the case of deterministic rule bases, the validity of multiple conclusions is a function of the situation. Again, however, realize that not all development packages permit multiple conclusions (see Exercise 4.26 at the end of this chapter).

Further Guidelines[6]

In the previous sections, we have focused primarily on the components and terminology of rule bases while presenting a few guidelines for the construction of rules. Here, let us summarize (and illustrate) certain additional policies that should normally be followed in the development of rule bases. The adherence to such guidelines, in fact, often separates the successful knowledge engineer from the ineffectual knowledge technician.

RULE GROUPING. Knowledge bases evolve over time. As a result, they may all too easily become a seemingly random collection of unrelated rules. Quite often, even those who developed such disorganized rule bases will have difficulty in understanding—and maintaining—such collections. Consequently, for the purpose of understanding, maintenance, validation, and ease of documentation, some degree of rule organization should be maintained. It is usually advisable to group rules together according to similar conclusion clause attributes. Thus, each rule in such a group will have a particular conclusion clause attribute in common. This

---

[6]Portions of this section of the text have been motivated and influenced by the works of Pederson [1989a, 1989b, 1989c, 1989d].

can make it much easier to appreciate the various factors (i.e., premises) that lead to the same conclusion clause attribute—as well as make it easier to catch errors in rule formulation or to make any additions or deletions to the knowledge base.

RULE GROUP ORDER. Once rules have been grouped, each group should be ordered. Typically, rule-group ordering is determined according to the conclusion attributes. That is, the highest ranked group is that one that concludes the final goal of the knowledge base. The next highest level is assigned to the rule group (or groups) that serve to conclude any premises in the higher ranked group. And the ordering process continues in this manner.

In order to further clarify the notions of rule grouping and rule-group ordering, consider the knowledge base listed below. Here, we have (evidently) a collection of rules used by an office manager in the assignment of jobs to clerical staff. For the sake of illustration, let us assume that we are still in the rule-base development phase. Thus the rules below may represent only a partial knowledge base.

Rule Mary1: *If* typing = technical
             and graphics = included
             Then assignment − Mary

Rule Susan: *If* requestor = Susan
             Then typing = technical
             and graphics = not included

Rule Mary2: *If* typing = nontechnical
             and graphics = included
             Then assignment = Mary

Rule Frank: *If* requestor = Frank
             Then typing = nontechnical
             and graphics = included

Rule John: *If* typing = nontechnical
             and graphics = not included
             Then assignment = John

Rule Fred: *If* requestor = Fred
             Then typing = nontechnical
             and graphics = not included

Let us begin by forming rule groups. The conclusion clause attributes of the various rules are *typing*, *graphics*, and *assignment*. Let us arbitrarily first form the group that has *typing* as a conclusion attribute. This group is listed below:

Group: *Typing*

Rule Susan: *If* requestor = Susan

> *Then* typing = technical
> *and* graphics = not included

Rule Frank: *If* requestor = Frank
> *Then* typing = nontechnical
> *and* graphics = included

Rule Fred: *If* requestor = Fred
> *Then* typing = nontechnical
> *and* graphics = not included

Next, let us form the group having *graphics* as a conclusion clause attribute. This so happens to result in the same group that we just formed for *typing*. Thus we may proceed to form the group having *assignment* as a conclusion clause attribute. The resulting group is listed below:

Group: *Assignment*

Rule Mary1:   *If* typing = technical
> *and* graphics = included
> *Then* assignment = Mary

Rule Mary2:   *If* typing = nontechnical
> *and* graphics = included
> *Then* assignment = Mary

Rule John: *If* typing = nontechnical
> *and* graphics = not included
> *Then* assignment = John

Observing this most recent group, it should now be more obvious that the rules Mary1 and Mary2 may (if we so desire, and if our software package permits the use of disjunctive clauses) be combined into a single rule as follows:

Rule Mary: *If* typing = technical
> *or* if typing = nontechnical
> *and* graphics = included
> *Then* assignment = Mary

Next, let us order these two groups. Obviously, the single conclusion to be derived from this knowledge base is the assignment of a word processing technician (Mary or John) to a given task. In other words, the goal of the knowledge base is assignment. Thus, the rules in the assignment group are ranked first. Following this, we find those groups whose conclusions appear as premises in the higher ordered group. Since there is but one group left, this is a trivial task. The final knowledge base is provided on the next page where the horizontal lines serve to indicate the rule group borders:

Rule Mary: *If* typing = technical
               or if typing = nontechnical
                      and graphics = included
               Then assignment = Mary

Rule John: *If* typing = nontechnical
               and graphics = not included
               Then assignment = John

Rule Susan: *If* requestor = Susan
               Then typing = technical
               and graphics = not included

Rule Frank: *If* requestor = Frank
               Then typing = nontechnical
               and graphics = included

Rule Fred: *If* requestor = Fred
               Then typing = nontechnical
               and graphics = not included

Observing the grouped and ordered rule base, it should now be easier to note that there is a definite gap in the assignment process. Specifically, if Susan is the person requesting word processing service, there is no assignment possible. Assuming that the office has been working without difficulty, we most likely have either missed a rule or need to modify a rule. For example, we might be able to assign nontechnical typing jobs to Mary, whether or not graphics are included. While this may result in a work overload for Mary, it will permit an assignment of Susan's tasks. Alternatively, we may have left out one or more existing clerical staff members to whom word processing assignments are made.

In the general case, a rule base may have more than one final conclusion clause attribute. In this instance, we would group according to each attribute and then order the resulting groups according to some measure of the importance of the respective group.

CONJUNCTIVE CONCLUSION CLAUSES: SHOULD THEY BE AVOIDED? Note the second group of rules as formed above. Recall that this group so happened to be based, simultaneously, on the conclusion clause attributes of *typing* and *graphics*. As such, one possible problem in rule grouping was avoided—the existence of overlapping groups. However, consider the following set of rules:

Rule 1:   *If* A = a   Rule 2:   *If* B = b   Rule 3:   *If* A = c
            Then R = W   *Then* R = U   *Then* S = Z
            and S = X

Thus, for rule grouping, we have the following dilemma: "Do we assign all three rules to one group, or do we assign each to an individual group?" The decision may be simplified if we replace rule 1 with the two equivalent rules listed below. Now, all we need do is to establish two groups, one for those concluding R and the other for those concluding S.

Rule 1*a:*   *If* A = a   Rule 1*b:*   *If* A = a
              Then R = W   *Then* S = X

Thus, one group consists of rules 1*a* and 2 while the other consists of rules 1*b* and 3.

Because of the problem that may be faced in rule grouping, as well as a desire to keep the knowledge base as straightforward as possible, it is often advocated to limit rules to a single conclusion attribute. In general, we agree with this advice. However, on occasion, we shall ignore this guideline for sake of illustration.

RULE SUBGROUPS. Earlier, we stated that rules should be grouped according to similar conclusion clause attributes. We can also form subgroups of rules within each grouping. Consider, for example, a rule base for the diagnosis of skin infections. Obviously, the goal of such a rule base is the determination of the disease encountered. Consequently, we would group all rules with the conclusion clause attribute *disease* together. However, there will be a number of different diseases identified. Thus, a natural subgrouping procedure is to group each of these together according to the specific disease identified. As an example, one subgroup under *disease* would be all rules that conclude that the disease is *psoriasis*.

RULE ORDER WITHIN RULE GROUPS. There are also certain guidelines that are normally followed for the placement of rules within a group (or subgroup). Generally, we prefer to list the *most probable* rules first. By *most probable,* we mean the rule that is most likely, in general, to be fired. Since many inference procedures move from top to bottom in a list of rules, placing those that are the most likely to be fired first may result in a savings in time as well as a reduction in the number of queries presented to the user.

Two other ways to impose a rule ordering are through the use of the rule *priority* and *cost* properties. That is, we may assign a priority to each rule and order them (within their group) accordingly. Alternatively, we may associate a cost with each rule (possibly, the cost of reaching the associated conclusion), and then order the rules—from lowest to highest cost. Some software packages include the provisions for one or both of these properties.

PREMISE CLAUSE ORDER. Sometimes, within a single rule, the *order* of the premise clauses is particularly important. Consider, for example, the rule listed on the next page :

below:

Rule 1:   *If* gas color = brown
        and sector no. 5 gas pipeline rupture = yes
        Then valve no. 2 action = close valve

This is part of a rule base for determining the actions that need to be taken, for various situations, at a chemical plant. In this rule, we may note that we close valve no. 2 if a gas pipeline has ruptured *and* the color of the escaping gas is brown. However, note what would happen if the user is queried according to rule premise order. That is, the user might *first* be asked: "What is the color of the escaping gas?" And it may be that there is no escaping gas as the pipeline has not ruptured. Obviously, in such an instance, this query would be disconcerting. The user may even infer that the pipeline has ruptured (otherwise, he or she might reason, why was I asked about the color of the gas?). Further, even if the rupture has occurred, we will first ask the user about the color of the escaping gas—and then ask him or her if the pipeline ruptured. Assuming that gas cannot escape unless there is a rupture, such a line of questioning appears *dumb*. Not only is such a line of questioning confusing and inefficient, it is also quite frustrating to the user. To avoid this, in rule 1, we would simply reorder the two premise clauses.

AVOID FALSE ECONOMIES. Those who have had previous experience in the development of other forms of models (e.g., mathematical models, network models, etc.) have most likely developed the habit of *cleaning up* such models by reducing unnecessary details and redundancies. Employment of such practices seems eminently logical and, as a consequence, one would guess that they may be carried over into expert systems. Unfortunately, such habits can lead to false economies in rule-base development. Consider, as an example, the rule base as developed earlier for the office word processing technician assignment problem, and as repeated below:

Rule Mary: *If* typing = technical
        or if typing = nontechnical
            and graphics = included
        Then assignment = Mary

Rule John: *If* typing = nontechnical
        and graphics = not included
        Then assignment = John

Rule Susan: *If* requestor = Susan
        Then typing = technical
        and graphics = not included

Rule Frank: *If* requestor = Frank
               Then typing = nontechnical
               and graphics = included

Rule Fred: *If* requestor = Fred
               Then typing = nontechnical
               and graphics = not included

Further, examine carefully the first rule, as listed below:

Rule Mary: *If* typing = technical
               or if typing = nontechnical
                   and graphics = included
               Then assignment = Mary

Now, looking at this rule, there may be a temptation on the part of some readers to eliminate the first two premise clauses. That is, the *improved* rule would be

Rule Mary-bad: *If* graphics included
                    Then assignment = Mary

This new rule (Mary-bad) may appear to be the same as the original, but it does not capture the same knowledge. First, we no longer can tell, from rule Mary-bad, that Mary can type (i.e., either technical or nontechnical documents). Second, should there exist another employee (or should one be hired in the future) who does nothing but graphics, both that individual and Mary will always be candidates for the same set of jobs.

Returning again to the rule base for the word processing assignments, there is yet another temptation that one might encounter. Specifically, why fool around with all these rules when one could just make the assignment directly—by matching the output of the requestor with the capabilities of the word processing technician. For example, we might develop the following rule:

Rule Frank/Mary: *If* requestor = Frank
                       Then assignment = Mary

This last rule is, in fact, a valid one. However, there is one major drawback of this rule as contrasted to the original set of rules. That is, *knowledge has been lost*. More specifically, if we were to construct these types of rules, we would lose all knowledge about the capabilities of each technician and the output of each requestor. Consequently, should we ever wish to revise this rule base (and we will!), we will most likely have to recapture this lost knowledge.

We shall discuss such matters as redundancies in rules and unnecessary premise clauses in Chap. 8. There, we shall present a formal procedure for rule base checking. In the meantime, we caution the reader to be particularly careful in eliminating premise clauses. Realize that rules are meant to be autonomous, each with their own particular knowledge content, and refrain from any actions that might purge (necessary) knowledge from a rule.

AVOID PROCEDURAL CONTENT. Recall that rules are meant to capture *what we know,* and should not include *what we do*. However, following this particular guideline can be difficult for those schooled in algorithmic methods, where relationships and procedures comingle. Rule bases should, *ideally,* contain no procedural content. Procedures (e.g., algorithms) are best left either as a part of the inference engine or as external routines, accessed when needed by the expert system. For sake of illustration, consider the rule base listed below, where the developer has implicitly incorporated certain control procedures within the rules. This rule base is, in fact, derived from that one used previously in Example 4.1, and should be compared with that formulation.

Rule C130: *If* engine type is prop
            Then plane is a C130

Rule B747: *If* wing position is low
            Then plane is a B747

Rule C5A
  or C141: *If* bulges on fuselage are absent
            Then plane is a C5A
            Else plane is a C141

The line of thought that was employed to produce this rule base, rather than the original rule base presented in Example 4.1, is as follows:

- If we determine that the plane has propellor-driven engines, we know that it is a C130. If not, we do not need to ask that question again.
- If the plane is not a C130, and its wing position is low, then it is a B747. Further, we need not ask about wing position again.
- If the plane is not a C130 or a B747, and it has no bulges, it is a C5A. Otherwise, it is a C141.

Having *bought* that line of thought, the developer of the most recent rule base might congratulate himself or herself for reducing the number of clauses and rules in the knowledge base. This is not only false economy, it is a failure to realize that the new rule base relies on certain implicit procedures. That is, for the rule base to reach a correct conclusion, it is necessary that

- The order of the rules remains unchanged.
- The order of the inferencing process is consistent with the *procedure* implied by the rules.
- Inferencing stops as soon as the first conclusion is reached.

To illustrate, let us assume that we now place rule C130 at the end of the knowledge base. Further, let us assume that the plane to be identified is a C130. A fairly typical inference scheme would begin with the first rule and ask the user about the wing position. Since the wing position of a C130 is high, rule B747 would be discarded and we would move to rule C5A/C141. Here, whether or not a C130 has bulges on its fuselage (it actually does, under the wings), we would conclude that the plane sighted was either a C5A or C141. We may summarize this particular guideline by noting that one should never construct a rule base so that the correct conclusion can only be reached when the rules are ordered in a specific manner.

USE SYMBOLS WHEREVER POSSIBLE. Expert systems are most appropriate for problems wherein the attribute values are symbols. This is not to say that they cannot be used in cases where attribute values are numbers; they can, but they tend to be less efficient. There is yet another reason to employ symbols rather than numbers: humans tend to relate more naturally to symbols. For example, we rarely describe a person by his or her precise weight (e.g., 145 pounds). Instead, we describe individuals using such symbols as thin, normal, or obese. Even in the case where the domain expert actually uses numbers, it is generally best to convert these numbers into symbols. Such a conversion generally makes the knowledge base more readable. Consider, for example, the following rule:

Rule Weight: *If* weight $\geq$ 1000 kilograms
                    Then additional braces = yes

This rule may be replaced by the following two rules:

Rule weight': *If* weight = heavy
                    Then additional braces = yes

Rule load: *If* load $\geq$ 1000 kilograms
                    Then weight = heavy

With the addition of rule heavy, we now know that weights above 1000 kilograms are considered heavy—something that may not be directly interpreted from the original rule.

AV-Pair Tables

In support of the rule-base development effort, it is strongly encouraged that one employ a systematic approach. The guidelines previously discussed provide a means to follow such a process. In addition, and from our experience, we believe that the development of AV-pair tables [Ignizio, 1988] will do much to further enhance the effort, and to reduce the development time. AV-pair tables are particularly useful in the transition from the written rule base to the representation employed by the software package utilized. If nothing else, AV-pair tables serve as a means for documentation of the components of a given rule base. Such tables are best explained by means of an example. To illustrate, we shall use the rule base given below:

Rule A–X: *If* waste origin = source A
           and waste amount = large
           Then waste destination = site X

Rule A-Y: *If* waste origin = source A
           and waste amount = modest
           Then waste destination = site Y

Rule B-Z: *If* waste origin = source B
           and waste content = toxic
           Then waste destination = site Z

Rule B-Y: *If* waste origin = source B
           and waste content = nontoxic
           Then waste destination = site Y

This particular rule base summarizes the procedure used to determine the routing of different types of solid waste to disposal sites. Observe that each rule has been named according to the specific source-destination pair of that rule. Now, let us establish the corresponding AV-pair table for this rule base. This is provided by Table 4.1 and the contents of the table should be relatively obvious.

Table 4.1 has been divided into two sections. The top section lists the attributes and values of the rule-base *premise* clauses. The bottom section lists those for the *conclusion* clauses. If the rule base has intermediate conclusions, as will be the case in most real problems, we may include a section for such clauses directly above the conclusion clauses. This may be illustrated by means of the rule base listed below:

Rule 1: *If* A = a
         and B = b
         Then R = r

Rule 2: *If* A = NOT a
         and C = c
         Then S = s

Rule 3: *If* S = s
         and B = b
         Then X = x

The AV-pair table associated with this rule base is presented in Table 4.2. The middle section of this table associated with rule 2, clause 3, represents the single intermediate conclusion clause of our latest rule base.

TABLE 4.1
Attributes and values for knowledge base

Rule	Clause	Attribute	Value
A–X	1	Waste origin	Source A
A–Y	1	Waste origin	Source A
B–Z	1	Waste origin	Source B
B–Y	1	Waste origin	Source B
A–X	2	Waste amount	Large
A–Y	2	Waste amount	Modest
B–Z	2	Waste content	Toxic
B–Y	2	Waste content	Nontoxic
A–X	3	Waste destination	Site X
A–Y	3	Waste destination	Site Y
B–Z	3	Waste destination	Site Z
B–Y	3	Waste destination	Site Y

Through an examination of the AV-pair table, such as the one exhibited in Table 4.2, we should be able to obtain a better appreciation of the contents of the rule base—as well as identify potential problems. For example, we may immediately note from the most recent table that there is one intermediate conclusion and two final conclusions.

Next, we may readily observe the position of each attribute within the rule base (e.g., attribute A appears in rule 1, clause 1 and in rule 2, clause 1). Thus, the AV-pair table serves to indicate the particular architecture of a rule base. Also, while not evident from this particular table, the AV-pair table is useful in observing whether or not rule grouping—and subgrouping—has been employed. For example, if we examine Table 4.1, we can observe that we have failed to group the rules according to subgroups in this rule base. That is, by examining the conclusion clauses of that table, we may observe that rules A–Y and B–Y have

TABLE 4.2
Attributes and values for knowledge base

Rule	Clause	Attribute	Value
1	1	A	a
2	1	A	NOT a
1	2	B	b
3	2	B	b
2	2	C	c
3	1	S	s
2	3	S	s
1	3	R	r
3	3	X	x

the same conclusion attribute value, and thus we may wish to place these two rules in a single subgroup. This may be accomplished by simply interchanging the positions of the third and fourth rules.

Further, we can observe, from the right-hand column, the complete list of values that each attribute can possibly either be tested against (if a premise clause attribute) or assigned (if a conclusion clause attribute). We may wish to compare these against the set of legal values for each attribute. If some value appears in the table that does not appear in the set of legal values, we may either have a problem (i.e., the rule base needs to be modified) or we may just have to enlarge the set of legal values. If the values in the table are but a subset of the legal values, this may indicate that we can reduce the set of legal values—*or* that we need to incorporate some additional rules. For example, observe the set of values for attribute B in Table 4.2. From the table, it may be noted that attribute B (which is a premise clause attribute) will never be tested against any value other than "b." A similar situation occurs for attributes C and S. Now, such a situation can occur and the rule base may still be valid. However, there are two *potential* problems associated with this case:

1. If we are employing a menu-driven expert system (i.e., one that employs prompts that are menus), the user may not be presented with a feasible selection. Consider, for example, the following prompt for the value of attribute B:

   ```
   The value of B is:
   1. b
   Type in number of selection: _____
   ```

   Now, even if B does not equal "b," the user will be unable to proceed unless he or she types in a "1."
2. We may have simply forgotten to include a rule that is triggered whenever the value of B is something other than "b." That is, our rule base is incomplete.

Generally, we should have at least two values for every premise clause attribute. For example, we may wish to use the values of *NOT b, NOT c,* and *NOT s* in our rule base. In fact, when using menu driven prompts, this will have to be done. And rules associated with these new values will have to be constructed and added to the knowledge base.

Inference Networks—and a Preview of Chaining

AV-pair tables present a useful means for documentation and evaluation of rule bases. There is also a graphical approach to rule-base documentation that can be quite useful. This is the use of inference networks.

We shall employ the following conventions in our inference networks. First, a box will represent an assertion, that is, the association of a premise attribute with a specific value. Circles will be used to represent conclusions. A circle

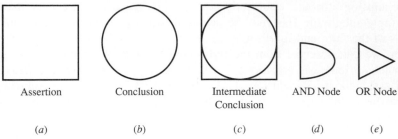

Assertion	Conclusion	Intermediate Conclusion	AND Node	OR Node
(a)	(b)	(c)	(d)	(e)

FIGURE 4.5
Inference network symbols.

enclosed by a box will represent an *intermediate* conclusion. Finally, assertions and intermediate conclusions may be combined by logical connectives, specifically AND or OR operators. The notation that we will employ in the development of our inference networks is depicted in Fig. 4.5.

As an extremely simple illustration of the use of such notation, consider the inference network of Fig. 4.6. Here, we have but four assertions and two conclusions (i.e., there are no intermediate conclusions).

Examining Fig. 4.6, we should be able to recognize that it represents two rules. These are

Rule 1 (R1): *If* assertion #1 is true
 and assertion #2 is true
 Then conclusion #1 is established

Rule 2 (R2): *If* assertion #3 is true
 or assertion #4 is true
 Then conclusion #2 is established

A somewhat more general inference network is exhibited in Fig. 4.7. In this figure, we have employed uppercase letters (i.e., A through H) to represent a complete clause. For example, A has been used to represent one complete premise clause of rule 2. Notice in particular the *intermediate* conclusions as represented by G and H.

The equivalent production rule set for the network of Fig. 4.7 is simply:

Rule 1: *If* B and C
 Then G

Rule 2: *If* A and G
 Then I

Rule 3: *If* D and G
 Then J

Rule 4: *If* E or F
 Then H

Rule 5: *If* D and H
 Then K

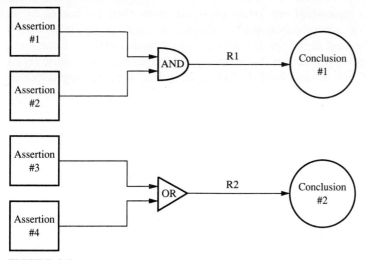

FIGURE 4.6
Simple inference network representation.

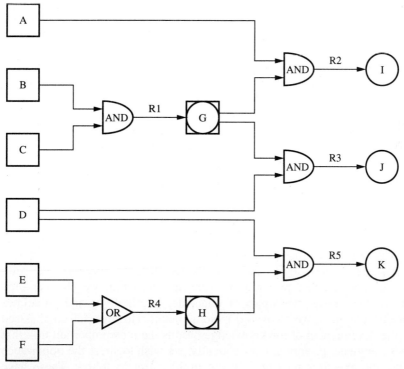

FIGURE 4.7
A more general inference network.

The inference procedure can often be made more clear by means of an inference network. As we have mentioned, inference normally proceeds in either a forward manner (forward chaining) or backward (backward chaining). Examining Fig. 4.7, note what would happen if we were provided with the following set of data: *A*, NOT *B*, *C*, *D*, *E*, and NOT *F*. Referring to Fig. 4.7, we may conclude that

- Rule 1 may be discarded (since we know NOT B).
- Rule 2 cannot be triggered since we cannot conclude G.
- Rule 3 cannot be triggered since we cannot conclude G.
- Rule 4 may be triggered (since we have E).
- H may be concluded if we fire rule 4.
- Rule 5 may be triggered since we know D, and H has been concluded.
- Firing rule 5, we may conclude K.

Observe that, in reaching our final conclusion (i.e., K), we moved from left to right in the inference network—the direction used by forward chaining. Further, our search was *data driven,* that is, we used the initial set of data to conduct the search.

We may also use an inference network to illustrate backward chaining. Again, using Fig. 4.7, let us assume that we are trying to determine the final conclusion "I." Now, to conclude "I," we must first trigger rule 2—which can occur only if both A and G are true. Further, to conclude G, we must first trigger rule 1—which can only occur if both B and C are true. Thus, moving backward through the network, from right to left, we know precisely what premises must be satisfied to reach the final conclusion of "I."

While inference networks provide a nice graphic illustration of chaining, a somewhat more formal procedure is necessary to actually conduct the inference process for larger, more realistic, problems. We shall present such procedures in Chap. 6.

SUMMARY

In this chapter, we have focused our attention on knowledge representation. After a brief introduction to alternative modes of representation (i.e., OAV triplets, semantic networks, frames, logic, and neural networks), we addressed the rule-based mode of knowledge representation—the approach that we shall employ throughout this text. Here, the topics of concepts, terminology, and guidelines for rule-base development were covered in some detail. However, as emphasized repeatedly, the development of the knowledge base is the most important task that the knowledge engineer performs. Consequently, we wish to stress the importance of addressing the exercises presented in the problem set to follow. These have been developed in an attempt to reinforce the material presented, and to develop a deeper appreciation of the need to employ a systematic, thoughtful procedure in knowledge-base construction.

In the next chapter, just how one might go about acquiring such rules, using either a domain expert or set of historical records as a source, will be discussed. Together, Chaps. 4 and 5 represent, we believe, the most critical facets of knowledge engineering. There may be, however, some question in the reader's mind as to just why the chapter on knowledge representation was presented *before* that on knowledge acquisition. This is because one might view the development of an expert system as a process that proceeds smoothly from one activity to the next—and it would appear that one must first acquire knowledge before one represents knowledge.

Actually, in the development of any real-world expert system, the process of knowledge acquisition and knowledge representation go hand in hand. That is, we acquire some knowledge (i.e., determine a few heuristics used by the domain expert) and then represent these heuristics by our specific choice of knowledge representation. Thus, and in an incremental manner, we gradually develop the complete knowledge base.

There is yet another, more pragmatic, reason why knowledge representation has been presented before knowledge acquisition. To discuss knowledge acquisition in any depth, one must first be aware of knowledge representation. Otherwise, our discussion of knowledge acquisition would be limited to one of primarily philosophical content.

EXERCISES

4.1. List, for our earlier discussion of the bank loan officer, the set of (*a*) a priori facts, (*b*) a priori rules, and (*c*) inferred facts and rules.

4.2. How might one actually verify the allegation that Dan Smith is, indeed, an *expert* loan officer?

4.3. Consider the following partial rule base. What problems do you see? Can you develop an improved rule base?

> R1: *If* car = red
>   *and* car = new
>   *and* car = convertible
>   *Then* car = mine
>
> R2: *If* car = old
>   *and* car = damaged
>   *and* car = blue
>   *Then* car = Fred's
>   *or* car = Sally's

4.4. A real estate company, with offices located across the nation, wishes to develop an expert system for customer consultation. Specifically, based upon data about a customer and his or her housing and location preferences, the expert system will list the houses on the market that are most likely to be of interest. Discuss the employment of multivalued attributes in such an expert system.

4.5. Construct an improved version of the rule listed below:

> Rule: *If* you must avoid whole milk
> *and* you dislike skim milk
> *or* you wish to reduce your cholesterol
> *Then* consider soy milk

4.6. Critique the following rule sets:

(*a*)   R1:   *If* temperature is high
and pressure is high
Then reduce flow
or terminate processing

(*b*)   R1:   *If* A = X   R2:   *If* B = Y
and B = Y   *and* A = X
Then C = Z   *Then* C = Z
                                   and D = W

(*c*)   R1:   *If* A = X   R2:   *If* A = X
and B = Y   *and* B = NOT Y
Then C = Z   *Then* C = Z

4.7. List the AV-pair properties for the following rules:

(*a*)   *If* subject's age ≥ 21
and subject's height ≤ 5'10"
and subject's weight = normal
Then candidacy = approved

(*b*)   *If* VSWR[7] ≤ 1.25
and pattern is approximately omnidirectional
Then antenna test is satisfactory

4.8. Given the rule base listed below:

(*a*) group and order all rules

(*b*) construct the associated inference network

(*c*) construct the associated AV-pair table

R1:   *If* order backlog = large   R2:   *If* order backlog = large
and labor supply = nil   *and* labor supply = adequate
Then bid decision = decline   *Then* tentative decision = consider

---

[7]VSWR stands for voltage standing wave ratio, the ratio of the reflected component of a waveform to the propagated component.

R3: *If* order backlog = minimal
or order backlog = modest
Then bid decision = yes

R4: *If* bid decision = yes
and historical data = available
Then bid methodology = GP

R5: *If* bid decision = yes
and historical data = not available
Then bid methodology = HRS

R6: *If* tentative decision = consider
and customer = important
Then bid decision = yes

R7: *If* tentative decision = consider
and customer = not important
Then bid decision = decline

4.9. The rule base of Exercise 4.8 is one associated with the procedure through which a certain firm decides on whether or not to bid on a project and, if so, the type of approach to be used to develop the cost estimates. One approach is GP (goal programming) while the other is HRS (a heuristic rule set), where the choice of approach is based upon the existence or lack of existence of appropriate historical data. Given a case where

Order backlog = large
Labor supply = adequate
Customer = important
Historical data = available

list the properties associated with each clause *and* rule in the rule base.

4.10. Given the following rule base plus the facts that

Sensor status = unknown
Warning light = on

list the properties associated with each clause *and* rule in the rule base.

R1: *If* sensor status = working
and warning light = on
Then terminate process

R2: *If* sensor status = not working
and warning light = on
Then repair sensors

R3: *If* sensor status = working
and warning light = off
Then continue process

4.11. How might you rewrite the rule set of Exercise 4.10 in order to reach a decision when the sensor status is unknown and cannot be determined by either a user prompt or inference? With this rewrite, list the properties associated with each clause *and* rule in the rule base.

4.12. Can an attribute appear in *both* a final conclusion clause and an intermediate conclusion clause? If so, provide an example rule base in which this happens. If not, explain why not.

4.13. Domain experts often employ rule bases that involve a number of intermediate conclusions, where these serve to essentially prune away (i.e., filter) the associated search space. An example of such usage is provided in the rule base listed below. Convert this rule base into one which employs no intermediate conclusions, and comment on the result.

R1: *If* publication date is recent
      and topic is western
      or topic is science fiction
  Then check further is yes

R2: *If* publication date is not recent
  Then check further is no

R3: *If* check further is yes
  and price is under $15
  and author is one of your favorites
  Then read review is yes

R4: *If* check further is yes
      and price is over $15
      or author is not one of your favorites
  Then read review is no

R5: *If* read review is yes
  and review is good
  Then buy book

R6: *If* read review is yes
  and review is not good
      or review is no
  Then don't buy book

4.14. Rewrite the (original) rule base of Exercise 4.13 to avoid disjunctive clauses.

4.15. Group and order the rule base developed for Exercise 4.14.

4.16. An (alleged) expert in the selection of personnel for assignment to knowledge engineering training programs has been consulted. The following is a summary of his opinions about trainees. Place these statements into a rule base.

When I look for a candidate for a knowledge engineering position, I first look for someone who can deal with ill-structured problems and who has a *broad* perspective. I have also noticed that the best candidates are those with a sense of organization and structure. I really think that messy, sloppy, unorganized people make lousy knowledge engineers. Another factor that is an absolute must, in my opinion, is that the individual be able to communicate with others—and be patient. Generally, I rule out computer "hackers" as knowledge engineering candidates. Hackers are more interested in playing with the software and hardware than in building rule bases. You can usually recognize a hacker by his or her extensive knowledge of about every software and hardware package in existence—or that has been or will be in existence. Hackers are often great people for knowledge engineering support but rarely are they effective as knowledge engineers. Engineers (i.e., traditional engineers such as civil engineers, electrical engineers, and industrial engineers) *may* make good knowledge engineers *if* they are practical and have a broader perspective than their more theoretical colleagues. The same is true of management scientists (MS) and operations research (OR) analysts. Those in MS and OR who are interested in real-world application tend to be excellent candidates; those who focus on narrow topics (e.g., the refinements of algorithms) make poor candidates. Finally, I have also noted that, with proper training, psychologists and art majors can be excellent candidates. I haven't quite figured out why, but many individuals within these groups seem to have the right perspective and abilities to satisfy most of the requirements that I personally believe are necessary to be a good knowledge engineer.

4.17. For the rule base developed for Exercise 4.16, group and order all rules and then develop the associated AV-pair table.

4.18. Examine the rule base developed for Exercise 4.16 (and summarized in the AV-pair table of Exercise 4.17). Is each rule representative of a single *chunk* of knowledge? Does the rule base describe what is known, rather that what to do? Will adding or dropping a rule have minimal impact on the validity of the resulting rule base? Why is it important to consider each of these factors?

4.19. Develop a rule base that reflects the manner in which you personally derive your class schedule for each school term.

4.20. Develop a rule base that reflects the approach you would employ in the search for, and selection of, a new job.

4.21. Develop the associated AV-pair table for the rule base of Exercise 4.19.

4.22. Develop the associated AV-pair table for the rule base of Exercise 4.20.

4.23. Develop an improved version of the following rule base:

R1: *If* problem involves predominance of symbols
    and environment is relatively stable
    and domain expert is available
    and alternative methods are less appropriate
    and expected benefits outweigh expected cost
    Then expert system recommendation = yes

R2: *If* problem involves predominance of numbers
    and exact solution is necessary
    and majority of operations are mathematical
    and quantitative model can be formed
    and problem is not combinatorially explosive
    Then algorithmic approach recommendation = yes

4.24. Let us assume that you have been discussing expert systems with a friend (don't laugh, it might happen). Further, your friend has never taken a course in expert systems but has used the personal computer quite extensively to develop software, employing either the BASIC, FORTRAN, or C programming languages. When you note that an expert system is made up of a rule base that is itself made up of IF-THEN rules, he appears bemused. As he states it: "What is the big deal about that. I have been using IF-THEN statements in all of my software development." Prepare an explanation that may clarify the matter.

4.25. Construct the inference network for the following rule base:

R1:   *If* T *or* Q
       Then C

R2:   *If* X
       Then 3

R3:   *If* S *and* W *and* G
       Then 1

R4:   *If* Q *or* V
       Then H

R5:   *If* S *and* U
       Then Q

R6:   *If* U *and* H
       Then 2

4.26. Refer back to the investment strategy rule base within the section titled "Multiple Conclusions." Let us assume that the software package upon which this rule base is to be implemented does not permit the use of multiple conclusions (i.e., multiple values for a conclusion clause attribute). Show how the rule base may be modified in order to work with such a software package and still arrive at multiple investment strategies.

CHAPTER

5

KNOWLEDGE ACQUISITION

INTRODUCTION

In the preceding chapter, we addressed the problem of knowledge representation. In this chapter, we focus on the companion problem of knowledge acquisition. As mentioned in the conclusion to Chap. 4, knowledge acquisition and knowledge representation are phases of expert systems development that proceed virtually hand in hand. And both phases are absolutely vital to the integrity of the rule base for the expert system ultimately constructed.

The knowledge acquisition phase of development is also one that can be extremely frustrating as well as time consuming. Some have termed it the *bottleneck* of expert systems development. Here, we are often dealing directly and intimately with domain experts. While dealing with people in general can be difficult, interfacing with domain experts can be many times as frustrating. Fortunately, there are certain guidelines that can ease the process, and these will be discussed in the sections to follow.

However, first we might point out that knowledge acquisition is not an entirely new endeavor. Those involved in the development of heuristic programs and decision-support systems have, for quite some time, faced very much the same problem as is now faced in expert systems. Specifically, how does one best elicit the facts and rules within the human expert's knowledge base? The conclusion has always been that this process is one that is mainly an art, rather than a science.

While guidelines can provide assistance, it is only through actual experience that one ultimately achieves a certain level of proficiency in this phase of knowledge engineering.

In this chapter, two different approaches to knowledge acquisition will be addressed. The first will be that of the acquisition of knowledge directly from the domain expert, or experts. The second will focus on this acquisition through the use of historical records, that is, by rule induction. Both approaches are widely employed, and sometimes used in combination in rule-based expert systems.

KNOWLEDGE ACQUISITION AND THE DOMAIN EXPERT

It would seem that the most obvious way in which one may acquire a knowledge base is to go directly to the human expert. However, there are at least four reasons why this may not work, or at least not provide totally satisfactory results.

- For some problems, there simply may not be an expert. One example that comes to mind is that of investing in the stock market. While some investors or investment advisory services do well for a relatively brief time, they typically have spells in which their performance is mediocre to terrible. One school of thought claims that there simply aren't any stock market experts while another claims that, if such experts exist, they would never be so foolish as to consider revealing their approach.

- The alleged experts may actually be exhibiting poor to mediocre performance. All too often, the term *expert* is loosely applied to anyone who simply *gets the job done*. Thus, while such *experts* may be available, it may well be wise to avoid building a knowledge base about (or, at least, solely about) their heuristic rules.

- The experts may not wish to reveal their *tricks of the trade*. In some cases, such individuals simply refuse to cooperate. In others, a potentially far more serious problem occurs—they appear to cooperate but intentionally provide false information.

- Finally, there are some experts who are just unable to articulate the approach that they use. Many experts, in fact, simply and honestly do not really understand how they actually make their decisions. As such, when asked to explain how they solve a problem, they respond with a description of their most recent *perception* of the process, which may or may not have any relationship to the procedure they actually employ.

However, based upon the assumption that a human is now performing the task that is to be performed in the future by an expert system, our first step is to identify that person (or persons). Once this person has been identified, the next step is to set up an initial meeting with the alleged domain expert. This meeting should be informal because you most decidedly want the inaugural meeting with this person to take place in a relaxed atmosphere. Consequently, in this meeting, one should avoid the use of audio/visual recording. In fact, it is generally wise to even avoid taking notes during this particular meeting. This first meeting establishes some rather critical first impressions. And first impressions in an atmosphere of tension and excessive formality are not likely to be favorable.

There are several purposes for the initial meeting. First, as mentioned, we wish to relax the individual. Second, we should attempt to explain (*simply* and *clearly*) to the individual just precisely what it is that we intend to accomplish. Typically, one emphasizes that our purpose (i.e., as knowledge engineers) is not to use and then discard the expert, but rather it is to provide him or her with a computerized assistant so that he or she can pursue *more interesting* work.[1] Here, we are addressing one of the most obvious fears of the expert. Third, we should lay out a proposed schedule for follow-on meetings, and describe their general form and purpose. If we intend to use audio/visual recorders during subsequent meetings, we should ask the expert's permission to do so.[2]

It should be stressed that each and every meeting with the designated expert should be carefully considered beforehand, and that a formal agenda should be drawn up. Any mistakes made during such meetings have the potential for destroying the often fragile working relationship between the knowledge engineers and the expert, and any time that is perceived to be wasted will not be well received by most busy experts. All meetings after the first should be conducted, in fact, in a businesslike atmosphere—although without excessive formalities or overt pressure.

In certain cases, the initial meeting should be followed, or even preceded by an on-site visit. For example, if the problem under investigation is related to some facility (e.g., a manufacturing facility, a power plant, or a farm), it is important to actually visit that site *as soon as possible*. There is simply no substitute for actually being able to view the problem in its physical context. The knowledge engineers who are reluctant to leave their offices or to get their *hands dirty* are not likely to be very successful.

There is yet another purpose to the initial meeting, as well as those that follow, that should not be openly discussed. Specifically, we should use this meeting, as well as follow-on meetings, to *attempt* to evaluate the true extent of the *expertise* of our expert. As mentioned before, all too often one is designated an expert as a result of simply getting the job done, even when the job is not be-

---

[1] As a point of information, it should be noted that, in some cases, firms have paid the domain expert for the knowledge cloned—*up front* or through some type of royalty arrangement.

[2] There are mixed feelings with regard to the benefits of audio/visual recordings. Some knowledge engineers believe that they are indispensable, others feel that they are not worth the effort—particularly the effort of sitting and listening to them (or having them transcribed). In our experience, we have used audio recordings as mainly a backup (e.g., in the event we needed to clarify some point made during the meeting).

ing performed all that well. Cloning the expertise of such individuals will result in nothing more than a perpetuation of this less-than-desirable performance, and this, in fact, is precisely all that a number of existing expert systems accomplish. However, if one is inclined toward cynicism, it may also be noted that few of the users of such expert systems actually appreciate this *problem*. That is, they suffer the same illusions about the expert system as they did about the human *expert*.

If and when you encounter a domain expert in whom you have no confidence, there are a number of alternatives that should be considered. First, and most obviously, you might try to find a replacement—someone else in the organization who seems reasonably competent in the domain under consideration. This option, of course, requires a certain amount of diplomacy. Second, you might consider learning enough about the problem at hand so that you can act as the domain expert. Third, you might wish to examine historical records of the decisions made. From these, you might be able to extract those decisions that appear to have been *good* ones—and use them to develop a set of rules. We will, in fact, cover the latter two alternatives in subsequent sections. As yet another alternative, you might wish to consider the employment of hybrid expert systems—the topic of Chap. 9. A hybrid expert system (by the definition employed in this text) combines an expert system with some other, more conventional method for decision analysis (e.g., heuristic programming, simulation, optimization). For example, the output of the expert system might serve as the input to another technique. In the right setting, and when the hybrid combination is properly designed and matched, the resultant solution can often represent a substantial improvement over that produced by the expert system alone. In the event that your expert is not all that exceptional, this is certainly an option to consider.

Returning to the primary point of our discussion, as long as we feel confident that the expert systems approach is indeed the most appropriate, and that the domain expert is reasonably competent, we may continue with the knowledge acquisition process. Thus, following our initial informal meeting with the domain expert, we should conduct a series of formal meetings designed to extract as much information, as painlessly as possible.

The conduct of the follow-on meetings is generally best handled through the employment of *two* knowledge engineers. *And at least one of these individuals should be experienced*. One knowledge engineer (typically, the more experienced of the two) should be given the primary responsibility for conducting the interview, while the other knowledge engineer listens carefully to both the questions and responses. The second knowledge engineer will also make sure that the meeting is being properly recorded and, when necessary, replace tapes and move microphones—and do this as discreetly as possible. Another role of the second knowledge engineer is to take notes concerning topic areas that should be explored in more depth—something that may be difficult for the knowledge engineer conducting the meeting to either observe or recall.

As mentioned above, at least one of the knowledge engineers should be experienced in knowledge acquisition and in the *successful* development of expert systems. One of the worst mistakes being made in the expert systems development area is the assignment of inexperienced personnel to the effort.[3] Such inexperience will generally be apparent, even to the domain expert with no familiarity whatsoever in expert systems.

There should hopefully develop, during the knowledge acquisition sessions with the domain expert, a sense of mutual respect. Ideally, the knowledge engineers should have a genuine respect for the domain expert, for the job that he or she does, and for the heuristics that have been developed through real-world experience. Even more important, the domain expert must develop a genuine respect for the knowledge engineers. Thus, while the knowledge engineers are obviously novices with respect to the specific domain of interest, it must be apparent that they have an ability to solve real problems and are knowledgeable and experienced in their own profession. One of the best ways in which to gain the respect of the domain expert is to make an all out attempt—well before the initial meeting—to understand the problem, the environment in which it exists, and the specific terminology and jargon employed by the expert. This last point cannot be emphasized enough. Contrast, if you will, the example of the American tourist in a foreign country who has made an effort to learn at least a few common words and phrases in the language of that country with the tourist who keeps asking "Why don't these ignorant people speak English?"—and I think you can guess how each will be perceived by the locals.

One must always keep in mind that the purpose of these meetings is to extract the knowledge base of the expert. While this sounds obvious, it has been observed that the discussions in some knowledge acquisition meetings meander off onto tangents as the discussants pursue points that have little if any bearing on the knowledge base. For example, one particular meeting that we attended quickly degenerated into a gossip session when the expert began to complain about the work habits of a fellow employee. These work habits, in turn, were so bizarre as to virtually end further discussion (for that session) of the original problem under consideration. While one cannot always avoid such excursions, and while you should obviously never abruptly change the subject, it is necessary to at least attempt to maintain the focus of the meeting.

There are several modes through which the knowledge base may be extracted during such meetings. One is to simply ask the expert to explain the procedure through which he or she arrives at a conclusion. Another approach is to conduct demonstration sessions wherein the expert is asked to proceed through the decision-making process for a series of examples. In general, the second approach lends itself better to knowledge acquisition. We can, however, combine both of these approaches, and such a combination seems to work particularly well in practice.

---

[3]Since expert systems is relatively new it may be argued that there is no pool of experienced expert system developers. However, even if that is true, one should take care to assign at least one person to the effort who has experience in the solution of real-world problems (i.e., via whatever methodology was appropriate), and in dealing with people with a non-technical background.

One of the practices that we have employed, with (we believe) consider-able success, is to ask the domain expert to go through a demonstration of the decision-making process—at *our* offices. We have found that this is an excellent way to determine just what data the domain expert actually requires for decision making. He or she will either bring this data or else discover, part way through the demonstration, that one or more sources of information are missing—and we ask the expert to list these sources and bring them to the next meeting. In this manner, in one or two meetings, the information needed in support of the expert's decisions can be identified.

At the conclusion of each session, the knowledge engineers will typically try to restate the responses of the expert in the production rule format. Thus, after a few sessions, it should be possible to develop a simple, prototype expert system. Once a prototype is available, we may use it to extract additional knowledge from the expert. That is, once the expert is introduced to the prototype, a fuller appreciation of the concept and purpose of an expert system is obtained. Further, the prototype provides a tangible basis for evaluation. In particular, the expert can note inconsistencies, limitations, and deficiencies of the prototype, which in turn may be used to refine and enhance the rule base.

A word of caution with regard to the prototype and the domain expert is in order. The domain expert should be warned that any conclusions reached by a prototype, in its earliest forms, are likely to be both naive in nature and limited in scope. We usually explain that the evolution of a prototype expert system is somewhat akin to, and can be just as dramatic as, the evolution of a human—from infancy to adulthood. Initially, the prototype is the proverbial empty slate. It knows nothing, including the fact that it does not even know what it does not know. As more and more heuristic rules are entered into the knowledge base, and as more demonstration examples are encountered, the *intelligence* of the prototype will grow—first slowly, then faster and faster until a reliable decision analysis assistant has evolved. The domain expert must be assured that, with patience and training, the prototype expert system will prove its worth.

There are a number of good papers that discuss the conduct of the knowledge acquisition phase. Prerau [1987] and Surko [1989], in particular, have provided some excellent guidelines for knowledge acquisition. Here, we have attempted to summarize our own thoughts on knowledge acquisition, where the influence of numerous other authors is acknowledged. These guidelines are presented in the list that follows. While it may not be possible to abide by each and every point, the guidelines should at least be considered in the planning of the knowledge acquisition effort.

Selection of the Domain

- The domain should be one for which the expert systems approach is truly appro-priate, and for which an expert system would provide some distinct advantage over any alternative methods.

- Good decision making within the domain should be of sufficient importance to management that they are willing to commit the time and resources necessary to support the development and implementation of the expert system.
- Management must recognize both the costs and risks of expert systems development. Any nontrivial expert system is going to require the employment of competent knowledge engineers, over a reasonable period of time. If management has totally unrealistic expectations (e.g, in terms of time, cost, or capabilities), make an attempt to clarify the process. If their expectations remain unrealistic, there is probably no point whatsoever in pursuing the project.[4]
- The domain should be relatively stable; in particular, dramatic changes over the period of the development effort should not be foreseen.

Selection of the Knowledge Engineers

- Ideally, two knowledge engineers should be used, where at least one of these is experienced in the development and implementation of (successful) expert systems.
- The knowledge engineers should not be *one-trick ponies*. That is, they should at the very least be aware of alternative approaches to decision analysis.
- The primary skills of the knowledge engineers should be in the areas of eliciting knowledge and in forming a model of that knowledge (i.e., the rule base).

Selection of the Expert

- Ask the organization to provide you with the names of candidate domain experts, that is, those individuals who are believed to have significant expertise within the domain in question.
- Select a domain expert whose performance is generally acknowledged to be *above and beyond* that of most others performing the same task.
- Select an expert with a successful *track record* over a period of time.

---

[4]In one recent instance, we were asked to construct an expert system for the automation of proposal response (e.g., the determination of time, labor, and material costs for custom-made equipment). The company had been responding to over one thousand requests for quotation each year, with each proposal taking about four or five weeks to complete. And their success rate was about 5 percent, less than half that of their major competitors. The firm's management was convinced that, without some form of automation for this process, they would fall further and further behind their competition (mainly, the Japanese). In fact, their share of the market had already fallen drastically over the past few years. We were assured of their support. Development of the system involved the need to deal with three different domains and domain experts, and of the construction and implementation of a rather massive computer-based database (those records that were being kept were being kept manually). And they wanted the expert system (and database) developed and implemented within the month, at a cost of under $1000. We politely declined.

- Select an expert who is both willing and able to communicate personal knowledge, and who is relatively articulate in doing so.
- Select an expert who is both willing and able to devote the time necessary to support the development effort.
- If no expert can be identified, or made available, consider the development of the rule base through alternative means (as will be discussed in the sections to follow).

The Initial Meeting

- Prior to this meeting, the knowledge engineers should make an all-out effort to familiarize themselves with the problem, the domain, and the terminology used within the domain.
- Locate this meeting in comfortable surroundings. Limit the duration of the meeting (in our opinion, such a meeting should last less than 2 hours).
- This meeting should be conducted in an informal, relaxed manner.
- Tell the expert what your plans and goals are, and explain just what an expert system is and what it can do (and cannot do) for the expert as well as the organization.
- Explain the evolution of the expert system (in particular, discuss how initial decisions, developed by early prototypes, are likely to appear naive).
- Reinforce your discussion of expert systems with the demonstration of the use of some existing expert system. However, avoid the demonstration of an expert system that is all too obviously a *toy* (e.g., one that selects the type of wine to go with a meal).
- If audio/visual recording is desired, ask the expert for permission to do so—and explain that these recordings will be for the private use of the knowledge engineering team.

Background

- Where appropriate, make a site visit—and do this as soon as possible.
- Determine the existence of any existing manuals, reports, or other written material that serve to describe the domain, the problem, and the terminology employed.
- Ask the expert to present an informal tutorial session on the subject. In this session, no questions are asked, the knowledge engineers merely listen and learn.

Organization of Follow-on Meetings

- Attempt to minimize the possibility of interruptions. Set aside meeting times during which the expert can devote his or her full attention to the effort.
- Establish a formal agenda for each meeting.

- Establish goals and objectives for each meeting.
- Once a prototype expert system has been developed, establish access to the supporting software and hardware (e.g., for prototype demonstrations and their critique).

Conduct of the Follow-on Meetings

- Elicit the rules through discussion and demonstration.
- Attempt to identify all external sources of data and information that are used by the expert.
- Be patient. Don't interrupt the domain expert.
- Avoid criticism—instead, focus on clarification.
- Always remember that you are building a model of the expert's rule base, *not* a model of *your* rule base.
- If you don't understand a point made by the expert, don't be afraid to admit it. Ask for clarification.
- Use test cases to both demonstrate the decision-making process and to identify the limits over which the rule base is valid.
- Acquaint the domain expert with production rules; this may encourage the expert to begin stating his or her rules in this format.
- *Always remember what you are there for.*

Documentation

- Document the results of the meeting as soon as possible after the meeting (preferably, immediately after the meeting).
- Documentation for each meeting should include such facts as:

  Date, time, and location of meeting
  Name of expert (i.e., if more than one expert is being used)
  List and description of the rules identified during the meeting
  Listing of any new objects, attributes and/or values encountered—and their properties
  Identification of any new outside sources and references
  Listing of any new terminology encountered, and associated definitions
  Listing and discussion of any gaps or discrepancies encountered
  Reminders (e.g., of points that need to be clarified)

- Documentation in support of *all* production rules thus far developed should include such facts as:

  A listing and description of all rules thus far developed
  A listing and description of all objects, attributes, and values thus far encountered

Source and reference list
Glossary of domain terminology
Listing and discussion of the test cases used to evaluate the prototype

While there will obviously be exceptions to such guidelines, they do provide an overall basis for the systematic and efficient conduct of the knowledge acquisition phase of expert systems development. Those who follow such a procedure will most likely reach their goal in less time and with greater efficiency.

However, one word of warning. Examine once again the last element of the list under the "Conduct of the Follow-on Meetings" guidelines. Sometimes, we can get so involved in the problem, and in procedures in general, that we lose sight of just what it is that we are there for. Remember, your goal is to develop a rule base that represents, as closely as possible the heuristics used by the domain expert. This must always be the focus of the effort.

In order to establish a (good) rule base, there are certain things to look for, in particular:

- The expert may not understand the concept of rules and heuristics. It is generally best to ask him or her about any neat little *tricks* or *shortcuts* that are employed to reach a decision. Usually, these represent pruning, or filtering, heuristics (i.e., heuristics used to reduce the number of paths searched in the inference network).

- Ask the expert about those instances in which he or she can come to a really quick conclusion. That is, just when are decisions easy? For example, the expert may note that, under certain circumstances he or she can immediately determine that outside help is needed, or that the problem is trivial. Associated with these situations are rules that need to be incorporated at the very top of the knowledge base. We want to identify these easy decisions as soon as possible in the inferencing process, in order to reduce both the time required to find a solution and the number of questions presented to the user.

- Is the problem one of *classification* or *construction*? If it appears to be a classification (i.e., diagnosis) problem, then try, in particular, to determine

   The *symptoms* the expert uses in classification
   The *treatments* that the expert recommends
   The relationships between symptoms and treatments (i.e., how are they matched)

If the problem faced is one of construction (e.g., the development of a schedule, loading scheme), then look especially for

   The data employed by the expert
   The list of alternatives (i.e., final conclusions), and whether or not this list may be preenumerated
   The heuristics used to prune the list of alternatives

- Determine, as soon as possible in the effort, the data used (and its source) by the expert in reaching his or her conclusions. If your expert system is to reach conclusions similar to those of the domain expert, it will require access to the same data.

Multiple Domain Experts

Surko addresses, in her paper [1989], one additional consideration in knowledge-base development: the existence of more than one domain expert. Some authors have noted that this situation can be particularly frustrating if not properly and delicately handled. Surko, however, advises that the knowledge engineer need not be particularly concerned about multiple experts. That is, using a rule base cloned from one expert, we build a prototype expert system and then let the other domain experts critique the results. She states: "If two experts disagree, then you'd probably better just quietly choose the version you feel is right. Your aim is to get the best knowledge, but you must do so with the minimum discord. . . . " In addition, she notes that getting experts together to air their differences is usually too risky for the potential gains.

Our own experiences in dealing with multiple experts has followed a similar approach. We have always selected one domain expert as the individual from whom the rules were to be acquired, that is, as the *key expert*. And, as Surko suggests, we have presented the prototypes to the remaining experts for a critique. In doing this, we have tried to discourage the key expert from attending such presentations. We feel that his or her attendance may cause the other experts to feel less free in making their comments and criticisms. From the suggestions, comments, and criticisms received from the other experts, we have then attempted to identify those that seemed to be both constructive and important. We then presented these to the key expert for his or her comments. Rather surprisingly, at least to us, we have almost always found the key expert to be responsive and objective in such situations.

There is yet another situation in which multiple experts may be encountered. However, rather than having mastery across the entire domain of interest, these experts may each have expertise in various portions of the domain. One approach to this situation is to develop a set of expert systems, one for each *subdomain*. Another is to utilize separate knowledge bases (i.e., one for each expert) and to coordinate these through a single expert systems package—by means of the *blackboarding* approach. And this is precisely the approach used in HEARSAY, an expert system described in Chap. 3.

KNOWLEDGE ACQUISITION:
AN EXAMPLE

In the August 12, 1988 issue of *The Wall Street Journal* [Rose, 1988], a front page story described one recent, and somewhat less than successful attempt at knowledge acquisition. Since there are lessons to be learned through both success and failure, let us consider this particular situation. While reading the summary of this effort (as extracted from the referenced article), compare if you will what was done with the guidelines that we have presented earlier.[5]

Mr. Thomas Kelly was, at the time of the publication of this story, a 55-year-old civil engineer working for Southern California Edison. Over a period of 20 years he had become an expert in the diagnosis and treatment of problems of a massive and troublesome earthen dam high in the Sierra Nevada mountain range of central California. For example, from supposedly just a trickle, Mr. Kelly is able to determine if the dam is bleeding—or just breathing.[6]

Mr. Kelly's expertise was considered too valuable to risk losing. Thus, the company spent two years and roughly $300,000 in an attempt to create an expert system that cloned his knowledge of earthen dams. One particular dam under consideration is named Vermilion. It is a trouble-plagued, 24-year-old dam that spans Mono Creek, a tributary of the San Joaquin River. The dam is 165 feet high and about a mile in length and can hold as much as 40 billion gallons of water. It is particularly prone to water seeping through its foundation and Mr. Kelly's expertise has been a major factor in keeping the dam—and the population in the valley below—safe. An expert system was proposed to achieve a level of problem solving comparable to the human expert, for application to *any* earthen dam.

Southern California Edison called upon Texas Instruments, Inc. for the development of an earthen dam expert system. Texas Instruments sent two *knowledge engineers* to proceed with the knowledge acquisition phase. One was a 30-year-old engineer and *computer whiz* and the other was a 30-year-old management systems expert. This was *the first assignment for both*.

To quote directly from *The Wall Street Journal* article[7]:

> The programmers were nervous when they first sat down with Mr. Kelly, who is almost twice their age. They had boned up on books about dam safety and construction, but that was just a start. Before them was a professional engineer, with decades of experience that they needed to dissect. . . .

> Their first meeting was a marathon, 7-hour session in a windowless room, where the programmers grilled Mr. Kelly. Tell us about Vermilion's earth and gravel interior, the pair asked. What about the drainage systems that riddle the dam? Every syllable was tape-recorded and transcribed for later study. When [they] left the meeting, everyone's eyes were bugging out. . . .

As the sessions continued, Mr. Kelly began to find them more and more troubling. The knowledge engineers were equally frustrated. Mr. Kelly would

---

[5]Reprinted by permission of *The Wall Street Journal*, ©Dow Jones & Company, Inc., 1988. All Rights Reserved Worldwide.

[6]We must admit to being skeptical of such statements. Our experience has been that the domain expert invariably understates the amount of data that he or she employs to actually reach a conclusion.

[7] The article serves, unfortunately, to perpetuate the perception of knowledge engineers as computer programmers, or "computer whizes."

listen to questions and reply with a simple "yes" or "no." When he provided explanations, which was evidently not too often, they were brief. While Mr. Kelly insisted that he was being cooperative, the knowledge engineers wondered if he was perhaps unconsciously reluctant to *part with his expertise*.

The knowledge engineers' first attempt at constructing a prototype expert system met with equally frustrating results. For almost every problem posed, the system would respond with the suggestion to pack the offending wet area of the dam with gravel and keep it under observation. Finally, the knowledge engineers visited the dam. And this visit provided them with what they termed a *visual breakthrough*, an enhanced appreciation of the elements of the problem.

However, even months later, the expert system's knowledge base consisted of only about 20 rules, and still produced trivial advice. After a year, and concerned now about the time and funds being expended, Southern California Edison decided to narrow the project's scope to consider only the Vermilion dam.

Texas Instruments assigned yet another employee to the project. Gradually, the group's insight into the problem grew, as did the knowledge base. The number of rules increased to about 80. However, confidence in the system, on the part of Mr. Kelly, seemed lacking.

Ultimately, the project *wound down*. The Texas Instruments' team moved on to other projects. The expert system developed is not in regular use and Southern California Edison officials state that the system "needs more work to be really useful."

The less than satisfactory results of this particular effort are hardly unique. A *formula* for the lack of success in developing the earthen dam expert system can be noted in a host of other expert systems efforts. Analyzing the situation, one may note a number of factors that may have diminished the project's chance of success. These include

- The assignment of inexperienced knowledge engineers (recall that this was the first assignment, in the construction of an expert system, for both)
- An initial knowledge acquisition session that immediately may have served to get things off on the wrong foot (i.e., the 7-hour marathon/grilling described above)
- A failure to make an on-site visit of the dam as soon as possible in the effort (such a visit should have ideally taken place *before* the acquisition sessions were actually initiated)
- An evident failure, on the part of the knowledge engineers, to overcome the domain expert's reluctance to provide detailed explanations
- An evident failure to achieve a relationship of *mutual* respect

To clarify the last point, it would seem, at least from reading the article, that the two novice knowledge engineers were in awe of the domain expert. On the other hand, Mr. Kelly may have perceived such a reaction as an indication of a lack of experience coupled with naiveté. Mr. Kelly could point to a successful career as an earthen dam expert; the two knowledge engineers had yet to accomplish their first successful expert systems project. And this was evidently still the case at the conclusion of this $300,000, two-year effort.

THE KNOWLEDGE ENGINEER AS THE DOMAIN EXPERT

As mentioned earlier, the selection of an expert may be a problem in the knowledge-acquisition phase. There may, in fact, be no practical access to a domain expert. The expert may have died, left the company, or may simply refuse to participate in the development of an expert system. In such an instance, we have at least two options. One is to use (or generate) historical data through which a rule-based expert system may be constructed. The second is to become our own expert. In this section we will focus on the latter approach.

Quite often in the expert systems literature there are warnings against becoming one's own expert. While we agree that such an approach is a matter of concern, we also believe that it may prove worthwhile if conducted properly. In fact, one of the most highly touted expert system, and one that is in actual use, is the R1 (or XCON) system for the configuration of VAX computers at the Digital Equipment Corporation (DEC)—and the initial prototype of this system was developed primarily by a knowledge engineer. Specifically, John McDermott of Carnegie-Mellon University began this project with a meeting of the VAX configuration experts at DEC. This meeting simply provided McDermott with an overview of the task. He then took two DEC configuration manuals back to his office and reviewed them in detail. Using this input, he developed a prototype expert system consisting of about 250 rules in about three person-months. The demonstration of the prototype indicated that it was able to satisfy all of the basic configuration problems provided by DEC. This success resulted in an extension of the project wherein an expert was used to evaluate the refined system. However, the initial prototype was developed primarily by the knowledge engineers. And this demonstrates that it is indeed possible to be one's own expert, at least in certain situations.

One argument against becoming one's own expert is the belief (as based on rather substantial empirical evidence) that it takes approximately 10 years to become an expert in a particular domain. Rather obviously, if this were the case, it would be very rare indeed to devote those 10 years to the development of our expertise simply for the purpose of constructing a single expert system. However, the 10-year figure is based on the belief, by psychologists, that a world-class expert (e.g., a chess grandmaster, a composer of classical music, or a Nobel laureate in a field of science) must store somewhere between 50,000 and 100,000 *chunks* of heuristic information prior to becoming an expert—and that it takes at least 10 years to acquire 50,000 chunks [Harmon and King, 1985]. Such an estimate may be reasonable. However, most often the types and breadth of expertise that a knowledge engineer is concerned with are hardly in the same league as that of a Nobel laureate.

In some respects, there are certain benefits to be gained by such an approach. That is, when knowledge engineers are their own experts, some aspects of the knowledge acquisition procedure are immediately simplified. For example, one can proceed directly to a statement of the rules in the IF-THEN format. Also, we do not have to worry about the expert's schedule, or the expert's reluctance (or inability) to describe the procedures used. Further, we may obtain a more objective view of the procedure since the knowledge engineer may have no vested interests in the organization under consideration. All too often, those who are dealing directly with a problem are *unable to see the forest for the trees*, and thus an outsider's perspective may represent a significant improvement. At any rate, there are advantages and disadvantages in being one's own expert and one should not dismiss, out of hand, such an approach.

THE DOMAIN EXPERT AS THE KNOWLEDGE ENGINEER

Now, if the knowledge engineer can act as the domain expert, why not employ the domain expert as the knowledge engineer? Actually, the ultimate goal of expert systems is to provide a development package that interacts directly with the domain expert, and thus supposedly eliminates the need for the knowledge engineer. However, this goal has yet to be reached and it is unlikely, despite certain claims in this area, that it will be—at least in the foreseeable future.

There has been, however, a certain amount of success in training domain experts to develop small-to-modest sized rule bases, and to implement these rule bases on various expert systems development packages—particularly on some of the more *user friendly* expert systems shells that are now available. Several firms have, in fact, reported significant improvement in overall productivity through the development of numerous small (rule-based) expert systems by their domain experts.

Our reaction to such an approach is mixed. We are all for the increased involvement on the part of everyone in the organization in expert systems development and implementation. Further, it is not inconceivable that many domain experts can, through proper training, become competent knowledge engineers. As just one example, in the case of Mr. Kelly, our expert on earthen dams for Southern California Edison, such an approach may have well proved fruitful. However, there are certain drawbacks. These include

- The time and funds required to train the domain expert
- The likelihood that any domain expert, so trained, will tend to *solve* any and all problems through expert systems—even when far more appropriate, effective, and efficient means exist

Each of these drawbacks may be alleviated, to some degree, by providing access to an in-house or external group of knowledge engineers. The knowledge engineers may provide the training and assist in rule base development. This is, in fact, the approach that has been taken by a number of firms.

We have had the opportunity to observe three firms that have employed domain experts as knowledge engineers. While some successful expert systems have indeed been developed, the overall picture has not been, in our opinion, entirely positive. In every instance that we encountered, the training of the (potential) domain expert has been through the immediate introduction of an expert systems shell. And the bulk of training has been dedicated to the *care and feeding* of that software, that is, to the input and output of the shells. Moreover, the training was provided by the software vendor (or representatives of that vendor). And once the training was *completed*, these people disappeared from the scene. Not surprisingly, this has resulted in the application, or attempted application, of expert systems to problems for which the approach is simply not suitable, or even reasonable. Further, even when a proper problem has been identified, some of the rule bases developed have left much to be desired.

However, it should be noted that all three of these organizations feel, at least thus far, that their approach has been successful. In particular, they have noted just how much they have saved by this approach versus that of adding a knowledge engineering group to the organizational structure. However, in doing so they have simply compared the costs devoted to training the domain experts to the costs of establishing a knowledge engineering group. Of course, following this line of reasoning, an even cheaper alternative (i.e., to the training of the domain expert in rule base development) would be to simply do nothing.

Another point that is made by the firms that encourage such a practice is that it takes less time to train a domain expert in expert systems than it does to train a knowledge engineer in the specific domain. This may or may not be true, based upon the particular situation. However, we question the basic premise of the argument. Knowledge engineers need to become only somewhat familiar with the domain, they do not have to be experts in that area in order to interface with the domain expert.

In our opinion, a better strategy is to provide a level of training in expert systems sufficient to allow the domain experts to be able to understand the purpose, scope, and limitations of expert systems—and to be better able to *identify* problems for which expert systems may be appropriate. Once this identification has been made, we believe that the domain expert should work with a trained, experienced knowledge engineer in the remaining phases of development and implementation. The knowledge engineer, if trained in decision analysis in general, may note that the problem identified can be solved more efficiently by some other approach. Or, if expert systems do apply, the knowledge engineer most likely brings a more objective, outsider's perspective to the development process.

However, from a less idealistic point of view, we must acknowledge that it is better to improve a situation than to leave it be—if the benefits of the improvement outweigh the costs. As such, and even in light of the problems cited, we would much prefer that a firm train its domain experts in knowledge engineering rather than ignore the benefits of the expert systems approach completely.

KNOWLEDGE ACQUISITION
VIA RULE INDUCTION

An alternative to the acquisition of knowledge through the interface with a human (i.e., an expert or a knowledge engineer assuming the role of the expert) is to convert an existing (*and appropriate*) database into a set of production rules [Carter and Catlett, 1987; Quinlan, 1983, 1987*a*; Thompson and Thompson, 1986]. The appropriate database, in turn, must consist of data that encompass examples pertaining to the type of problem under consideration—*where the examples selected should represent desirable outcomes* (i.e., it makes little sense to use examples which reflect poor judgment). More specifically, one needs examples of *good* decision making. In some cases, this approach may provide adequate results while, in others, it may at least lead to the development of a credible prototype system. Several commercial expert systems shells, in fact, incorporate means (actually, supporting programs) for the accomplishment of such a process. We will, in fact, present the results of the application of two commercial software packages (i.e., for the development of rules from a database) later in this chapter. However, one should most definitely first understand precisely how this is done, as well as the scope and limitations of this approach.

Before describing one popular approach to rule generation from data, let us first reflect upon the components of a knowledge base consisting of production rules. To clarify our discussion, let us return to the simplified aircraft identification problem as introduced in Chap. 4. Recall that our problem is concerned with the identification of various types of aircraft, where we will limit the number of aircraft under consideration to just the C130 (Lockheed Hercules), C141 (Lockheed Starlifter), C5A (Lockheed Galaxy), and B747 (Boeing Jumbo Jet). Note carefully that, for purpose of illustration, these four aircraft will be the *only* inhabitants of our universe of airplanes.

Identification of Objects, Attributes, and Values

Rather obviously, the objects in our knowledge base are the four different types of aircraft. Our next step is to consider the attributes that serve to distinguish these aircraft from one another. For example, some of the many attributes exhibited by these aircraft include

- Number of engines
- Type of engines (e.g., jet or propellor)
- Wing position (i.e., high on the fuselage or low on the fuselage)
- Wing shape (i.e., swept back or conventional)
- Tail shape (e.g., T-shaped or conventional)
- Bulges on the fuselage (e.g., aft of the cockpit, aft of the wing, under the wing, or none)
- Size and dimensions

- Color and markings
- Speed and altitude

Having identified a candidate set of attributes and their values, we should next consider filtering out those attributes that do not serve to support our decision-making process. For example, the number of engines is obviously not necessary since each of the four aircraft under consideration has exactly four engines. (However, in a more realistic aircraft identification model, we would extend our concern to all possible aircraft and there the number of engines does, of course, serve to distinguish one aircraft from another.)

We might also be able to drop the last three attributes since, at the distance we expect to see the aircraft, we will not be able to distinguish colors and markings or estimate size and dimensions, speed and altitude. We are then left with the second through the sixth set of attributes, which may or may not be enough to permit precise aircraft identification. Actually, as we will see, these five attributes (i.e., engine type, wing position, wing shape, tail shape, and bulges) will be more than sufficient—under the rigid assumptions of a strictly deterministic rule base and a completely stable system.

Before we proceed, let us repeat our warning, from Chap. 4, with regard to the avoidance of false economies. Realize that, by eliminating attributes, we are eliminating premise clauses in the rule base, as well as knowledge about the situation. Thus, one must always be careful that the attributes dropped from consideration are not a part of the *necessary* knowledge for the problem under consideration. This guideline can, however, create a dilemma. That is, should we seek the minimal set of attributes or, to be safe, should we try to incorporate every conceivable attribute? Unfortunately, there is no clear-cut answer. Too many attributes may make the knowledge base unwieldy—and require an inordinate amount of data and/or responses from the user. As a result, the expert system will appear to be plodding, if not outright *dumb*. Too few attributes can limit the usefulness of the expert system, as well as make future modifications more difficult. Thus, when all is said and done, we normally seek some acceptable compromise.

Assuming that we feel relatively confident with the filtered set of attributes, we may list these and their associated values. The resulting five attributes are presented in Table 5.1.

The Establishment of a Decision Tree

Recall, from Chap. 4, our discussion of inference networks. These are networks that serve to illustrate the rule base and inference process. Another, and generally more simple network is that of a *decision tree*. It also provides an alternative representation of rule bases (and, as such, is actually another mode of knowledge representation). Such a network may be best explained by means of an example. Consequently, we shall construct a decision tree for the aircraft identification problem of Table 5.1. Each node in the decision tree will represent either a question about the value of an attribute, or a conclusion. Each branch that emanates

TABLE 5.1
Aircraft attribute listing

Attribute	Aircraft type			
	C130	C141	C5A	B747
Engine type	Prop	Jet	Jet	Jet
Wing position	High	High	High	Low
Wing shape	Conventional	Swept-back	Swept-back	Swept-back
Tail	Conventional	T-tail	T-tail	Conventional
Bulges	Under wings	Aft wings	None	Aft cockpit

from a node pertaining to a question will represent one of the possible values of the associated attribute. Nodes pertaining to questions will be represented by boxes, while those depicting conclusions will be represented by circles.

Now, for our problem, let us arbitrarily use the node *engine type* as the *root node* of the tree, that is, the node at the highest level in the tree and which includes, under it, all of the possible solutions (i.e., conclusions) contained within the tree. At the next levels in the tree let us use the attributes *wing shape*, then *wing position*, then *tail shape*, and finally *bulges*. Our corresponding tree is then shown in Fig. 5.1. Notice, from Fig. 5.1, that our conclusions are shown at various levels in the tree. For example, as soon as we determine that the engine type is propellor, then we know (at least for our limited model) that the plane must be a C130. Also, we can theoretically arrive at a conclusion wherein we simply do not know what type of aircraft has been sighted. Such events are noted by the question marks as, for example, shown beneath and to the left of the tail shape attribute node. That is, the question mark refers to the fact that, to reach this event, we must have sighted a plane with jet engines, swept-back wings, high wing position, and a conventional tail shape. In our limited universe of planes, there is no such type of aircraft.

We may develop an alternative decision tree by simply reordering the nodes *and* eliminating any branches leading to impossible conclusions. One such ordering is demonstrated in Fig. 5.2. Notice in particular that the decision tree of Fig. 5.2

- Requires fewer attributes to arrive at the identification process (i.e., the attributes *wing shape* and *tail shape* are unnecessary)
- Never results in an impossible event (i.e., the question marks of the previous figure)

Generally speaking, we would prefer to develop a tree with as few levels (i.e., attributes) as possible and with all events representing known conclusions. The fewer the number of attributes, the fewer the data and/or user responses that typically will be needed to arrive at a conclusion. As such, it would appear that the second tree is preferred to the first. And this is essentially true as long as we are dealing with deterministic rule bases.

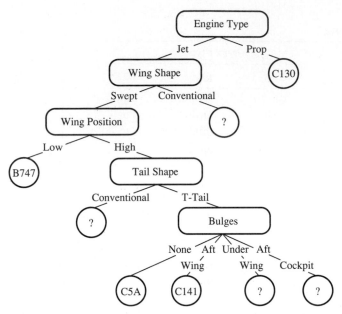

FIGURE 5.1
Possible decision tree for aircraft identification.

Generating Rules from Trees

At this point, it should be clear that we can proceed through a systematic aircraft identification process by simply proceeding down the decision tree according to the values of the attributes for the sighted plane. For example, using Fig. 5.2, we may determine that a sighted aircraft is a B747 by first noting that it has jet engines and then that its wings are positioned low on the fuselage. We can, in fact, do something more interesting, as well as more relevant to rule-based expert systems. Specifically, we can convert the decision tree into a set of production rules. To illustrate this process, we will use the tree of Fig. 5.2. However, first

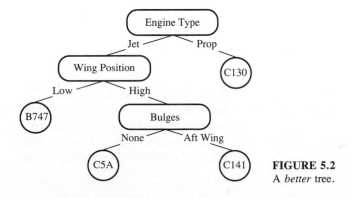

FIGURE 5.2
A *better* tree.

note that this tree does not contain any intermediate nodes. The absence of such nodes will, in fact, be a prerequisite to the use of the approach listed below:

- *Definition*. A *chain* is defined as a path from one node in the tree to another where we transverse the branches in only one direction.
- *Step one*. Identify any conclusion node that has not yet been dealt with.
- *Step two*. Trace the chain from the conclusion node backward through the tree (i.e., *up* the tree) to the root node (i.e., the highest level node in the tree).
- *Step three*. In the chain, as identified in step 2, circled nodes are considered THEN nodes, or conclusion clauses, while boxed nodes are considered IF nodes, or premise clauses.
- *Step four*. Construct the corresponding production rule set for the chain under consideration, and repeat this process for every conclusion node.

To demonstrate, one conclusion node in Fig. 5.2 is the node concluding that the aircraft is a C130. Tracing back up the tree, we return to the root node which pertains to the question, "What is the engine type?" Consequently, our production rule set for this node is

Rule 1:   *If* engine type is prop
           Then plane is C130

Similarly, for the remaining three conclusion nodes, we have:

Rule 2:   *If* engine type is jet
           and wing position is low
           Then plane is B747

Rule 3:   *If* engine type is jet
           and wing position is high
           and bulges are none
           Then plane is C5A

Rule 4:   *If* engine type is jet
           and wing position is high
           and bulges are aft of wing
           Then plane is C141

And we have, in fact, constructed a knowledge base for the aircraft identification problem that is complete and consistent.

However, before becoming too satisfied with our results, let us consider whether or not our complete and consistent rule set is also efficient. One possible definition for an efficient rule set is one that contains the fewest attributes (and thus, the fewest potential questions). While the rule set constructed from Fig. 5.2 is certainly more efficient than that which would be constructed from Fig. 5.1, it is not as efficient as the one for Fig. 5.3 shown on the next page.

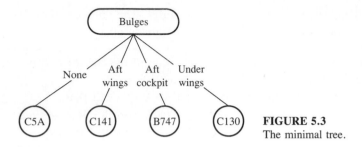

FIGURE 5.3
The minimal tree.

Specifically, the rule set for Figure 5.3 would simply be

Rule 1:   *If* bulges are none
           Then plane is C5A

Rule 2:   *If* bulges are aft of wings
           Then plane is C141

Rule 3:   *If* bulges are aft of cockpit
           Then plane is B747

Rule 4:   *If* bulges are under wing
           Then plane is C130

The only question that needs ever be asked of the user is the one concerning the character of the bulges on the plane's fuselage. For this example, we should by now realize that the most important attribute (the one containing the most information) is that of fuselage bulges. In fact, for this contrived, highly simplified problem, it is the *only* attribute of importance.

Before proceeding further, let us first consider whether or not it is always best to develop a decision tree (and corresponding production rule set) having a minimal number of attributes. If the situation is *deterministic*, and this has been our assumption thus far, then a tree with the minimum number of attributes is, *in theory—and for the static case—*optimal. That is, if an observer is absolutely sure that the aircraft sighted has bulges aft of the wing, and if it is true that the only aircraft in our universe with bulges aft of the wing are C141s, then we have established the identity of the plane with complete certainty. On the other hand, if the observer is not absolutely positive that he or she saw bulges aft of the wings (perhaps it was a cloudy day, or the perspective was less than ideal), it could prove extremely beneficial to offer supporting information on other aircraft attributes (e.g., wing position, wing shape). In the case of *uncertainty*, the case most typical of real-world problems, it is generally best to reinforce our partial information.

However, even with a deterministic rule base, it may still prove unwise to minimize the number of attributes (i.e., recall our discussion on false economies from Chap. 4). That is, if we design a rule base to minimize the number of attributes (i.e., premise clauses) for the existing situation, that rule base may have to be extensively modified in the near future—and may require the inclusion of attributes that were not previously necessary. The reader should bear this point in mind in our discussion of various approaches to the development of rule bases from examples.

Returning to our aircraft identification example, we may note that we develop different trees, and thus different production rule sets, depending upon the ordering of the attributes (and, particularly, with respect to the choice of the root node or attribute). In this example, we have generated the decision trees in an unstructured manner. In any real problem, with many attributes and nodes, one would like to have a much more systematic way in which to develop an efficient production rule set. One approach that has been developed to handle this problem is available through Quinlan's ID3 algorithm [Quinlan, 1983]. This algorithm, or variations on the algorithm, may be found in a number of commercial programs. We will now present our version of the ID3 algorithm by means of an illustrative example.

The ID3 Algorithm for Rule Generation

To demonstrate the ID3 algorithm, let us consider a hypothetical problem with regard to the generation of a production rule set for a simplified investment scenario. Specifically, since we may be dubious as to the existence of any true stock market experts, let us attempt to construct an expert system based solely on historical performance. The investment opportunities that we will consider will be limited to

- A mutual fund that invests solely in blue chip stocks
- A mutual fund that invests solely in North American gold mining stocks
- A mutual fund that invests solely in mortgage-related securities

Our goal will be to determine, at any given time and under any given set of conditions, just which *one* of these mutual funds in which to place *all* of our cash holdings. For convenience, let us classify the mutual fund's expected value into three classes: high, medium, or low. These three values then correspond to the conclusions that are to be reached by our expert system.

A much more difficult, and far more critical problem is one associated with the identification of an associated set of attributes. In our aircraft identification example, attribute identification was relatively trivial. Unfortunately, in most problems this factor is far more complex. Further, selection of attributes is primarily an art, coupled with a certain amount of luck, and enhanced by one's insight and experience. For purpose of discussion, we will assume that we have noted that the following factors seem to have the most apparent impact on the values of the three types of mutual funds:

- Interest rates
- Amount of cash available in Japan, Western Europe, and the United States
- The degree of international tension (e.g., prospects for military operations, incidents of terrorism, etc.)

Based on our attribute list, we next peruse historical data to generate the results exhibited in Table 5.2. Each row in Table 5.2 is said to represent a specific example. We will leave, as an open question, just how far back in time one should go in assembling such a data set. Note in particular that the attribute *fund value* represents the conclusion that is sought by the expert system. In order to be consistent with the terminology of the ID3 methodology, we will also term this attribute the *classification*, or *class*. That is, given any scenario (i.e., combination of interest rates, cash availability, and tension), we would wish to classify the mutual fund selections into high, medium, or low expected fund values.

Our next step is to develop a decision tree from this data set. However, rather than employ the arbitrary procedure used in our aircraft identification example, let us use a more systematic approach based on the measure of the entropy of each attribute. Entropy, in turn, is a measure commonly used in information theory. The higher the entropy of an attribute, the more uncertainty there is with respect to its outcomes (or values). Thus, we would wish to select attributes in order of increasing entropy, where the root node of our tree would correspond to the attribute with the lowest entropy value.

The formula for the entropy of any given attribute, A_k, is given as

$$H(C|A_k) = \sum_{j=1}^{M_k} p(a_{k,j}) \cdot \left[-\sum_{i=1}^{N} p(c_i|a_{k,j}) \cdot \log_2 p(c_i|a_{k,j}) \right] \qquad (5.1)$$

TABLE 5.2
Investment data set (examples)

Mutual fund-type	Interest rates	Cash available	Tension	Fund value (*class*)
Blue chip stocks	High	High	Medium	Medium
Blue chip stocks	Low	High	Medium	High
Blue chip stocks	Medium	Low	High	Low
Gold stocks	High	High	Medium	High
Gold stocks	Low	High	Medium	Medium
Gold stocks	Medium	Low	High	Medium
Mortgage-related	High	High	Medium	Low
Mortgage-related	Low	High	Medium	High
Mortgage-related	Medium	Low	High	Low

where

$$H(C|A_k) = \text{entropy of the classification property of attribute } A_k$$

$$p(a_{k,j}) = \text{probability of attribute } k \text{ being at value } j$$

$$p(c_i|a_{k,j}) = \text{probability that the class value is } c_i \text{ when attribute } k \text{ is at its } j\text{th value}$$

$$M_k = \text{total number of values for attribute } A_k; j = 1, 2, \ldots, M_k$$

$$N = \text{total number of different classes (or outcomes);}$$
$$i = 1, 2, \ldots, N$$

$$K = \text{total number of attributes; } k = 1, 2, \ldots, K$$

In Table 5.2, we have

- Four attributes (mutual fund-type, interest, cash, and tension); thus $K = 4$
- Three classes (i.e., the fund value is either high, medium, or low); thus $N = 3$
- Three values for the attribute *mutual fund-type* (blue chips, gold, or mortgage); thus $M_1 = 3$
- Three values for the attribute *interest* (high, medium, or low); thus $M_2 = 3$
- Two values for the attribute *cash* (high or low); thus $M_3 = 2$
- Two values for the attribute *tension* (high or medium); thus $M_4 = 2$

and we may compute the values of the entropy for each of our attributes by means of Eq. (5.1). For example, to compute the entropy for the attribute *cash*, we proceed as follows:

$$p(a_{3,1}) = \text{probability that cash is high} = 6/9 \text{ (i.e., from Table 5.2, cash availability is high 6 out of 9 times)}$$

$$p(a_{3,2}) = \text{probability that cash availability is low} = 3/9$$

$$p(c_1|a_{3,1}) = \text{probability that a fund value is high when cash is high} = 3/6$$

$$p(c_2|a_{3,1}) = \text{probability that a fund value is medium when cash is high} = 2/6$$

$$p(c_3|a_{3,1}) = \text{probability that a fund value is low when cash is high} = 1/6$$

$$p(c_1|a_{3,2}) = \text{probability that a fund value is high when cash is low} = 0/3$$

$$p(c_2|a_{3,2}) = \text{probability that a fund value is medium when cash is low} = 1/3$$

$$p(c_3|a_{3,2}) = \text{probability that a fund value is low when cash is low} = 2/3$$

and substituting into Eq. (5.1), we have

$$H(C|\text{cash}) = \left(\frac{6}{9}\right) \cdot \left[-\frac{3}{6} \cdot \log_2\left(\frac{3}{6}\right) - \frac{2}{6} \cdot \log_2\left(\frac{2}{6}\right) - \frac{1}{6} \cdot \log_2\left(\frac{1}{6}\right)\right]$$

$$+ \left(\frac{3}{9}\right) \cdot \left[-\frac{0}{3} \cdot \log_2\left(\frac{0}{3}\right) - \frac{1}{3} \cdot \log_2\left(\frac{1}{3}\right) - \frac{2}{3} \cdot \log_2\left(\frac{2}{3}\right)\right] = 1.2787$$

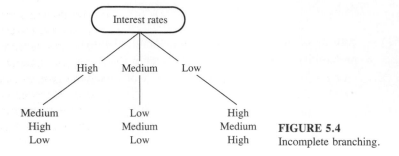

FIGURE 5.4
Incomplete branching.

Similarly, for our remaining attributes, we have

$$H(C|\text{interest}) = 1.140333$$
$$H(C|\text{tension}) = 1.2787$$
$$H(C|\text{mutual fund-type}) = 1.140333$$

There is a tie for the lowest entropy value between *interest* and *mutual fund-type*. Breaking this tie arbitrarily, and constructing a decision tree with a root node as *interest*, we may develop the result shown in Fig. 5.4. Listed below each attribute value in Fig. 5.4 are the classes (i.e., mutual fund values) associated with that particular branching of the decision tree. For example, under *interest* equals *high*, we may note (from Table 5.2) that the associated classes are *medium, high,* and *low*. Since all three branches result in an inconclusive classification, we must branch on yet another attribute.

Since we have already branched on the attribute *interest*, we must partition Table 5.2 according to the attribute *interest*. This results in three subtables (one for interest equals high, one for interest equals medium, and one for interest equals low). The subtable for interest equals high is shown in Table 5.3. Once again, we may compute the entropy for each attribute (i.e., *mutual fund-type, cash,* and *tension*), but now only for the examples provided in the subtable. The result is that the attribute *mutual fund-type* has the lowest entropy (a value of zero). In fact, for all partitions of *interest* (i.e., the other subtables formed when interest equals medium and interest equals low), this will be the case. Thus, we next branch according to *mutual fund-type*, resulting in the tree shown in Fig. 5.5. In this figure, BC, GS, and MR refer to the blue chip fund, the gold stock fund, and the mortgage-related fund, respectively. As may be seen, each branch leads

TABLE 5.3
Classifications when interest equals high

Mutual fund	Cash	Tension	Fund value
Blue chip stocks	High	Medium	Medium
Gold stock	High	Medium	High
Mortgage-related	High	Medium	Low

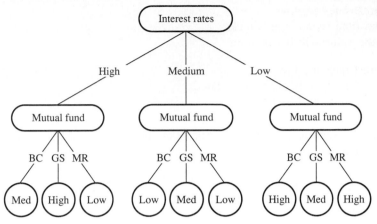

FIGURE 5.5
Final tree.

to a unique conclusion (i.e., class) and thus we are finished with the construction of the tree. Specifically, only two attributes are necessary to classify all results.

The associated (and unordered and ungrouped) rule set for Figure 5.5 is then

Rule 1: *If* interest rates are high
       and the fund type is blue chip
       Then the value will be medium

Rule 2: *If* interest rates are high
       and the fund type is gold stocks
       Then the value will be high

Rule 3: *If* interest rates are high
       and the fund type is mortgage-related
       Then the value will be low

Rule 4: *If* interest rates are medium
       and the fund type is blue chip
       Then the value will be low

Rule 5: *If* interest rates are medium
       and the fund type is gold stocks
       Then the value will be medium

Rule 6: *If* interest rates are medium
       and the fund type is mortgage-related
       Then the value will be low

Rule 7: *If* interest rates are low
       and the fund type is blue chips
       Then the value will be high

Rule 8:   *If* interest rates are low
          and the fund type is gold stocks
          Then the value will be medium

Rule 9:   *If* interest rates are low
          and the fund type is mortgage-related
          Then the value will be high

We could also easily combine certain rules. For example, rules 5 and 8 and rules 3 and 6 may be combined as

Rule 5/8:   *If* fund type is gold stocks
            and interest rates are medium
            or interest rates are low
            Then the value will be medium

Rule 3/6:   *If* the fund type is mortgage-related
            and interest rates are high
            or interest rates are medium
            Then the value will be low

We may, in fact, go even further and, for example, combine rules 7 and 9. However, again recall that the use of disjunctive premise clauses may not be supported by your particular choice of expert systems development package. Further, as more and more rules are joined, the rule base becomes more difficult to read—as well as to maintain.

DON'T GENERALIZE! Now, before we go any further, let us note that the decision tree formed in Fig. 5.5 may lead one to conclude that, at each level in the tree, the same attribute is encountered. For example, at the second level of the tree in Fig. 5.5, we find the attribute mutual fund-type. This, however, is only a coincidence. Typically, there are different attributes at different levels—as well as a different number of attributes in the various chains.

Some Warnings

The conversion of data into production rules is seductively simple, and particularly so in small examples of the type shown above. However, even the above example should indicate some of the problems that are associated with this approach. First, note carefully that while we have nine lines of data (or *examples*) in Table 5.2, they really only represent three distinct data points. That is, we have only listed three possible combinations of interest, cash, and tension scenarios. Thus, the number of examples provided in Table 5.2 is woefully small if we intend to actually construct a reasonably efficient expert system.

We may also encounter conflicts in the database, that is, two examples with identical attribute values may lead to different classes. In such an instance, we most likely have not identified a sufficient number of attributes.

Another difficulty may occur when one deals with attributes having continuous values. If you note our example of Table 5.2, only examples with discrete values (e.g., high, medium, and low) were dealt with. However, it should also be recognized that these values are actually discrete representations of (arbitrary) intervals for continuous values. For example, we noted that interest rates could take on the values of high, medium, or low. In reality, interest rates take on continuous values. Thus, to represent these continuous values by discrete values, we might let

low = interest rates from 0 to 5 percent

medium = interest rates from greater than 5 and up to 9 percent

high = interest rates greater than 9 percent

While this approach certainly eliminates continuous values, it raises yet another question, that is, what ranges of values should one employ for each discrete representation?

The ID3 algorithm will, for problems of any realistic size and complexity, develop very large trees in terms of the number of nodes and branches needed for representation. This results in equally large production rule sets. Unfortunately, large tree sizes do not necessarily result in better expert systems. In fact, it may well be that the effectiveness of the expert system (or classification scheme) is impacted negatively by large trees. Quinlan has developed various tree pruning methods to reduce tree sizes [Quinlan, 1983, 1987a, 1987b; Carter and Catlett, 1987]. One such approach is contained within Quinlan's C4 algorithm (an enhanced version of ID3 that includes *pessimistic* pruning). Using such an approach, a problem requiring 109 attribute nodes and 58 conclusion nodes was reduced to one with but 39 attribute nodes and 17 conclusion nodes. Not only was the size of the tree reduced, but the *hit ratio* (i.e., the proportion of cases correctly identified) was also significantly improved (from about 80 percent for ID3 to over 85 percent for C4).

Yet another problem may occur when one is dealing with stochastic events. Using ID3 for such data may result in trees that are enormously *bushy*. However, Quinlan has also developed extensions of ID3 to at least partially alleviate this situation. Additional comments and criticisms of the ID3 methodology may be found in the references [see, in particular, Mingers, 1986, 1987].

As a further comment on the use of data to generate rule-based expert systems, it should be noted that such an approach is only appropriate when one is faced with a diagnostic type of problem, that is, a problem in which, given certain symptoms (data), we wish to find an appropriate diagnosis (classification). Such a problem is also known by such names as classification analysis, discriminant analysis, or pattern recognition—and there are a number of conventional approaches to this type of problem. Some, based upon conventional statistical analysis [Johnson and Wichern, 1988; Lachenbruch, 1975], require that the attributes have multivariate normal densities with common covariance matrices (a requirement that is often not met in actual data) while others, based upon mathematical programming methods [Cavalier et al., 1989; Freed and Glover, 1981a, 1981b; Ignizio, 1987a] are not burdened by such restrictive assumptions. A fair portion of the artificial intelligence community seems, however, to be unaware of the mathematical programming-based approaches to discriminant analysis and all too often justify their use of the ID3 (or equivalent) algorithm primarily (if not solely) on the limitations of statistically based discriminant analysis methods. As a result, a number of rule-based expert systems have been built via approaches such as ID3 when a more appropriate alternative existed.

It may be observed that the ID3 algorithm attempts to minimize the number of attributes in the decision tree. As discussed, this has both good and bad points. One must weigh, based upon each individual case, whether or not the advantages of such an approach outweigh the disadvantages.

Finally, rule induction processes develop what are termed *flat rules*. That is, each rule results in one or more *final* conclusions. This is just another way of saying that rules with intermediate conclusions are not generated. For example, each of the rules of the rule set developed for the examples of Table 5.2 results in a final conclusion (i.e., choice of investment strategy). Intermediate conclusions, however, often serve to document the filtering or pruning process used by most domain experts to reduce the effort they expend in the search for a conclusion. As an illustration, in order to avoid searching through all stocks listed on a stock exchange, the domain expert may first filter out those that do not pay dividends above a certain amount. This results in a new, reduced set of alternatives. Next, the domain expert may rank the remaining set of candidate stocks according to the price-to-earning ratios (PE ratios), and eliminate those stocks that exceed some certain figure (e.g., a PE ratio of 12). A rule set developed by a knowledge engineer, in conjunction with such an expert should then contain a number of rules with intermediate conclusions—and, as a consequence, the pruning process employed by the domain expert is clearly documented. A rule set developed by means of induction from examples of stock selections will only implicitly include such a process.

Neural Networks and Rule Induction

In Chap. 4, we touched briefly on neural networks as a means of knowledge representation. Recently, neural networks have become a topic of considerable interest within the AI community. They have been found to be particularly effective in dealing with certain types of problems within the general realm of classification (e.g., discriminant analysis and grouping). And one of their primary advantages is that one need not develop production rules, instead neural networks develop (internally) their knowledge through training—on example problems.

However, and as we mentioned earlier, one of the biggest drawbacks of neural networks is that their knowledge base is almost totally opaque. Recently, a fairly extensive effort has been devoted to the induction of rules from neural networks. That is, we may use neural networks to derive the examples that are then presented to rule induction procedures, such as ID3. As a result, a set of rules corresponding to the internal knowledge base of the neural network may be acquired and established—in a transparent mode.

SOFTWARE FOR RULE INDUCTION

Many commercial expert systems shells provide supporting routines for the development of production rules from examples. This serves to reduce, considerably, the labor involved in such an effort. We will now describe, briefly, the use and evaluation of two particular packages for rule induction.

One of the most popular expert systems shells now on the market is the VP-Expert package available through Paperback Software. All remarks to follow are based on the results that we obtained when employing versions 1.2 and 2.0 of this package. A somewhat less well-known package (at least in the United States) is Xi Plus, available through Expertech Ltd., out of Berkshire, England. The Xi Plus package used was release 2.0. The evaluation itself was conducted during the summer of 1988.

A discussion of the use of VP-Expert and Xi Plus should help to indicate just how commercial packages employ rule induction systems. In addition, some potentially valuable insight into just how one might approach the evaluation of a proprietary (i.e., closed) software routine might be gained.

The Xi Plus software has a separate routine titled Xi Rule (version 1.00 was employed) that is used to derive rules from data. The method employed by Xi Rule is ID3, our old friend from the previous section.

VP-Expert uses a command termed INDUCE to develop rules from data. However, no explanation is given (at least in the manual that we were provided) concerning the underlying algorithm employed. Evidently, in such an instance, one is expected to accept the methodology employed strictly on faith, an approach that we personally find unacceptable. However, in fairness, it must be noted that VP-Expert is hardly alone in this approach.

There is a real problem in the purchase of software that is probably not discussed in the detail that it deserves. That is, since the source code for the software is usually proprietary (and thus unavailable for examination by the buyer/user), one is left with only three alternatives in regard to the use of such software. First, we might accept the validity of the software on faith. After all, we might think, if the package has sold hundreds or thousands of copies, must it not be all right? Second, we may peruse any existing reviews of the software, or find someone who is using the package and ask his or her opinion. Alternatively, and for the more skeptical, we could develop examples that are useful in exploring the limits of such software, and for detecting problems and inconsistencies. Our personal view, reinforced by experience, is that the third course is the only rational one. Consequently, after receiving copies of VP-Expert and Xi Plus (and Xi Rule), the first thing we did was to test each package on the set of examples provided

TABLE 5.4
Aircraft attribute listing—rule induction data set 1

	Aircraft type			
Attribute	C130	C141	C5A	B747
Engine type	Prop	Jet	Jet	Jet
Wing position	High	High	High	Low
Wing shape	Conventional	Swept-back	Swept-back	Swept-back
Tail	Conventional	T-tail	T-tail	Conventional
Bulges	Under wings	Aft wings	None	Aft cockpit

in Tables 5.4 to 5.7. Although each data set is small and evidently simple, each permits the quick determination of certain features (or lack thereof) of the rule induction software. Such testing is also invaluable, as we will see, in determining the specific mode of operation (and, in particular, of the inference process) of the package.

We have dealt with the aircraft identification example previously and, if you recall, the *only* attribute needed to classify the four aircraft is that of *bulges*. Thus, one might expect that a rule-induction procedure would develop the same set of rules as depicted earlier in Fig. 5.3. The medical diagnosis example is included to determine just how the software deals with examples that are in conflict. Specifically, if the eyes are bloodshot, temperature is high, and nose is runny, then one example states that the disease is a cold while the other states that it is the flu. One would hope that such a conflict, or the potential for such a conflict, would be identified in the rule-induction program. Finally, the investment data set is employed simply to determine if the results we obtained manually using ID3 will be the same as those obtained by the software packages.

The first data set was initially input into VP-Expert. The software includes an editor and screen into which you type the rows of the example data. The rows that should be typed in for the aircraft identification database are shown in Table 5.8. Note that the first row is simply a list of attribute names while the

TABLE 5.5
Aircraft attribute listing—rule induction data set 2

	Aircraft type			
Attribute	C130	C141	C5A	B747
Engine type	Prop	Jet	Jet	Jet
Wing position	High	High	High	Low
Wing shape	Conventional	Swept-back	Swept-back	Swept-back
Tail	Conventional	T-tail	T-tail	Conventional
Bulges	Under wings	Aft wings	None	Aft cockpit
Engines	4	4	4	4

TABLE 5.6
Medical diagnosis—rule induction data set 3

	Disease		
Attribute	None	Cold	Flu
Eyes	Clear	Bloodshot	Bloodshot
Temperature	Normal	High	High
Nose	Clear	Runny	Runny

TABLE 5.7
Investment data—rule induction data set 4

Mutual fund-type	Interest rates	Cash available	Tension	Fund value (*class*)
Blue chip stocks	High	High	Medium	Medium
Blue chip stocks	Low	High	Medium	High
Blue chip stocks	Medium	Low	High	Low
Gold stocks	High	High	Medium	High
Gold stocks	Low	High	Medium	Medium
Gold stocks	Medium	Low	High	Medium
Mortgage-related	High	High	Medium	Low
Mortgage-related	Low	High	Medium	High
Mortgage-related	Medium	Low	High	Low

TABLE 5.8
Induction table—aircraft identification via VP-Expert

Engine	Wing position	Wing shape	Tail shape	Bulges	Plane
Prop	High	Conv.	Conv.	Under wing	C130
Jet	High	Swept	T-tail	Aft wings	C141
Jet	High	Swept	T-tail	None	C5A
Jet	Low	Swept	Conv.	Aft cockpit	B747

remainder are the set of example data (i.e., as derived by observing the four different aircraft). Further, observe that this table is a *transposition* of Table 5.4.

The rules, as developed by VP-Expert for this example, are listed below:

Rule 0:   *If* engine = prop
        and wing position = high
        and wing shape = conv
        and tail shape = conv
        and bulges = under wings
        Then plane = C130

Rule 1:   *If* engine = jet
        and wing position = high
        and wing shape = swept
        and tail shape = T-tail
        and bulges = aft wings
        Then plane = C141

Rule 2:   *If* engine = jet
        and wing position = high
        and wing shape = swept
        and tail shape = T-tail
        and bulges = none
        Then plane = C5A

Rule 3:   *If* engine = jet
        and wing position = low
        and wing shape = swept
        and tail shape = conv
        and bulges = aft cockpit
        Then plane = B747

Thus, despite the fact that we only need one attribute (bulges) to classify the four aircraft, the software develops one rule for each aircraft, where the value for *every attribute* must be determined. Rather obviously, VP-Expert is not using ID3.

But, one may ask, is such a rule base *wrong*? The answer is, quite simply, no. Either this rule base or the one employing bulges only will reach the same conclusion. Further, if uncertainty exists (e.g., with regard to the attribute values input to the system), then the rule base developed by VP-Expert might actually be preferred. However, it is still a bit disconcerting to note that, no matter how many attributes are selected, all of these will be employed in the rule bases developed by VP-Expert.

When Xi Rule is applied to the same set of data, we do achieve the set of rules that would be expected, rules employing only a single attribute as shown on the next page:

Rule 1: *If* bulges are none
 Then plane is C5A

Rule 2: *If* bulges are aft of wings
 Then plane is C141

Rule 3: *If* bulges are aft of cockpit
 Then plane is B747

Rule 4: *If* bulges are under wing
 Then plane is C130

We then modified the aircraft data set to include the attribute *engines* (i.e., number of engines) to correspond to the data set of Table 5.5. For VP-Expert, this is accomplished by simply adding a new column to Table 5.8.

Since there are four engines on all of the planes, one would most likely expect that any package would easily recognize that the attribute engine is unnecessary. Xi Rule once again had no trouble with this example, and the set of rules did not change. However, VP-Expert proceeded to add a new premise clause to each of its four rules: *and* engines = 4. Now, one can argue that this is inefficient or one may argue that we need such knowledge (e.g., in the event that a new plane, with other than four engines, is added to our universe). While it is much easier said than done, we would have liked to see the rules developed by VP-Expert displayed with those representing unnecessary (or potentially unnecessary) premise clauses highlighted. In this manner, the knowledge engineer can decide which attributes (of those highlighted) to include and which to discard.

Next, we entered the third data set into both VP-Expert and Xi Rule. One would expect the conflict between the examples (i.e., for a diagnosis of cold or flu) to be identified. Xi Rule immediately detected the problem and flashed a warning message on the screen. In addition, the set of rules developed by Xi Rule, as listed immediately below, further served to indicate the existence of the problem:

Rule 1: *If* nose is clear
 Then disease is none

Rule 2: *If* nose is runny
 Then CLASH

Thus, Xi Rule correctly determined that only one attribute is necessary for classification (the value for nose) *and* that there is a potential conflict, or clash, whenever the value of nose is *runny*. However, note that if there really is no conflict (i.e., in the case of multivalued conclusions), then this feature can be more of a nuisance than a help.

When VP-Expert is used to develop the knowledge base from the same set of data, the result is the set of production rules listed on the next page:

Rule 1:   *If* nose = clear
          and temperature = normal
          and eyes = clear
          Then disease = none

Rule 2:   *If* nose = runny
          and temperature = high
          and eyes = bloodshot
          Then disease = cold

Rule 3:   *If* nose = runny
          and temperature = high
          and eyes = bloodshot
          Then disease = flu

When this knowledge base is consulted (i.e., via VP-Expert), it will provide the following answers:

- nose = clear, temperature = normal, eyes = clear: **disease = none**
- nose = runny, temperature = high, eyes = bloodshot: **disease = cold**

Thus, even though rule 3 should be triggered (by the latter set of attribute values), it is not fired. Again, there is nothing intrinsically wrong with this result. Rather, it simply tells us something about the mode of control in the inference engine of VP-Expert. Evidently, the first rule that is fired for a given conclusion clause attribute will terminate the inference procedure. However, VP-Expert does have a command statement (PLURAL) that may be employed whenever such situations are encountered. Using PLURAL, multiple conclusions may be reached.

Finally, both packages were tested with the fourth data set. VP-Expert continued to produce rules that included all of the attributes, even though *cash* and *tension* are unnecessary. However, in this instance, it did note the *frequent occurrence* (this is the terminology employed in VP-Expert, not our terminology) of both cash equals high and tension equals medium, and asked if we wished to combine these two into a single occurrence. That is, if one examines the fourth set of data, it may be noted that whenever cash equals high, tension equals medium. (Oddly enough, it failed to notice the *frequent occurrence* of both cash equals low and tension equals high.)

When Xi Rule was applied to this last example, precisely the same tree (i.e., Fig. 5.5) and associated set of production rules were obtained as had been developed by hand in the previous section. This is as expected since Xi Rule employs the ID3 algorithm.

At the conclusion of these four tests, it was clear that the methodologies employed by the two software packages were based on two totally different philosophies. Personally, we would like to see a package that combines both approaches, that is, that (as discussed earlier) lists all attributes but highlights those that are not necessary for the minimal decision tree. In any event, it was only through such an evaluation that the scope, limits, and operating philosophies of the two approaches (and particularly that of VP-Expert) could be ascertained.

SUMMARY

Knowledge acquisition is that phase of expert systems development dedicated to the identification of the rules and facts that comprise the knowledge base. In some cases, such acquisition may be accomplished through interviews with human experts. In others, human experts either do not exist or are unavailable. In this latter instance, one may either attempt to be one's own expert or utilize, if possible, historical data to construct a set of production rules. Yet another alternative is that we might consider training the domain expert to at least recognize the existence of problems that might be approached by expert systems.

Knowledge acquisition through a human expert is a delicate task that needs to be well thought out and carefully and deliberately conducted. Guidelines to this approach exist as a result of the observations of those who have used such a procedure in earlier efforts. However, such guidelines are obviously subjective and incomplete. One can only truly appreciate this task through actual experience.

Acting as one's own expert is generally frowned upon. However, in certain instances there is no other reasonable choice. Further, successful expert systems have been developed (at least in part) through such an approach and thus it is an alternative that should, at least, be considered.

Another alternative to knowledge acquisition may, in some cases, be achieved through the conversion of historical data (or examples) into production rules. The ID3 algorithm, and its various extensions, provides one means for accomplishing this when faced with a problem of classification. However, there may also exist more conventional alternatives [Cavalier et al., 1989; Freed and Glover, 1981a, 1981b; Ignizio, 1987a; James, 1985; Johnson and Wichern, 1988; Lachenbruch, 1975] to such problems that should also be considered prior to making a decision as to the approach to be used.

EXERCISES

5.1. You have been hired to construct an expert system for a firm. They have identified the domain expert for you. However, after a few sessions with this expert, you discover that he has been taking credit for the decision-making abilities of a subordinate. What should you do?

5.2. You have been hired to construct an expert system for a firm. They have identified the domain expert for you. This person is an acknowledged master of a certain task, but gets along poorly with her supervisors. The firm wants to *capture* her knowledge and then terminate her, and have asked that you keep this plan in confidence. What should you do?

5.3. Recall Exercise 4.16 from Chap. 4. Let us assume that you are faced with *two* alleged experts in the area of personnel selection for knowledge engineering positions. One holds the views presented in Exercise 4.16, the other has a diametrically opposed perspective. He, in fact, believes that only computer science graduates with training in LISP should ever be considered as knowledge engineers. How do you resolve this dilemma?

5.4. There is some debate as to whether or not it is easier, and more cost effective, to train a domain expert as a knowledge engineer than to hire a knowledge engineer to extract the rule base from the domain expert. Outline a systematic, formal plan of research that might resolve this issue. How might the choice of expert systems development tools affect the results of such an investigation?

5.5. A firm has asked you to develop an expert system for them. However, the domain expert from whom the knowledge base is to be acquired has orders to stay at the plant during working hours and to respond, immediately, to any requests (i.e., from the firm) for assistance. He, in fact, must wear a pager at all times. As a result, you find that your sessions with the domain expert are repeatedly interrupted and, consequently, your progress toward knowledge-base development is proceeding at a snail's pace. What might you do?

5.6. Why is it important to try to determine the set of conditions that leads the domain expert to *quickly* reach a conclusion (i.e., the so-called easy decisions)? Provide some examples of such conditions.

5.7. In building an expert system, you have asked that the domain expert meet with you at your office (i.e., to avoid interruptions). The company involved agrees to such an arrangement but the domain expert is reluctant. He notes that he is most comfortable in his own surroundings. What might be the implications of such a statement? How might you deal with such a situation?

5.8. Consider our earlier discussion of Mr. Kelly, the earthen dam expert for Southern California Edison. Further, assume for sake of discussion that this expert simply had difficulty in articulating his rule set. What approach do you feel might then be most appropriate for the conduct of the knowledge acquisition task?

5.9. Referring again to the earthen dam expert, let us assume that you are considering using Mr. Kelly as *both* domain expert *and* knowledge engineer. However, a representative from an AI consulting firm advises you to not try this approach. He notes that expert systems should be constructed using the most powerful and sophisticated tools available. These range from AI languages (e.g., LISP, PROLOG) to extremely sophisticated frame- and object-based expert systems development packages. Moreover, he notes that, to gain competence in the use of such tools takes years of study and experience. Thus, it would simply be impractical to train Mr. Kelly. Do you agree with this argument? Why or why not?

5.10. The acquisition of production rules via induction is only advised for problems of classification. Explain, clearly and concisely, just why this is so.

5.11. A marketing firm has been asked to develop a new advertising campaign for an automobile manufacturer. The company makes two cars: the Scatabout and the Hotshot. The Scatabout is a no-frills subcompact. The Hotshot is a *loaded* muscle car. They wish to determine the customer type most likely to buy each car, and then selectively address these potential customers through advertisements. List the attributes that you believe are most likely to discriminate between the car buyers.

5.12. Ten people have been surveyed concerning their car preference in support of the objectives of Exercise 5.11. For each type of car (subcompact or muscle car), they have been asked to list their preference as either high, medium, or low. The results are summarized in the table on the next page. Develop the associated rule set by means of the ID3 algorithm.

Car type	Age	Income	Job type	Preference
Subcompact	≤30	Low	Blue collar	Medium
Subcompact	≤30	Medium	White collar	High
Subcompact	>30	Medium	Blue collar	Medium
Subcompact	>30	High	Blue collar	Medium
Subcompact	>30	High	White collar	Low
Muscle car	≤30	Low	White collar	High
Muscle car	≤30	Medium	White collar	Low
Muscle car	>30	Low	Blue collar	High
Muscle car	>30	Medium	White collar	High
Muscle car	>30	High	White collar	Low

5.13. The mechanical properties of some heavy nonferrous alloys are presented in the table below. Use the ID3 algorithm to develop a rule set that will serve to identify the specific metal/alloy under investigation.

Density	Tensile yield strength	Tensile strength	Tensile modulus elasticity	Elongation in 50 mm	Metal or alloy
High	Low	Medium	High	High	Annealed copper
High	High	High	Medium	High	Annealed brass
High	High	High	Medium	Medium	Annealed bronze
Low	Low	Low	Medium	Medium	Aluminum (1100–0)
Low	Medium	Medium	Low	Low	Magnesium (AM100A)
Low	Medium	Medium	Medium	Medium	Aluminum (5452)

5.14. Consider the professors within your major department, or your coworkers if you are not in school. Develop a rule base, via ID3, that may serve to identify each of these individuals where the attributes used are restricted to *visual* clues (e.g., color of hair, amount of hair, height, etc.).

5.15. Apply the ID3 algorithm to the following example sets within Chap. 5:
   (*a*) Table 5.4
   (*b*) Table 5.6

5.16. The *Red Cross First Aid Guide* presents a listing of symptoms, or signs, to look for to identify various illnesses. The following table is a simplification of one such list. Use the ID3 algorithm to induce the associated rule base.

Sign	Illness				
	Fainting	**Stroke**	**Heart attack**	**Hyper-glycemia**	**Hypo-glycemia**
Loss of consciousness	Sometimes	Sometimes	Sometimes	Sometimes	Sometimes
Skin color	Pale	Pale	Normal or bluish	Normal	Pale
Sweating	Sometimes	None	Sometimes	None	Cold sweat
Paralysis	None	Often	None	None	None
Breathing	Normal	Difficult	Difficult	Deep/rapid	Normal
Speech	Normal	Normal or slurred	Normal	Confused	Normal
Pain	None	Sometimes	In chest	None	None

5.17. For Exercise 5.16, can you reduce the problem size by simple inspection? If so, show your results and discuss the approach.

5.18. Using the information presented in the table below, develop a rule base via the ID3 algorithm for the determination of the choice of the solution approach to various problems of the classification category. (Please note that the results presented in this table are for illustration only and are not meant to represent the optimal, or only, choices for the conditions shown.)

Rule base transparency required?	Disjoint data sets required?	Normality required?	Value type	Most appropriate approach
No	Yes	No	Numeric and continuous	Single-layer perceptron
No	No	Yes	Numeric and continuous	Statistical-based
Yes	No	No	Symbols	Expert system
No	No	No	Numeric or symbolic	Math programming

5.19. Discuss, for the rule base developed above, the potential problems associated with any reduction in the number of attributes.

5.20. Tom and Betty are both physicians with a specialty in allergies. Both have been highly recommended as the domain expert candidates for a medical expert system (for the diagnosis and treatment of allergies) that you have been funded to develop. Both physicians started their practice about the same time and in the same city. However, Tom's practice has flourished while Betty's is but marginally successful. Watching both at work, you note that Tom is pleasant and understanding, and spends a surprising amount of time with each patient. Whenever he is trying to identify the patient's affliction, he goes through a carefully planned, predetermined, and lengthy question-and-answer session before deciding on the set of tests to run on the patient. Betty, however, is considerably more reserved and, in fact, almost brusque in her doctor-patient relationships. Her sessions with each patient are short and to the point, and the few questions she asks seem to vary widely from patient to patient (as contrasted to Tom's questions which hardly ever seem to change). Which candidate would you expect to be the best choice as your domain expert? Why?

5.21. Most literature on expert systems notes that domain experts are *smarter* than expert systems. However, there are certainly a number of expert systems that outperform the experts—even the expert from whom the rule base may have been acquired. In light of such results, who do you believe is smarter, the expert or the expert system?

CHAPTER
6

THE
INFERENCE
ENGINE

THE ROLE OF THE INFERENCE ENGINE

The inference engine serves as the inference and control mechanism for the expert system and, as such, is an essential part of the expert system as well as a major factor in the determination of the effectiveness and efficiency of such systems. Inference, in turn, is the process of drawing a conclusion (either intermediate or final) by means of a set of rules, for a specific set of facts, for a given situation. Inference is thus the *knowledge processing* element of an expert system.

The most common inference strategy employed in expert systems is known as *modus ponens*. Simply stated, *modus ponens* means that if the premise of a rule is true, then its conclusion is also true. Thus, if A infers B and A is true, then B is true. This may be represented as: A $\rightarrow$ B. Notice, however, that if B is true, we cannot say that A is true. For example, we can say

> *If* an animal gives live birth
> *Then* it is a mammal

However, we cannot say that if an animal is a mammal, it gives live birth (e.g., the platypus, which lays eggs, is a mammal).

The control responsibilities of the inference engine are those used to determine such matters as

- How to start the inference process
- Which rule to fire if more than one is triggered
- The manner in which the search for a solution is conducted

Like the knowledge base, the inference engine contains rules and facts. However, the rules and facts of the knowledge base pertain to the specific domain of expertise while the rules and facts of the inference engine pertain to the more general control and search strategy employed by the expert system in the development of a solution. These two sets of facts and rules are purposely kept separate in the typical expert system. This, as we have noted earlier, is one of the key features of expert systems that serve to differentiate it from heuristic programming.

This separation results in several advantages. First, it permits one to make changes in the knowledge base with minimal impact on the inference engine, and vice versa. Second, it provides for the development and use of expert systems shells. Expert systems shells, as discussed, contain all of the necessary components of an expert system with the exception of the knowledge base. Should the inference strategy of such a shell be compatible with that required by a given knowledge base, then the shell may be used to accommodate that knowledge base. This approach results in a type of *plug-in* architecture where we insert various knowledge bases and run these via a host expert systems shell. In doing so, we should be able to avoid much of the time-consuming effort required to construct our own expert system from scratch. Given the capabilities of existing expert systems shells, there is, in fact, little need for the knowledge engineer to ever have to deal with the construction of an inference engine—or any part of an expert system outside that of the knowledge base. However, any successful knowledge engineer must at least understand the fundamentals of the inference process.

UNDERSTANDING WHAT'S INSIDE
THE INFERENCE ENGINE

Many expert systems shells are restricted in terms of the mode of inference used. Thus, while it may be possible to use a particular shell for a given knowledge base, it may also be the case that substantially improved performance could be achieved using a more compatible shell. All too often, however, a user becomes enamored with a particular shell (or perhaps it is simply the only shell to which he or she has easy access, or familiarity) and uses it indiscriminately.

In far too many cases (and this, unfortunately, is becoming more true with the increase in popularity of expert systems—and the resultant proliferation of courses in the subject matter), formal courses in artificial intelligence/expert systems are so focused on the implementation of a specific expert systems shell that one is never exposed to just exactly what goes on *inside* this shell. Instead, the shell is treated as a combination *universal problem solver* and mysterious *black box* where one accepts, evidently on faith, that it does indeed perform whatever operations are required. Those *educated* in this manner are likely to be manifestations of the old saying, "a little knowledge is a dangerous thing." This type of approach to expert systems is both unfortunate and unnecessary as it requires relatively little time to develop a reasonably sufficient understanding and appreciation of the workings of the inference engine—and it is to this topic that the remainder of this chapter is devoted.

SEARCH STRATEGIES

The purpose of an expert system is to develop and recommend a proposed solution (or set of alternative solutions) to a given problem. To accomplish this task, the expert system must conduct a *search* for the solution; and it is the responsibility of the inference engine in particular to perform this search in an efficient and effective manner. In the search process, we are faced with a number of alternatives (i.e., potential solutions) and, typically, a variety of constraints. For example, when faced with the problem of determining just what automobile to purchase, our alternatives theoretically include literally all of the different automobiles in existence. However, we are also faced with certain constraints, including

- Budgetary limitations: we must restrict our attention to automobiles that are priced less than or equal to a certain amount
- Automobile availability (i.e., not all cars in existence are for sale, or at least not for sale within a reasonable distance): we might then restrict our attention to only those cars that are for sale within, say, a 50-mile radius
- Style: for example, if we have a family, we must have a car with four doors and a large storage area
- Time: we must make our decision within a certain amount of time

Such constraints serve to filter out the number of potential automobiles from which we will make our selection. Other factors that we will typically use to reduce the number of alternatives to a reasonable level include age of car, mileage on car, preferred manufacturer, preferred dealers, color, and so on. Ultimately, we will then focus on but a few cars from which we make our final selection.

The search strategy implied in the selection of a car may be described in more technical terms as a *forward chaining* search with *pruning*. That is, we begin the process with certain data concerning the type of automobile desired, its style, cost, age, mileage, and so on. These data, along with our constraints, serve to filter out the majority of the potential alternatives and thus we ultimately arrive at only a few automobiles from which we make our final selection. From the point of view of the expert system, the pruning process reduces the size of the associated inference network, resulting in a reduction in search requirements. The inference network of Fig. 6.1 serves to depict a simplified version of our process. The assertions in the boxes to the left represent

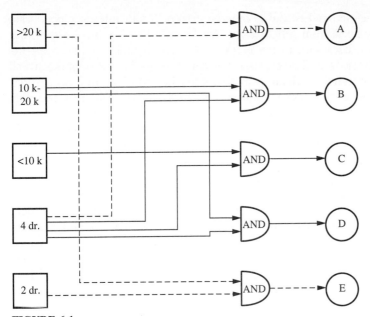

FIGURE 6.1
Inference network for automobile selection.

- Cost of automobile (i.e., greater than $20,000; between $10,000 and $20,000; and less than $10,000)
- Number of doors (i.e., 2 door or 4 door)

The conclusions, to the right, represent automobile choices (A, B, C, D, and E). Now, if we restrict ourselves to cars that are no more than $20,000 in cost and that have four doors, we can reduce the inference network in size. Specifically, we may eliminate the dashed branches as well as all assertions, conclusions, and logical connectives associated solely with these branches. Such a pruning process often enables one to develop solutions to otherwise unsolvable problems.

We could approach our automobile selection problem from an entirely different direction by first specifying a particular car for purchase, and then determining whether or not it meets our needs. When using this approach, we are said to be employing *backward chaining*. For example, we might first consider the purchase of a Porsche 944. Having established this as a tentative decision, we then determine whether or not it is feasible (i.e., does it satisfy the associated premise clauses of our knowledge base?). If we have a large family and require a car with four doors and substantial storage area, the Porsche obviously does not qualify. However, if we want a sporty, fast car that serves to inflate our self-image, the Porsche may be the most attractive candidate.

The two fundamental search strategies employed by an expert system are then forward and backward chaining. Forward chaining proceeds from premises (or data) to conclusions, and is said to be *data driven*. In our inference networks, we would then proceed from left to right. Backward chaining proceeds from a tentative conclusion backward to the premises to determine if the data supports that conclusion. Backward chaining is often called a *goal-driven* approach and proceeds from right to left. Ultimately, both approaches will lead to a conclusion, but their search efficiency is dependent on the nature of the problem faced, that is, on the nature of the inference network associated with the problem.

Specifically, if one has a few premises and many conclusions, then forward chaining is generally the best search strategy. Otherwise, with many premises and relatively few conclusions, we should normally employ backward chaining. The inference networks associated with forward chaining are said to *fan outward* while those for backward chaining *fan inward*. The (original) automobile selection example is then one with many conclusions and (relatively) few premises. As such, we would most likely prefer to approach this problem with a forward chaining search strategy.

We do not always employ just backward chaining or just forward chaining for a specific problem. Rather, there are instances in which we may employ *both* approaches. As an example, consider the problem of trying to locate a fugitive. The particular fugitive in question, Mr. Z. Worst, has absconded with the money entrusted to him by some rather naive investors. When last seen, Mr. Worst was boarding a plane bound for Buenos Aires. The detective hired by several of his victims has discovered the following facts about Mr. Worst:

- He speaks only English and German, and is more comfortable with German.
- His father was an infamous Nazi war criminal.
- He suffers a serious disease that requires periodic medical attention.
- He despises large cities, crowds, and traffic.
- He is a habitual gambler and loves, in particular, horse races.

Using these facts, the detective decides that Mr. Worst is not going to remain in Buenos Aires and, in fact, is unlikely to remain in Argentina. He feels that, because of his background and languages, Mr. Worst will most likely head for a community in which there is a large German population. Further, since he avoids large cities, the detective decides that he will probably locate in one of several small villages in Paraguay; each of which is known to have a substantial German population and even some neo-Nazi inclinations. Notice that, so far, the detective has used forward chaining to focus in on a limited set of candidate hiding spots for our fugitive.

At this point, the detective shifts to backward chaining. In particular, he conjectures that one small community is a likely spot. Based upon this belief, he then determines what information needs to be collected in order to corroborate his guess. That is, does the town have access to good medical care? Is the town reasonably close to a horse-racing track? If these conditions are indeed true, then our detective will start his search in this town.

As may be seen, a combination of forward and backward chaining may indeed prove to be better suited to a problem than just a single mode of chaining. Some commercial packages do, fortunately, permit the use of such combinations. However, before proceeding further, let us describe in detail the processes involved in both backward and forward chaining, beginning our discussion with the forward chaining approach.

THE FORWARD CHAINING ALGORITHM

In Chap. 2, we noted that an algorithm is a step-by-step procedure which ultimately converges to an optimal solution. Further, it was noted that neither expert systems nor heuristic programming methods are considered algorithmic as they cannot, in general, guarantee optimality. As such, the reader may think that the phrase "forward chaining *algorithm*" is a bit of a contradiction in terms. It is not; the algorithm to be introduced for forward chaining merely provides a step-by-step procedure that guarantees only the *correct performance of the inference process* of an expert system that employs data driven search. Since the expert system itself is made up of heuristic rules, there is still no guarantee that the conclusion ultimately reached by this process is itself optimal.

Fundamentally, there are two different philosophies of forward chaining. In one, we deliver all known data to the expert system at the very onset of the consultation session. In the second approach, we provide the expert system with only those specific elements of data that it requests during consultation. Both approaches have their advantages and disadvantages. The first approach often works well when the expert system is embedded in some automated process and receives its data directly from a computer resident database—or from a set of sensors. The second approach seeks to reduce the amount of data requested to that absolutely necessary to reach a conclusion. In the event that the generation of data requires extensive time and cost, this particular philosophy can be particularly appealing. Our procedure, to be described, shall use the latter approach.

Let us now consider just one algorithm by means of which the forward chaining search procedure may be conducted. In order to more clearly understand and appreciate the algorithm, we shall present it in such a form as to permit *manual* implementation. Of course, before implementation of the algorithm, it is assumed that the rules, rule numbers, and the associated prompts (i.e., user queries) for each attribute have been listed. Further, remember that we are still dealing with deterministic rule bases.

Algorithm for Forward Chaining: Baseline Version

1. Initialization. Establish three empty tables, the Working Memory table, the Attribute-Queue table, and the Rule/Premise Status table. The Working Memory table will be used to record all assertions (i.e., all the facts deduced during the consultation). The Attribute-Queue table records, in order, all attributes for

which a value has been assigned or is being sought. The attribute at the top of this table is the attribute presently under consideration. The Rule/Premise Status table records the rule status (i.e., active, marked, unmarked, discarded, triggered, or fired) as well as the status of each premise clause (i.e., free, true, or false). All premise clauses are initially unspecified (i.e., free) and all rules are initially unmarked and active. The notation employed to represent the rule and clause states is

A	=	active rule	FD =	fired rule
D	=	discarded rule	FR =	free clause
U	=	unmarked rule	FA =	false clause
M	=	marked rule	TU =	true clause
TD	=	triggered rule		

2. **Start inference.** Assign a value to a specific premise attribute, where this attribute must not appear in any conclusion clause.[1] Record this attribute at the *top* of the Attribute-Queue table. Also record this attribute, *and its associated value*, at the *bottom* of the Working Memory table.

3. **Rule scan and check for convergence.** Examine the Rule/Premise Status table. If no rules are active, STOP. Otherwise, scan the *active* rule-set premise clauses for *all* occurrences of the attribute on the top of the Attribute-Queue table, and record any changes in status of the premise clauses of the active rule set.

   (a) If the premise[2] of any rule is false then mark the associated rule as being discarded. Repeat this for all rules having a false premise. When complete, proceed to step 3b.

   (b) If the premise[3] of any rule is true then mark the associated rule as being triggered and place its conclusion attribute (and rule number) at the *bottom* of the Attribute-Queue table. Repeat this for all rules having a true premise. When complete, proceed to step 3c.

   (c) If no rules are presently in the triggered state (i.e., via a check of the Rule Status table), go to step 5. Otherwise (i.e., if one or more rules are triggered), go to step 4.

4. **Rule firing.** Cross out the topmost attribute on the Attribute-Queue table. Change the status of the rule associated with the new topmost attribute from triggered to fired. Place the conclusion associated with the fired rule at the bottom of the Working Memory table. Return to step 3.

5. **Queue status.** Cross out the topmost attribute on the Attribute-Queue table and proceed to step 6.

---

[1] The choice, as far as this algorithm is concerned, of the premise attribute selected is arbitrary. We simply must volunteer at least one piece of data before the inference process may be started.

[2] By "premise," we are referring to the entire premise of the rule, which may be composed of several premise clauses.

[3] Same comment as for footnote 2, above.

6. Convergence check and rule marking. Scan the active rule set for any un-marked, active rule. If no such rules can be found, STOP. Otherwise, mark the first such rule found (i.e., in the Rule/Premise Status table) and go to step 7.

7. Query. For the most recently marked rule, query the user for the value of an attribute in any of the rule's free premise clauses. If the user has a response then go to step 8. Otherwise (i.e., if the user either has no response *or* if the attribute has no associated prompt), continue this step for all remaining free premise clauses of the marked rule. If all such clauses have been examined without a user response, return to step 6.

8. Rule unmarking. Place the associated attribute (and rule number) on the top of the Attribute-Queue table. Also place this attribute, plus its value, at the bottom of the Working Memory table. Unmark the most recently marked rule and return to step 3.

Example 6.1 The forward chaining algorithm is best understood by means of il-lustration. To demonstrate its implementation, let us first consider the rule set from Chap. 5, Fig. 5.2. The rules associated with this figure are repeated below[4]:

Rule 1:   *If* engine type is prop
          Then plane is C130

Rule 2:   *If* engine type is jet
          and wing position is low
          Then plane is B747

Rule 3:   *If* engine type is jet
          and wing position is high
          and bulges are none
          Then plane is C5A

Rule 4:   *If* engine type is jet
          and wing position is high
          and bulges are aft of wing
          Then plane is C141

We also need to associate prompts for all attributes that do not appear in a conclusion clause. These prompts are listed below:

- Engine type: "Are the aircraft engines jet or propellor driven?"
- Wing position: "Are the wings positioned high or low on the fuselage?"
- Bulges: "Where, on the fuselage, are there pronounced bulges?"

There are thus four rules with one, two, three, and three premise clauses per rule, respectively. The initial tables associated with this production rule set are shown in Table 6.1.

---

[4] We shall use this relatively inefficient rule set (recall Fig. 5.3) simply to illustrate the algorithm.

TABLE 6.1
Initial tables for Example 6.1

Rule/Premise table			
Rule number	**Rule status**	**Premise clause number**	**Premise clause status**
(1)	A, U	(1)–1	FR
(2)	A, U	(2)–1 (2)–2	FR FR
(3)	A, U	(3)–1 (3)–2 (3)–3	FR FR FR
(4)	A, U	(4)–1 (4)–2 (4)–3	FR FR FR

Attribute-Queue table (attribute listing)	**Working Memory table (attribute = value)**

As may be seen, all rules are active (A) and unmarked (U) while all premise clauses are unspecified or free (FR); and thus we have completed step 1 of the forward chaining algorithm. Moving to step 2, we must initiate the inference process by assigning a value to a specific premise attribute, where the attribute selected does not appear in any conclusion clause. Step 2 is somewhat akin to turning the key in a car's ignition; it serves to start the inference process. Examining the rule set, we see that the attributes *engine type*, *wing position*, and *bulges* are all possibilities. Let us arbitrarily select *engine type* to begin the process. We must then assign a value to the attribute *engine type*. Only two values exist, *jet* or *propellor*. To determine which value to assign, we must assume a specific aircraft type. Let us assume that the user has sighted a C5A. In that case, the value for *engine type* will be *jet*. Thus, we place the attribute *engine type* at the top of the Attribute-Queue table and also place this attribute, plus the value *jet*, on the Working Memory table. The results are shown in Table 6.2.

We now proceed to step 3. Since there are active rules (all of them are in fact active) we cannot terminate the algorithm. Thus we scan the premise clauses of the active rule set (rules 1 to 4) looking for all occurrences of the attribute *engine type*. This attribute appears in the premise clauses of all rules wherein, for rule 1, the first premise clause is now false and, for the remainder of the rules, their first premise clause is true. We enter this information in the premise clause columns of the Rule/Premise Status table. Moving to step 3*a*, we see that rule 1 is false

TABLE 6.2
Second set of tables for Example 6.1

Rule/Premise table			
Rule number	**Rule status**	**Premise clause number**	**Premise clause status**
(1)	A, U	(1)–1	FR
(2)	A, U	(2)–1 (2)–2	FR FR
(3)	A, U	(3)–1 (3)–2 (3)–3	FR FR FR
(4)	A, U	(4)–1 (4)–2 (4)–3	FR FR FR

Attribute-Queue table (attribute listing)	Working Memory table (attribute = value)
Engine type	Engine type = jet

TABLE 6.3
Third set of tables for Example 6.1

Rule/Premise table			
Rule number	**Rule status**	**Premise clause number**	**Premise clause status**
(1)	~~A~~, U, D	(1)–1	~~FR~~, FA
(2)	A, U	(2)–1 (2)–2	~~FR~~, TU FR
(3)	A, U	(3)–1 (3)–2 (3)–3	~~FR~~, TU FR FR
(4)	A, U	(4)–1 (4)–2 (4)–3	~~FR~~, TU FR FR

Attribute-Queue table (attribute listing)	Working Memory table (attribute = value)
Engine type	Engine type = jet

161

and may be discarded. The remaining rules are neither true nor false and thus remain active. These results are entered as shown in Table 6.3. Note the use of the overstrike to signify a change in status. Since no rules are presently triggered, we proceed to step 5.

We now cross out the topmost attribute (i.e., engine type) on the Attribute-Queue table and proceed to step 6. In step 6, we first scan the active rule set for any unmarked rule. Rules 2, 3, and 4 are all active and unmarked. We mark the first of these (i.e., rule 2) and proceed to step 7. The results of the actions involved in steps 5 and 6 are indicated in Table 6.4.

At step 7, we must ask for input from the user. The most recently marked rule is rule 2 and the only free premise clause is the second. Since there is a prompt for this clause, we ask the user: "Are the wings positioned high or low on the fuselage?" Since we have assumed that the aircraft is a C5A, the user will respond with the answer, "high." We thus move to step 8.

In step 8, we place the attribute *wing position* (and specify that this is from rule 2) on the top of the Attribute-Queue table. We place the same attribute and its value (i.e., *high*) at the bottom of the Working Memory table. We then unmark rule 2 and return to step 3. At step 3, we note that there are still active rules and thus we proceed to scan the active rule set (rules 2 to 4) for all occurrences of the attribute *wing position*. Since wing position is known to be high, we note that premise 2 of rule 2 is false while the second premise of rules 3 and 4 are true. Moving to step 3*a*, we see that rule 2 is false and may be discarded while the remaining two rules

TABLE 6.4
Fourth set of tables for Example 6.1

Rule/Premise table			
Rule number	**Rule status**	**Premise clause number**	**Premise clause status**
(1)	~~A~~, U, D	(1)–1	~~FR~~, FA
(2)	A, ~~U~~, M	(2)–1	~~FR~~, TU
		(2)–2	FR
(3)	A, U	(3)–1	~~FR~~, TU
		(3)–2	FR
		(3)–3	FR
(4)	A, U	(4)–1	~~FR~~, TU
		(4)–2	FR
		(4)–3	FR

Attribute-Queue table (attribute listing)	Working Memory table (attribute = value)
~~Engine type~~	Engine type = jet

TABLE 6.5
Fifth set of tables for Example 6.1

Rule/Premise table			
Rule number	**Rule status**	**Premise clause number**	**Premise clause status**
(1)	~~A~~, U, D	(1)–1	~~FR~~, FA
(2)	~~A~~, ~~U~~, ~~M~~, U, D	(2)–1 (2)–2	~~FR~~, TU ~~FR~~, FA
(3)	A, U	(3)–1 (3)–2 (3)–3	~~FR~~, TU ~~FR~~, TU FR
(4)	A, U	(4)–1 (4)–2 (4)–3	~~FR~~, TU ~~FR~~, TU FR

Attribute-Queue table (attribute listing)	Working Memory table (attribute = value)
~~Engine type~~ Wing position (rule 2)	Engine type = jet Wing position = high

TABLE 6.6
Sixth set of tables for Example 6.1

Rule/Premise table			
Rule number	**Rule status**	**Premise clause number**	**Premise clause status**
(1)	~~A~~, U, D	(1)–1	~~FR~~, FA
(2)	~~A~~, ~~U~~, ~~M~~, U, D	(2)–1 (2)–2	~~FR~~, TU ~~FR~~, FA
(3)	A, ~~U~~, M	(3)–1 (3)–2 (3)–3	~~FR~~, TU ~~FR~~, TU FR
(4)	A, U	(4)–1 (4)–2 (4)–3	~~FR~~, TU ~~FR~~, TU FR

Attribute-Queue table (attribute listing)	Working Memory table (attribute = value)
~~Engine type~~ ~~Wing position (rule 2)~~	Engine type = jet Wing position = high

are still active. The results of all of the actions in this paragraph are summarized in Table 6.5. Since no rules are in the triggered state, we proceed to step 5.

At step 5, we cross out the topmost attribute (wing position) of the Attribute-Queue table and proceed to step 6. In step 6, we scan for unmarked, active rules. Rules 3 and 4 are both active and unmarked. We mark rule 3 and proceed to step 7. The results of our actions during this go-around of steps 5 to 7 are shown in Table 6.6.

We are once again at step 7. The only free premise of the marked rule (rule 3) is the third premise, wherein the associated attribute is *bulges*. The prompt for bulges is, "Where on the fuselage are there pronounced bulges?" And the answer should be "none." We next move to step 8 wherein we place the attribute bulges (and its associated rule number) on the top of the Attribute-Queue table and place the same attribute, and its value, at the bottom of the Working Memory table. Rule 3 is then unmarked. The results of the actions of steps 7 and 8 are shown in Table 6.7. We return to step 3.

Rules 3 and 4 are still active so we cannot yet terminate the search. We then scan the premise clauses of rules 3 and 4 for the attribute *bulges*, which may be found in both rules. From this scan, we see that the third premise of rule 3 is now true while the third premise of rule 4 is false. Further, we may discard rule 4 and trigger rule 3. In triggering rule 3, we place its conclusion attribute (note that we are now at step 3*b*) and rule number on the bottom of the Attribute-Queue table.

TABLE 6.7
Seventh set of tables for Example 6.1

Rule/Premise table			
Rule number	**Rule status**	**Premise clause number**	**Premise clause status**
(1)	~~A~~, U, D	(1)–1	~~FR~~, FA
(2)	~~A~~, ~~U~~, ~~M~~, U, D	(2)–1 (2)–2	~~FR~~, TU ~~FR~~, FA
(3)	A, ~~U~~, ~~M~~, U	(3)–1 (3)–2 (3)–3	~~FR~~, TU FR, TU FR
(4)	A, U	(4)–1 (4)–2 (4)–3	~~FR~~, TU ~~FR~~, TU FR

Attribute-Queue table (attribute listing)	Working Memory table (attribute = value)
~~Engine type~~ ~~Wing position (rule 2)~~ Bulges (rule 3)	Engine type = jet Wing position = high Bulges = none

TABLE 6.8
Eighth set of tables for Example 6.1

Rule/Premise table			
Rule number	**Rule status**	**Premise clause number**	**Premise clause status**
(1)	~~A~~, U, D	(1)–1	~~FR~~, FA
(2)	~~A, U, M~~, U, D	(2)–1 (2)–2	~~FR~~, TU ~~FR~~, FA
(3)	A, ~~U, M~~, U, TD	(3)–1 (3)–2 (3)–3	~~FR~~, TU ~~FR~~, TU ~~FR~~, TU
(4)	~~A~~, U, D	(4)–1 (4)–2 (4)–3	~~FR~~, TU ~~FR~~, TU ~~FR~~, FA

Attribute-Queue table (attribute listing)	Working Memory table (attribute = value)
~~Engine type~~ ~~Wing position (rule 2)~~ Bulges (rule 3) Plane (rule 3)	Engine type = jet Wing position = high Bulges = none

The results are shown in Table 6.8. From 3c, we note that since we have a rule in the triggered state, we proceed to step 4.

This is the first time that we have reached step 4. Proceeding per directions, we cross out the topmost attribute (i.e., bulges) on the Attribute-Queue table and change the status of the rule associated with the new topmost attribute (i.e., plane and rule 3) from triggered to fired. We then place the conclusion of the fired rule (i.e., that the plane is a C5A) on the bottom of the Working Memory table. We then return to step 3. Further, since no rules are now active, we may terminate the search—with the results summarized in Table 6.9.

The primary results of interest, from Table 6.9, are those shown in the Working Memory table, and particularly the last entry in that table. That is, we have finally determined that the plane sighted was a C5A. In addition, we have a collection of all the facts concerning this sighting in the Working Memory table (i.e., the engine type is jet, the wing position is high, there are no bulges on the fuselage, and the plane is a C5A), whether provided by the user or developed by the inference process.

This example has not, however, by any means indicated the power of the forward chaining approach, which is to be expected whenever small examples are used. That is, to finally identify the plane, we had to ask the user about each and every premise attribute. Typically, we can do much better than this when dealing with more realistic problems (and, in particular, more realistic rule bases).

TABLE 6.9
Final set of tables for Example 6.1

Rule/Premise table			
Rule number	**Rule status**	**Premise clause number**	**Premise clause status**
(1)	~~A~~, U, D	(1)–1	~~FR~~, FA
(2)	~~A~~, ~~U~~, ~~M~~, U, D	(2)–1 (2)–2	~~FR~~, TU ~~FR~~, FA
(3)	~~A~~, ~~U~~, ~~M~~, U, ~~TD~~, FD	(3)–1 (3)–2 (3)–3	~~FR~~, TU ~~FR~~, TU ~~FR~~, TU
(4)	~~A~~, U, D	(4)–1 (4)–2 (4)–3	~~FR~~, TU ~~FR~~, TU ~~FR~~, FA

Attribute-Queue table (attribute listing)	Working Memory table (attribute = value)
~~Engine type~~ ~~Wing position (rule 2)~~ ~~Bulges (rule 3)~~ Plane (rule 3)	Engine type = jet Wing position = high Bulges = none Plane = C5A

Enhancements to the Forward Chaining Algorithm

The forward chaining algorithm presented and demonstrated above represents a most rudimentary form of this search procedure. In actual practice, one would augment the scheme by means of additional heuristic rules in the inference engine. For example, refer back to step 3 of the algorithm where we were asked to scan the active rule-set premise clauses for *all* occurrences of the attribute on the top of the Attribute-Queue table. Since we have a large number of rules and attributes, it may take a considerable amount of time to accomplish this scan. Consequently, we might modify the algorithm in order to scan the active premise clauses for just the *first* instance (or first few instances) of the attribute under consideration. In addition, as discussed in Chap. 4, it is often possible to group and order the rules and premise clauses to obtain increased search efficiency.

As another example, consider step 4. Note that it is quite possible, at this step, to have several rules in the triggered state. Step 4 will fire, however, only that rule associated with the topmost attribute in the Attribute-Queue table. This is a very arbitrary policy that can certainly be improved upon. That is, we need a good rule for determining which knowledge base rule to fire when more than one is triggered. Several heuristic firing rules have been proposed including the following:

- *Rule ordering*. Arrange the rules according to some priority and fire only that rule (from among those triggered) with the highest priority, or ranking.
- *Recency ordering*. Select the rule to be fired as the one most recently dealt with or, alternately, as the rule with which we have least recently dealt.
- *Rule partitioning*. Separate the rules into groups of which only some are active at a given time. Deal only with active groups. Note that this approach also requires that we provide some means and rules for activating and deactivating rule groupings.
- *Specificity ordering*. If the premise clauses of one rule are a superset of another, then fire that rule first. For example, if we have two rules such as the following:

R1. *If* A = X   R2. *If* A = X
    and B = Y   *Then* R = YY
    and C = Z
    Then Q = XX

and if both are triggered, we would fire rule 1 (R1).

We can, in fact, apply these enhancements to either forward or backward chaining search. Such enhancements may be demonstrated with yet another example of the forward chaining algorithm.

Example 6.2. X-007 is a robot owned by Slapdash Landscaping, Inc. The Slapdash Landscaping Company, in turn, provides landscaping services to tract homes in southern California. Slapdash has recently received a contract to provide minimal landscaping services to a block of five new tract homes constructed by Brown Box Housing. Each house is virtually identical and each sits on *ranchette* size lots (of 35 by 50 feet). Slapdash Landscaping is to simply provide trees for each lot. Their contract specifies that each lot must have at least two trees (counting any existing trees on the lot). Thanks to Brown Box's big bulldozers, some lots are completely bare of trees, while others have one or two trees still standing. Slapdash will be paid a fixed amount per lot regardless of the number of trees it unloads—as long as each lot has at least two trees.

Slapdash realizes that this job is simple enough for X-007 and reprograms the robot's knowledge base to conduct the task of simply scanning each lot, determining the number of existing trees on the lot, and then unloading the appropriate number of trees. Since there are five lots under consideration, the truck is initially loaded with ten trees. Any trees that are not needed will be returned to the company.

Before proceeding any further, the reader should take careful note of the rule base for this problem, as presented below. Specifically, this rule base *does not capture any heuristic rules*. Rather, all it does is describe a quantitative procedure. That is, all one has to do is to count the number of trees on a given lot and subtract this value from two. If the result is positive, we simply drop off the corresponding number of trees. Thus, virtually anyone who can add and subtract will arrive at the same conclusion—the same tree unloading scheme. As such, it must be emphasized that the single purpose of this example is to illustrate the use of the forward chaining algorithm as modified for an iterative consultation session.

Rule 1: *If* PARTITION is scan lot
         and number of trees on present lot is zero
         Then unload two trees on the present lot
         and revise database

Rule 2: *If* PARTITION is scan lot
         and number of trees on present lot is one
         Then unload one tree on the present lot
         and revise database

Rule 3: *If* PARTITION is scan lot
         Then let PARTITION be status check
         and revise database

Rule 4: *If* PARTITION is status check
         and last lot was not lot five
         Then let PARTITION be scan lot
         and revise database (including moving truck to next lot)

Rule 5: *If* PARTITION is status check
         Then STOP

In a more realistic version of this problem we might augment this rule base with some actual heuristics. For example, if our landscaping service cloned the expertise of an expert landscape architect, they might discover that certain heuristic rules may be applied to the choice of trees to be dropped off at each lot. For example, if there is a single existing tree, and this tree is short and slow growing, then drop off a fast growing tree. Or, if all the trees on the street are of the same type, then drop off trees of that type to maintain a consistency of foliage. However, simply for the sake of clarity in the discussion to follow, we shall restrict our attention to the set of five rules presented above.

Let us then employ the forward chaining algorithm to conduct this task. However, we shall make a few modifications in the algorithm. First, we shall maintain a database that may be directly accessed by the expert system. Second, we shall incorporate a few of the enhancements as listed earlier. And third, we will adjust the algorithm to permit an iterative process. Specifically, the expert system must be used to determine a *sequence* of decisions, one for each lot. The database for this problem is provided below. The last five rows of Table 6.10 indicate the number of existing trees on each lot. These are actually not known until the robot scans the lot in front of which the truck is parked. Let us assume, for sake of illustration, that the delivery truck is presently parked in front of the first lot. Thus, when X-007 scans this lot, it will note the presence of precisely one tree.

Notice that rules 1 to 3 are *partitioned* from rules 4 and 5 by means of the setting for the attribute *partition* (i.e., either *scan lot* or *status check*). Further, note that we shall use *specificity* in the determination of rules to be fired. For example, while rule 3 will always be triggered whenever the partition is *scan lot*, it will never be fired if either rule 1 or 2 is triggered, as the premise of rule 3 is a subset of

TABLE 6.10
Initial database for
Example 6.2

Attribute	Value
Partition	Scan lot
Last lot	Lot 0
Present lot	Lot 1
Load	10 trees
Lot 1 trees	1
Lot 2 trees	0
Lot 3 trees	2
Lot 4 trees	0
Lot 5 trees	1

that of rules 1 and 2. Similarly, rule 5 will be triggered whenever the partition is *status check* but it will not be fired as long as rule 4 is triggered, as the premise of rule 5 is a subset of that of rule 4.

Another unusual aspect of this rule set may be seen in the conclusion clauses of rules 1 to 4. Whenever these rules are fired, the existing database will be revised. This requires that one be able to access an associated database from the expert system—an ability that most expert systems shells now have, and that any useful shell should have.

The revised version of the forward chaining algorithm that will be employed is provided below.

Algorithm for Forward Chaining: Revised Version (for Database Access and Partitioned Rule Base)

1. Initialization. Establish the initial database and three empty tables, the Working Memory table, the Attribute-Queue table, and the Rule/Premise Status table. All premise clauses are initially free and all rules are initially unmarked and *inactive*. The only new notation associated with the Rule/Premise Status table is that the letter "I" will be employed to represent the fact that a rule is inactive.

2. Start inference. Access the *database* and update the Rule/Premise Status table accordingly (e.g., activate the specified rule set and update the status of the premise clauses for the activated rule set). List the attribute *partition* at the top of the Attribute-Queue table and also list this attribute, plus its value, in the Working Memory table. Proceed to step 4*b*.

3. Rule activation. Examine the attribute set at the top of the Attribute-Queue table to determine if the active/inactive status of the rule partitions must be changed (i.e., has there been a change in the partition setting?). If not, proceed to step 4. If so, clear the status of all rules and premises in the Rule/Premise Status table and activate those rules associated with the active partition while deactivating all others. Proceed to step 4.

4. Rule scan and check for convergence. Examine the active portion of the Rule/Premise Status table. If no rules are active, STOP. Otherwise, proceed to step 4a.

(a) Access both the present database and the Working Memory and record any changes in status to the premise clauses of the active rule set. When finished, go to step 4b.

(b) If the premise of any rule is false, mark the associated rule as inactive. Repeat this for all rules having a false premise. When complete, proceed to step 4c.

(c) If the premise for any rule is true, mark the associated rule as triggered. If no rules are in the triggered state, proceed to step 6. Otherwise, go to step 4d.

(d) If only one rule is triggered, term this the *selected rule* and proceed to step 4e. If more than one rule has been triggered, use a tie breaking heuristic (specificity) to select one of these triggered rules and proceed to step 4e.

(e) For the selected rule, place its conclusion attribute (and rule number) on the bottom of the Attribute-Queue table. Proceed to step 5.

5. Rule firing. Cross out the topmost attribute set on the Attribute-Queue table and change the status of the rule associated with the new topmost attribute from triggered to fired. Place the attribute(s) and associated value(s) related to the conclusion clause(s) of the fired rule at the bottom of the Working Memory table. If firing of the rule requires a revision to the database, perform this revision. Return to step 3.

6. Queue status. Cross out the topmost attribute set on the Attribute-Queue table and proceed to step 7.

7. Convergence check and rule marking. Scan the active rule set for any un-marked, active rule. If no such rules can be found, STOP. Otherwise, mark the first such rule found (i.e., in the Rule/Premise Status table) and go to step 8.

8. Query. For the most recently marked rule, query the user for the value of an attribute in any of the rule's free premise clauses. If the user has a response then go to step 9. Otherwise (i.e., if the user either has no response or if the attribute has no associated prompt), continue this step for all remaining free premise clauses of the marked rule. If all such clauses have been examined without a user response, return to step 7.

9. Rule unmarking. Place the associated attribute (and rule number) on the top of the Attribute-Queue table. Also place this attribute, plus its value, at the bottom of the Working Memory table. Unmark the most recently marked rule and return to step 3.

We may now continue Example 6.2. The modifications made to the forward chaining algorithm should then permit us to deal with our landscaping problem. Our initial set of tables is provided in Table 6.11.

Table 6.11 represents the completion of step 1 of the forward chaining algorithm. Notice that we have added the database for this example to our set of

TABLE 6.11
Initial set of tables for Example 6.2

Rule/Premise table			
Rule number	**Rule status**	**Premise clause number**	**Premise clause status**
(1)	I, U	(1)–1 (1)–2	FR FR
(2)	I, U	(2)–1 (2)–2	FR FR
(3)	I, U	(3)–1	FR
(4)	I, U	(4)–1 (4)–2	FR FR
(5)	I, U	(5)–1	FR

Attribute-Queue table (attribute listing)	Working Memory table (attribute = value)	Database	
		Attribute	**Value**
		PARTITION	Scan lot
		Last lot	Lot 0
		Next lot	Lot 1
		Load	10 trees
		Lot 1 trees	1
		Lot 2 trees	0
		Lot 3 trees	2
		Lot 4 trees	0
		Lot 5 trees	1

tables. We now move to step 2 of the algorithm. Here, we access the existing database. The attributes that pertain to the rule base from the existing database are *partition*, *last lot*, *next lot*, and *lot1 trees*. The present values of these attributes are then used to activate all rules in the partition *scan lot* and to update the status of the premise clauses for these newly activated rules. Let us also add the attribute *partition* to our Attribute-Queue table and this attribute, plus its value, to the Working Memory table. The results of implementing step 2 are summarized in Table 6.12.

Upon completion of step 2, we proceed directly to step 4b. Here, we note that rule 1 is set to inactive (its premise is false since there is one tree on the lot) and we move to 4c where both rules 2 and 3 are triggered. Moving to step 4d, we need to break the tie between rules 2 and 3 (in order to fire just one of these rules). Using the specificity heuristic, we will break the tie in favor of rule 2 as rule 3 is simply a subset of rule 2. This is an important point and the reader should be sure

TABLE 6.12
Second set of tables for Example 6.2

Rule/Premise table			
Rule number	**Rule status**	**Premise clause number**	**Premise clause status**
(1)	~~I~~, U, A	(1)–1 (1)–2	~~FR~~, TU ~~FR~~, FA
(2)	~~I~~, U, A	(2)–1 (2)–2	~~FR~~, TU ~~FR~~, TU
(3)	~~I~~, U, A	(3)–1	~~FR~~, TU
(4)	I, U	(4)–1 (4)–2	FR FR
(5)	I, U	(5)–1	FR

Attribute-Queue table (attribute listing)	Working Memory table (attribute = value)	Database	
		Attribute	**Value**
Partition	Partition = scan lot	PARTITION Last lot Next lot Load Lot 1 trees Lot 2 trees Lot 3 trees Lot 4 trees Lot 5 trees	Scan lot Lot 0 Lot 1 10 trees 1 0 2 0 1

that it is understood. We thus select rule 2 and proceed to step 4*e* where we place the attribute *unload (on) lot 1*, and the rule number (i.e., rule 2) at the bottom of the Attribute-Queue table. We then proceed to step 5 where the topmost attribute of the Attribute-Queue table is crossed out and we fire the rule associated with the new topmost attribute. We next place the attribute associated with the conclusion clause of the fired rule at the bottom of the Working Memory table (i.e., unload lot1 = 1 tree). Also, according to one of the conclusion clauses of rule 2, we must revise the database. The results of steps 4 and 5 are summarized in Table 6.13. The most recent change in the database is shown in boldface. That is, in Table 6.13, the "Load" value has been changed to 9.

Moving to step 3, we note that the activated partitions have not changed and thus we proceed to rule 4. Since there are still active rules in the active partition (specifically, rule 3), we go to step 4*a*. Accessing the present database (i.e., as

TABLE 6.13
Third set of tables for Example 6.2

Rule/Premise table			
Rule number	**Rule status**	**Premise clause number**	**Premise clause status**
(1)	~~I~~, U, ~~A~~, I	(1)–1 (1)–2	~~FR~~, TU ~~FR~~, FA
(2)	~~I~~, U, ~~A~~, ~~TD~~, FD	(2)–1 (2)–2	~~FR~~, TU ~~FR~~, TU
(3)	~~I~~, U, A, TD	(3)–1	~~FR~~, TU
(4)	I, U	(4)–1 (4)–2	FR FR
(5)	I, U	(5)–1	FR

Attribute-Queue table (attribute listing)	Working Memory table (attribute = value)	Database	
		Attribute	**Value**
~~Partition~~ Unload-lot 1 (rule 2)	Partition = scan lot Unload-lot 1 = 1 tree	PARTITION Last lot Next lot Load Lot 1 trees Lot 2 trees Lot 3 trees Lot 4 trees Lot 5 trees	Scan lot Lot 0 Lot 1 **9 trees** 1 0 2 0 1

revised in Table 6.13), we note that it has no impact on the Rule/Premise Status table and we move on to 4*b*. Rule 4*b* results in no change and we move to 4*c*. In step 4*c*, we do not trigger any new rules, but rule 3 has already been triggered from an earlier operation. Consequently, we proceed to step 4*d* and then onto 4*e* where the attribute *partition* and the rule number 3 is added to the bottom of the Attribute-Queue table. Now, since the partition attribute is just a device for the activation of the various portions of the rule base, let us cross out, in the Working Memory table, its previous value. That is, cross out "Partition = scan lot" at the top of the Working Memory table (and note that the database should be revised accordingly). Moving to step 5, we fire rule 3, place the associated conclusion attribute and value at the bottom of the Working Memory table, revise the database, and return to step 3. These actions are summarized in Table 6.14.

Returning to step 3, we note that the activated partition has now changed. We thus clear the status of all rules and premises and set the rules associated

TABLE 6.14
Fourth set of tables for Example 6.2

Rule/Premise table			
Rule number	**Rule status**	**Premise clause number**	**Premise clause status**
(1)	~~I~~, U, ~~A~~, I	(1)–1 (1)–2	~~FR~~, TU ~~FR~~, FA
(2)	~~I~~, U, ~~A~~, ~~TD~~, FD	(2)–1 (2)–2	~~FR~~, TU ~~FR~~, TU
(3)	~~I~~, U, ~~A~~, ~~TD~~, FD	(3)–1	~~FR~~, TU
(4)	I, U	(4)–1 (4)–2	FR FR
(5)	I, U	(5)–1	FR

Attribute-Queue table (attribute listing)	Working Memory table (attribute = value)	Database	
		Attribute	**Value**
~~Partition~~ ~~Unload-Lot 1 (rule 2)~~ Partition (rule 3)	~~Partition = Scan Lot~~ Unload-Lot 1 = 1 tree Partition = status check	PARTITION Last lot Next lot Load Lot 1 trees Lot 2 trees Lot 3 trees Lot 4 trees Lot 5 trees	Status ck Lot 0 Lot 1 9 trees 1 0 2 0 1

with the partition *status check* to active. We then move to step 4. Since there are active rules in the active partition (rules 4 and 5), we proceed to step 4*a*. In step 4*a*, we access the latest version of the database where we note that the last lot is now lot 1 and the next lot is lot 2, and the number of trees on lot 2 (obtained through our robot's vision system) is zero. Proceeding to 4*b* and 4*c*, we determine that both rules 4 and 5 are triggered and thus we move to 4*d* where the tie is broken in favor of rule 4. Thus, we place the attribute *partition* at the bottom of the Attribute-Queue table. Through step 5, the topmost attribute on the Attribute-Queue table is crossed out, rule 4 is fired, and the attribute *partition* plus its value (scan lot) are placed at the bottom of the Working Memory table. In the Working Memory table, we may also cross out the previous value of the partition attribute. The database is also revised (and note that we now move the truck to the next lot). We then return to step 3. The results of all the operations in this paragraph are summarized in Table 6.15.

TABLE 6.15
Fifth set of tables for Example 6.2

Rule/Premise table			
Rule number	**Rule status**	**Premise clause number**	**Premise clause status**
(1)	I, U	(1)–1 (1)–2	FR FR
(2)	I, U	(2)–1 (2)–2	FR FR
(3)	I, U	(3)–1	FR
(4)	~~I~~, U, ~~A~~, ~~TD~~, FD	(4)–1 (4)–2	~~FR~~, TU ~~FR~~, TU
(5)	~~I~~, U, A, TD	(5)–1	~~FR~~, TU

Attribute-Queue table (attribute listing)	Working Memory table (attribute = value)	Database	
		Attribute	**Value**
~~Partition~~ ~~Unload-lot 1 (rule 2)~~ ~~Partition (rule 3)~~ Partition (rule 4)	~~Partition = Scan Lot~~ Unload-lot 1 = 1 tree ~~Partition = Status Check~~ Partition = scan lot	PARTITION Last lot Next lot Load Lot 1 trees Lot 2 trees Lot 3 trees Lot 4 trees Lot 5 trees	**Scan Lot** **Lot 1** **Lot 2** 9 trees 1 0 2 0 1

Returning to step 3, we note that we must revise the active and inactive status of our rule set and clear the status of the rules and premises. Completing this, we move to step 4 where we access the present database and change the status of our premise clauses. Moving on to steps 4*b* and 4*c*, we note that rule 1 and rule 3 may be triggered while rule 2 is set inactive. Breaking the tie between the triggered rules, we fire rule 1, revise the database and return to step 3. These actions are summarized in Table 6.16.

We may continue in this manner until we have completed the delivery of the trees and then return the truck, plus four unneeded trees, back to the firm. As an exercise, the reader is invited to complete the remainder of the example. At this point, however, the reader may be wondering just why such a seemingly involved process is necessary to solve this problem. For example, why not just solve the problem for the first lot and then *restart* the inferencing process. Actually, that is in essence just what was accomplished. However, through the partitioning concept,

TABLE 6.16
Sixth set of tables for Example 6.2

Rule/Premise table			
Rule number	**Rule status**	**Premise clause number**	**Premise clause status**
(1)	~~I~~, U, ~~A~~, ~~TD~~, FD	(1)–1 (1)–2	~~FR~~, TU ~~FR~~, TU
(2)	~~I~~, U, ~~A~~, I	(2)–1 (2)–2	~~FR~~, TU ~~FR~~, FA
(3)	~~I~~, U, A, TD	(3)–1	~~FR~~, TU
(4)	I, U	(4)–1 (4)–2	FR FR
(5)	I, U	(5)–1	FR

Attribute-Queue table (attribute listing)	Working Memory table (attribute = value)	Database	
		Attribute	**Value**
~~Partition~~ ~~Unload-Lot 1 (rule 2)~~ ~~Partition (rule 3)~~ ~~Partition (rule 4)~~ Unload-lot 2 (rule 1)	~~Partition = Scan Lot~~ Unload-lot 1 = 1 tree ~~Partition = Status Check~~ Partition = scan lot Unload-lot 2 = 2 trees	PARTITION Last lot Next lot Load Lot 1 trees Lot 2 trees Lot 3 trees Lot 4 trees Lot 5 trees	Scan lot Lot 1 Lot 2 **7 trees** 1 0 2 0 1

we are able to automate the methodology. While it may be easier to restart the process at each new lot, when solving the problem by hand, we may want the computer to do this for us when faced with a real-world problem.

Some Reflections on Example 6.2

The original algorithm presented for forward chaining works fairly well for certain classes of problems; typically those which are static in nature and in which a single decision is to be made. The problem of Example 6.2, however, even though simplistic in form, represents a more difficult problem for solution by expert systems. And it also represents a problem of, quite often, far more interest and importance.

If we refer back to Example 6.1, we may note that the fundamental problem involved is to *classify* various entities (i.e., aircraft). Recall from the discussion in Chap. 3 that classification (or discriminant analysis or pattern recognition) is, by far, the major application of expert systems to date. It is employed, for example, for medical diagnosis (i.e., given a set of symptoms, classify the disease), for equipment maintenance diagnosis (e.g., given recent operational data on a system, determine if it is performing according to specifications and, if not, classify the specific source of the problem), for credit scoring (given the data on a credit application, classify the applicant according to credit risk), and a host of other instances. The search process for problems of classification may be of either the forward or (and more typically) backward chaining type.

The problem of Example 6.2 is not, however, a classification problem. Rather, it is a simplified version of a type of *packing* problem [Kaufmann and Henry-Labordère, 1977; Murty, 1976; Taha, 1975]. More generally, it is a problem of *construction*.[5] Here, we are constructing an unloading scheme. The specific example is simplistic and was selected merely to illustrate the general approach for dealing with such problems. However, realistic versions of this problem type include

- Cargo loading (e.g., the loading of planes, ships, and trucks)
- Equipment configuration (e.g., the fabrication of special-purpose equipment, as depicted in the well-known R1, or XCON project)

Construction problems are, unless relatively simple, generally addressed much more appropriately by forward chaining than by backward chaining. Typically, there is a massive, if not near infinite number of possible conclusions. To illustrate, just consider how many different ways in which one might place equipment onboard a plane, or configure the components of a computer, or sequence jobs through processing operations.

THE BACKWARD CHAINING ALGORITHM

As for the case of the forward chaining algorithm, the search algorithm to be presented for backward chaining is a simplified version, and designed for manual calculations for small knowledge bases. With the assumption that the knowledge base has been constructed and the user prompts developed, we may proceed with the search using the steps listed below.

Algorithm for Backward Chaining: Baseline Version

1. Initialization. Establish three empty tables, the Working Memory table, the Goal table, and the Rule/Premise Status table. The Working Memory table will

---

[5] The reader may wish, at this time, to refer back to our previous discussion of classification and construction, as presented in Chapter 3, under the section: "An Evaluation of Problem Types."

be used to record all assertions. The Goal table records, in order, those attributes for which a value is sought. The Rule/Premise Status table records the rule status (i.e., active, discarded, triggered, or fired) as well as the status of each premise clause (i.e., free, true, or false). All premise clauses are initially free and all rules are initially active.

2. Start inference. Specify a final goal (i.e., a conclusion clause attribute). Place the associated goal attribute at the *top* of the Goal table.

3. Rule scan and check for convergence. Scan the *conclusion clauses* of the *active* rules (i.e., rules that have not yet been fired or discarded) to find any occurrence of the goal attribute presently on the top of the Goal table.

   (*a*) If the Goal table is empty, STOP.

   (*b*) If only one such rule may be found, go to step 6. If several such rules may be found, and any of these are triggered, select any one of the triggered rules and proceed to step 6. Otherwise, arbitrarily select one rule from among the rules found that contains the subject goal attribute in its conclusion clause set, and go to step 6.

   (*c*) If no active rules are found that contain the subject goal attribute in their conclusion clause set, then go to step 4.

4. Query. For the goal attribute on top of the Goal table, find the associated query if one exists. If there is no query associated with this goal attribute, then STOP. Otherwise, query the user, record his or her response, remove the top goal attribute from the Goal table and place it, plus its value (as supplied by the user), in the Working Memory table. Go to step 5.

5. Rule/premise status update. Using the contents of the Working Memory table, update the Rule/Premise Status table. Specifically, if the premise of any rule is false, discard that rule, and if the premise is true, trigger that rule. Return to step 3.

6. Rule evaluation. For the rule found in step 3:

   (*a*) If this rule is triggered, then remove the current topmost goal attribute from the Goal table and place it, plus its value, in the Working Memory table. Change the status of this rule from triggered to fired. Go to step 5. Otherwise (i.e., if this rule is not triggered) proceed to step 6*b*, below.

   (*b*) If this rule is not triggered, then select the first unknown premise attribute of the rule and place it, plus the rule number, at the top of the Goal table. Return to step 3.

   Example 6.3 As before, we shall attempt to clarify the steps of the algorithm through its demonstration on a small-example knowledge base. Since we should now be thoroughly familiar with the aircraft identification knowledge base, as repeated below, we shall use the backward chaining algorithm for solution. For the purpose of discussion, this time let us assume that our novice observer has sighted a B747.

Rule 1:   *If* engine type is prop
          Then plane is C130

Rule 2:   *If* engine type is jet
          and wing position is low
          Then plane is B747

Rule 3:   *If* engine type is jet
          and wing position is high
          and bulges are none
          Then plane is C5A

Rule 4:   *If* engine type is jet
          and wing position is high
          and bulges are aft of wing
          Then plane is C141

Through the initialization phase of step 1, our first table is that shown as Table 6.17.
    Proceeding to step 2 of the algorithm, we must specify a final goal. Rather obviously, we are attempting to determine the type of plane and thus our final

TABLE 6.17
Initial set of tables for Example 6.3

Rule/Premise table			
Rule number	**Rule status**	**Premise clause number**	**Premise clause status**
(1)	A	(1)–1	FR
(2)	A	(2)–1	FR
		(2)–2	FR
(3)	A	(3)–1	FR
		(3)–2	FR
		(3)–3	FR
(4)	A	(4)–1	FR
		(4)–2	FR
		(4)–3	FR

Goal table	Working Memory table

goal is simply the class of plane. In fact, in this simple knowledge base, this is the only possible goal. We thus place the goal attribute *plane* at the top of the Goal table. Moving to step 3, we note that there are active rules and that the attribute *plane* appears in the conclusion clause of all four rules. For step 3*b*, we note that none of the rules are triggered and we arbitrarily select rule one. We then proceed to step 6. From 6*b*, the first unknown premise attribute of rule one is *engine type*. We place this attribute, and the rule number, at the top of the Goal table and return to step 3. The results thus far are shown in Table 6.18, below.

At step 3, we now scan the conclusion clauses of the active rules to find any occurrence of the attribute *engine type*. From 3*c*, we note that no active rules contain this attribute in their conclusion clauses and we proceed to step 4, the query of the user. Since there is a query associated with *engine type*, we query the user, who should respond with *jet* if he or she did indeed observe a B747. We thus remove the top goal attribute from the Goal table and place it, and its value, in the Working Memory table. We then proceed to step 5. In step 5, we use the contents of the Working Memory table to update the Rule/Premise Status table. Our results at the conclusion of step 5 are summarized in Table 6.19.

Returning to step 3, we now search the active rule set (rules 2 to 4) for any occurrence of *plane* in the conclusion clauses. Since all three rules contain

TABLE 6.18
Second set of tables for Example 6.3

Rule/Premise table			
Rule number	**Rule status**	**Premise clause number**	**Premise clause status**
(1)	A	(1)–1	FR
(2)	A	(2)–1	FR
		(2)–2	FR
(3)	A	(3)–1	FR
		(3)–2	FR
		(3)–3	FR
(4)	A	(4)–1	FR
		(4)–2	FR
		(4)–3	FR

Goal table	Working Memory table
Engine type (rule 1)	
Plane	

TABLE 6.19
Third set of tables for Example 6.3

Rule/Premise table			
Rule number	**Rule status**	**Premise clause number**	**Premise clause status**
(1)	~~A~~, D	(1)–1	~~FR~~, FA
(2)	A	(2)–1 (2)–2	~~FR~~, TU FR
(3)	A	(3)–1 (3)–2 (3)–3	~~FR~~, TU FR FR
(4)	A	(4)–1 (4)–2 (4)–3	~~FR~~, TU FR FR

Goal table	Working Memory table
~~Engine type (rule 1)~~ Plane	Engine type = jet

this attribute, we will arbitrarily select rule 2, and proceed to step 6. Rule 2 is not triggered so we select the first unknown premise attribute (*wing position*) and place it and the rule number at the top of the Goal table, and return to step 3. Table 6.20 summarizes the results to this point.

At step 3, we now search for *wing position* in the conclusion clauses of the active rules. Since it does not appear there, we proceed to step 4 and query the user, who should respond with *low* as the value for wing position. We remove wing position from the Goal table and place it, and its value, in the Working Memory table and proceed to step 5. In step 5, we again update the Rule/Premise Status table. Our results are summarized in Table 6.21.

Returning to step 3, we note that there is still one active rule (rule 2) and that the attribute *plane* appears in its conclusion clause. We thus proceed directly to step 6. From 6*a*, we remove plane from the Goal table and place it, plus its value (i.e., from the conclusion clause of rule 2) in the Working Memory table. Rule 2 is changed from triggered to fired. We then proceed to step 5 and update the Rule/Premise Status table. The results thus far are summarized in Table 6.22.

Returning to step 3, we note that the Goal table is now empty and thus we may terminate the algorithm. That is, we have determined that the plane sighted was a B747.

TABLE 6.20
Fourth set of tables for Example 6.3

Rule/Premise table			
Rule number	**Rule status**	**Premise clause number**	**Premise clause status**
(1)	~~A~~, D	(1)–1	~~FR~~, FA
(2)	A	(2)–1 (2)–2	~~FR~~, TU FR
(3)	A	(3)–1 (3)–2 (3)–3	~~FR~~, TU FR FR
(4)	A	(4)–1 (4)–2 (4)–3	~~FR~~, TU FR FR

Goal table	Working Memory table
Wing position (rule 2) ~~Engine type (rule 1)~~ Plane	Engine type = jet

TABLE 6.21
Fifth set of tables for Example 6.3

Rule/Premise table			
Rule number	**Rule status**	**Premise clause number**	**Premise clause status**
(1)	~~A~~, D	(1)–1	~~FR~~, FA
(2)	A, TD	(2)–1 (2)–2	~~FR~~, TU ~~FR~~, TU
(3)	~~A~~, D	(3)–1 (3)–2 (3)–3	~~FR~~, TU ~~FR~~, FA FR
(4)	~~A~~, D	(4)–1 (4)–2 (4)–3	~~FR~~, TU ~~FR~~, FA FR

Goal table	Working Memory table
~~Wing position (rule 2)~~ ~~Engine type (rule 1)~~ Plane	Engine type = jet Wing position = low

TABLE 6.22
Sixth and final set of tables for Example 6.3

Rule/Premise table			
Rule number	**Rule status**	**Premise clause number**	**Premise clause status**
(1)	~~A~~, D	(1)–1	~~FR~~, FA
(2)	~~A~~, ~~TD~~, FD	(2)–1	~~FR~~, TU
		(2)–2	~~FR~~, TU
(3)	~~A~~, D	(3)–1	~~FR~~, TU
		(3)–2	~~FR~~, FA
		(3)–3	FR
(4)	~~A~~, D	(4)–1	~~FR~~, TU
		(4)–2	~~FR~~, FA
		(4)–3	FR

Goal table	Working Memory table
~~Wing position (rule 2)~~	Wing position = low
~~Engine type (rule 1)~~	Engine type = jet
~~Plane~~	Plane = B747

Example 6.4 As our final example of the backward chaining algorithm, we address the knowledge base provided below. A, B, C, and D simply represent some hypothetical attributes and their only values will be true or false. For example, they might represent the state of a set of switches, either on or off. X and Y, however, represent conclusions (e.g., the system is working properly, has malfunctioned, etc.). While these rules should really be simplified, grouped, and ordered, let us use them as is in the inference demonstration.

Rule 1: *If* B is true
Then Y

Rule 2: *If* C is true
Then B is true

Rule 3: *If* C is false
Then B is false

Rule 4: *If* A is true
Then X
and Y

Rule 5: *If* C is true
Then D is true

For purpose of demonstration, let us assume that, for the system under consideration, C is true while A is false. And let the single goal to be determined be Y. Following the steps of the backward chaining algorithm, we will ultimately develop Table 6.23 as the final result of the search. That is, we have concluded

TABLE 6.23
Final set of tables for Example 6.4

Rule/Premise table			
Rule number	**Rule status**	**Premise clause number**	**Premise clause status**
(1)	A, ~~TD~~, FD	(1)–1	~~FR~~, TU
(2)	A, ~~TD~~, FD	(2)–1	~~FR~~, TU
(3)	A, D	(3)–1	~~FR~~, FA
(4)	A	(4)–1	FR
(5)	A, TD	(5)–1	~~FR~~, TU

Goal table	Working Memory table
~~C (rule 2)~~ ~~B (rule 1)~~ ~~Y~~	C = true B = true Y

Y, the desired goal. As an exercise, the reader should be able to develop the same results.

While this is but a simple problem, it does demonstrate some interesting results. Specifically, our algorithm stops as soon as Y is determined. Thus, the status of X (which, one will recall, is also a conclusion) remains unknown, and the rule that would determine this status (rule 4) is still active at the conclusion of the search. Since all that we wanted was the status of Y, there is nothing intrinsically wrong with this result. If, however, we are also interested in X, one would need to revise our baseline backward chaining algorithm to continue the search. In fact, we could revert to forward chaining to determine the status of rule 4. As an exercise, we advise the reader to solve this same problem by means of (only) forward chaining and compare the results. First, however, we advise that the rule base be properly established. That is, through rule simplification, grouping, and ordering, we have the equivalent rule base presented below:

Group 1: goal = Y

        Rule B/Y: *If* B is true
                             Then Y

        Rule A/Y: *If* A is true
                             Then Y

Group 2: support for group goal = Y

        Rule C/B/T: *If* C is true
                             Then B is true

        Rule C/B/F: *If* C is false
                             Then B is false

Group 3: conclusion = X

        Rule A/X: *If* A is true
                                  Then X

Group 4: conclusion = D

        Rule C/D: *If* C is true
                                  Then D is true

It should now be somewhat more obvious that rule C/D is possibly extraneous (this depends upon whether or not we are interested in the result that D is true) and that there are no rule groups to support the goal of X (thus, the only way to determine X is to ask the user for the value of A).

MIXED MODES OF CHAINING

The discussion of Example 6.4 serves to indicate that one might, on occasion, wish to combine forward and backward chaining. Such mixed modes are available through many existing commercial expert systems shells. However, while they represent a potentially significant source of computational improvement, they do require a thorough understanding of the procedure, the knowledge base, and of the specific shell used.

To demonstrate a simplified approach to the mixed mode of chaining, we shall employ the following knowledge base [Benchimol et al., 1987, chap. 4]. Note carefully a fundamental assumption which will be used is that all known data are presented at the start of the consultation session.

Rule 1:   *If* F and H
          Then K

Rule 2:   *If* E and A
          Then K

Rule 3:   *If* E and B
          Then H

Rule 4:   *If* A and G
          Then B

Rule 5:   *If* B and D
          Then H

Rule 6:   *If* G and D
          Then E

Rule 7:   *If* A and B
          Then D

Rule 8:   *If* A and C
          Then G

For the purpose of discussion, we will classify the eight rules as follows: rules 1 to 3 will be backward chaining rules, while rules 4 to 8 will be forward chaining rules. That is, for rules 1 to 3, the *normal* mode of search will be backward chaining, while for the rest, the *normal* mode is forward chaining.

Example 6.5: Mixed Chaining—Priority to Backward Chaining. Given the above knowledge base, let us first solve our problem by initially employing backward chaining, *and only resorting to forward chaining when we are unable to backward chain*. Further, let us assume that our working memory contains A and C (i.e., A and C are given) while our goal is to determine K.

The mixed chaining process may be summarized using Table 6.24. The first column of the table simply lists the step of the search process. The second column indicates the rule number and type under consideration. Initially, we encounter the null set, ϕ, wherein we know only the facts A and C, and our goal is to determine K. In the second line (step 2), and using the fact that priority is given to backward chaining rules, we encounter rule 1. The designation R1(B) then indicates the fact that we are dealing with rule 1 and that this is (normally) a backward chaining rule. The third column lists the presently known facts and the goals under consideration. The fourth column then indicates the type of chaining actually used for the rule under consideration, and whether or not it was fired at that step.

Let us now go through the steps of the mixed-mode chaining summarized in Table 6.24. First, we are given A and C, and our goal is to determine K. This is summarized in step 1, the first line of the table. Next, we proceed to the backward chaining rules and seek one having our desired goal (K) in the conclusion. The first such rule we come to is rule 1. From rule 1, we note that K is determined whenever F and H are true. Thus, in the third column of step 2 we replace K with F and H. In the fourth column we simply denote the fact that we employed backward chaining for rule 1 and that it was not fired at this step.

Still using backward chaining, we now seek a rule that has either F or H in its conclusion. The only such rule is rule 3. Thus, we move to rule 3 (and step 3 of the table). Here, we see that H is determined by E and B and thus we replace the goal H by E and B. In the fourth column we note that backward chain-

TABLE 6.24
Mixed chaining search for Example 6.5

Step	Rule (rule type)	Facts (goals)	Chaining (firing)
1	ϕ	AC(K)	
2	R1(B)	AC(FH)	B
3	R3(B)	AC(FEB)	B
4	R8(F)	ACG(FEB)	F(fired)
5	R4(F)	ACGB(FE)	F(fired)
6	R7(F)	ACGBD(FE)	F(fired)
7	R5(F)	ACGBDH(FE)	F(fired)
8	R6(F)	ACGBDHE(F)	F(fired)
9	R2(B)	ACGBDHEK(F)	F(fired)

ing was used and that the rule was not fired. We now are no longer able to use any of the backward chaining rules. Consequently, we move to the set of forward chaining rules (rules 4 to 8) and find that rule 8 is triggered (i.e., by the set of facts thus far known). We may then fire rule 8 which gives us the new fact, G, and G is added to our set of known facts in the third column of the table. In the fourth column we note that forward chaining was used and that the rule was fired (and may now be discarded). Knowing A, C, and G, we may now fire rule 4, to derive B. And this action is listed in step 5 of the table. Knowing A,C,G and B, we may fire rule 7 (see step 6) to derive D. Knowing the facts indicated in step 6, we may fire rule 5 (see step 7) to derive H. Knowing the facts listed in step 7, we may next fire rule 6 (see step 8) to derive E, and knowing E, we may remove E from our list of desired goals.

At this point, we have fired all of our forward chaining rules and we still seek goal F. Returning to our set of backward chaining goals, we note that F does not appear as the conclusion of any rule. Consequently, our only move is to use forward chaining on the remaining rule set, and this may be done for rule 2. Specifically, and as summarized in step 9, we may fire rule 2 to derive K. And since K is the goal that was originally desired, we may terminate the process.

Example 6.6: Mixed Chaining—Priority to Forward Chaining. Using the same knowledge base, let us now consider the solution to our problem where priority is given to forward chaining. Table 6.25 summarizes the search process. As may be seen, with priority given to forward chaining, the search process converges in two fewer iterations than in the case where priority is given to backward chaining. The procedure itself should be self-explanatory.

While it so happened that mixed chaining with emphasis given to forward chaining worked best for our simple knowledge base, one should not attempt to generalize these results. The particular form of mixed chaining that will work best for a given knowledge base is dependent on both the structure of that knowledge base and the order of the rules. And there are also a number of enhancements to the mixed mode of chaining that can result in increased efficiency. An alternative approach to mixed chaining is presented in Exercise 6.16, at the end of the chapter.

TABLE 6.25
Mixed chaining search for Example 6.6

Step	Rule (rule type)	Facts (goals)	Chaining (firing)
1	ϕ	AC(K)	
2	R8(F)	ACG(K)	F(fired)
3	R4(F)	ACGB(K)	F(fired)
4	R7(F)	ACGBD(K)	F(fired)
5	R5(F)	ACGBDH(K)	F(fired)
6	R6(F)	ACGBDHE(K)	F(fired)
7	R2(B)	ACGBDHEK	F(fired)

ON CERTAIN ILLUSIONS
ABOUT EXPERT SYSTEM MODULARITY[6]

One of the main advantages of the employment of expert systems is achieved through the separation of the knowledge base (i.e., what we know) from the inference engine (i.e., what we do). This separation lends itself to rule-base transparency, ease of maintenance, and the employment of expert systems shells. And each of these is a significant result. Unfortunately, there has been a certain degree of exaggeration concerning such features. In particular, it is sometimes either stated or implied that one can, at any time, add or delete rules in the knowledge base, *without any impact of the validity of the conclusions reached.* For example, we may be told that as long as the rule added to the knowledge base is valid, the entire rule base will remain valid.

While it is certainly far easier to make changes to the rule base of an expert system than it is for either algorithmic approaches or heuristic programming, one should not draw the conclusion that such changes have no impact on the system's performance. In fact, even minor changes, such as the addition or deletion of a single rule, have the potential for causing problems. While these problems may be alleviated through the guidelines for rule-base development as presented in Chap. 4, they cannot be totally eliminated by such means alone.

To illustrate just what may happen, consider a rule base that is still in development. The rule base will contain those rules used by a physician in the diagnosis of minor ailments. The resulting expert system may then be used, at home, by the layperson. Our illustrative medical expert system will be limited to simply determining whether or not the patient is *sick* or *well,* and then to making a corresponding prescription. In developing our knowledge base, let us begin with the first rule as shown below. Further, for the purpose of discussion, let us assume that we will use backward chaining in the search process and that, to break any ties in the conduct of the algorithm, we simply select the lowest numbered rule.

Rule 1:   *If* diagnosis = sick
          Then prescription = aspirin

Our goal is to determine whether or not the patient is ill and, if so, to provide him or her with a prescription (aspirin, if sick). Consequently, this goal appears as the conclusion clause attribute in rule 1. However, use of rule 1 would require that we ask the users (i.e., the patients) whether or not they are actually sick, and if they knew this, an expert system would hardly be needed. So let us add some rules that will more precisely determine the condition of the patient.

Rule 2:   *If* temperature $\geq$ 100 degrees
          Then diagnosis = sick

---

[6] The example employed in this section is based on a model that appears in the manual for the EXSYS expert systems software (Huntington, 1985, sec. G).

Rule 3:   *If* severe stomach cramps = yes
           Then diagnosis = sick

With the addition of rules 2 and 3, our expert system would perform as follows. First, our goal is given in the conclusion attribute of rule 1. Since the premise clause attribute of rule 1 (i.e., diagnosis) appears in the conclusion clause of at least one other rule, we do not query the user about whether or not he or she is sick. Instead, we move to rule 2. The patient is sick if his/her temperature is over 100 degrees and, since the temperature attribute does not appear in the conclusion clause of any other rule, we must ask the patient what his/her temperature is. If the patient states that his/her temperature is over 100 degrees, we conclude that the patient is sick. If not, we move to rule 3 and query the patient about the existence of severe stomach cramps.

Before proceeding further, note that if rule 2 is true we have determined that the patient is sick. However, depending upon the specific form of the backward chaining algorithm employed, we might either move on to rule 3 (which would be redundant if all we are trying to determine is whether or not the patient is sick) or simply ignore rule 3. In fact, in the version of the backward chaining algorithm provided in this chapter, we will do the latter.

Thus, if the premise of either rule 2 or rule 3 is true, the patient is sick rule (rule 1) will fire. However, what happens if neither rule 2 nor 3 is true? In the case of the three rules thus far given, no conclusion can be reached. Rather obviously, we need one or more rules to determine what to do if the patient is not sick. Let us add rules 4 and 5.

Rule 4:   *If* temperature < 100 degrees
           and severe stomach cramps = no
           Then diagnosis = well

Rule 5:   *If* diagnosis = well
           Then prescription = none

Notice that our goal attribute (prescription) now appears in both rules 1 and 5. Now, if rules 2 and 3 are both false, then rule 4 is triggered. Firing rule 4 causes rule 5 to be triggered. If it is fired, we have a prescription for the well patient.

Now, examine each of the existing five rules. While the problem is obviously grossly simplified, none of the rules are *wrong*. Using these rules, we will either prescribe aspirin to the sick patient or nothing to the well one.

However, let us assume that a new heuristic is identified. That is, some rather brilliant researcher has concluded that, if a person has a sore throat, then he or she is sick. Since this represents a valid rule, let us see what would happen if we simply append this rule to our knowledge base—as rule 6:

Rule 6:   *If* throat = sore
           Then diagnosis = sick

Now, if the concept of knowledge base modularity is indeed true, we may simply add this rule to our knowledge base (and, in fact, anywhere in the knowledge base) and have a new, improved expert system. However, note precisely what will happen if

- The patient's temperature is under 100 degrees
- The patient does not have severe stomach cramps
- And the patient has a sore throat

Given such a situation, we *should* arrive at the intermediate conclusion that the patient is sick, and a final conclusion to prescribe aspirin.

However, using the backward chaining algorithm of this chapter (try it!), and always breaking any ties by selecting the lowest numbered rule, we will conclude that the patient is *well* and that we should prescribe *nothing*. Further, at the conclusion of the chaining process, rule 6 will still be active. That is, we will never get the opportunity to ask about the condition of the patient's throat. Thus, even though the patient has a sore throat, and thus is sick, we reach the wrong conclusion.

However, instead of adding rule 6 to the end of the rule set, try making it the first rule. Now, if we again perform backward chaining, we will determine that the patient is now *sick* and that we should prescribe *aspirin*. That is, a simple reordering of the rules has led to an opposite conclusion. Rather obviously, one cannot simply add or delete rules from a knowledge base without a consideration of the impact of such actions on the final conclusion.

As our knowledge base is small, it is relatively easy to determine just what happened to cause this situation. Specifically, once rule 6 (i.e., the rule for a sore throat) was added to the rule base, rule 4, which was *previously* legitimate, is no longer valid. While rule grouping and ordering will make it easier to discover this problem, it will not eliminate it.

As suggested in Chap. 4, the use of an AV-pair table can sometimes be of assistance in spotting potential trouble spots. The AV-pair table for this rule base is shown in Table 6.26. We may observe at least two things from this table. First, the rules need to be grouped and ordered. As an exercise, the reader may do this. Second, the premise clause attribute *throat* has only a single value (*sore*). This should cause us to consider the other possible values for *throat*, and their impact on the rule base. In particular, what happens if the throat is *not* sore?

If we wish to add a rule 6 to our knowledge base and still retain its validity, *independent of the rule order,* we must modify rule 4. Specifically, we should note that a well person exhibits the absence of a high temperature, stomach cramps, *and* a sore throat. Thus, our modified version of rule 4 is:

Rule 4 (revised)   *If* temperature < 100 degrees
           and severe stomach cramps = no
           and throat = not sore
           Then diagnosis = well

TABLE 6.26
Attributes and values
for medical diagnosis rule base

Rule	Clause	Attribute	Value
1	1	Diagnosis	Sick
5	1	Diagnosis	Well
2	1	Temperature	≥ 100
4	1	Temperature	<100
3	1	Severe stomach cramps	Yes
4	2	Severe stomach cramps	No
6	1	Throat	Sore
2	2	Diagnosis	Sick
3	2	Diagnosis	Sick
4	3	Diagnosis	Well
6	2	Diagnosis	Sick
1	2	Prescription	Aspirin
5	2	Prescription	None

Using the knowledge base with the revised version of rule 4, the results will be reasonable—regardless of any permutation of the rule set.

SUMMARY

The inference engine contains, like the knowledge base, a set of rules. However, the rules in the inference engine are those used specifically to control the inference process, the process of determining a conclusion from the knowledge base, given a certain set of facts.

The reader has been provided a brief introduction to algorithms for backward, forward, and even mixed modes of chaining. The algorithms presented were purposely designed for manual use, and the reader is encouraged to use these methods to solve the problems at the end of the chapter. We are absolutely convinced that this is the only way that the novice can gain a full appreciation of expert systems—and the insight necessary to intelligently apply and evaluate the expert systems approach. However, remember that any results obtained by means of the inference engine are only as good as the rule base being employed. Further, our intent should be to *understand*, rather than to develop inference engines, as there are a sufficient number of commercial software packages available to perform this task. Our primary goal should always be to develop good rule bases, from which good conclusions may be drawn.

EXERCISES

Note carefully that, unless otherwise directed, you should use the manual methods of inference to solve the problems described in the exercises below.

6.1. When dealing with a real-world expert system, of any realistic size, it would obviously be impractical to consider solution by hand. Why then should one be exposed to the manual solution of such problems?

6.2. On what basis might one choose forward chaining over backward chaining? Backward chaining over forward chaining? Describe the advantages and disadvantages of both forward and backward chaining.

6.3. In discussions with the domain expert, you learn that the number of conclusions (i.e., solutions) for a given problem cannot be preenumerated. What should this tell you about the choice of chaining? The class of problem?

6.4. Solve the rule base of Exercise 4.9 using
   (a) Forward chaining (start inference with order backlog = large)
   (b) Backward chaining (use *bid methodology* as the goal)
   Comment on any differences in results obtained by the two approaches.

6.5. For the case in which the publication date is recent, the topic is science fiction, price is under fifteen dollars, author is not one of your favorites, and the review is good:
   (a) Solve the rule base of Exercise 4.13 by forward chaining (start the inference with topic = science fiction).
   (b) Solve the rule base of Exercise 4.13 by backward chaining (let goal = book-purchase decision).

6.6. What search strategy (i.e., chaining process) would you expect to be the most appropriate for the earthen dam expert system discussed in Chap. 5? Why?

6.7. Solve Example 4.2 by means of both forward and backward chaining.

6.8. Solve the rule base associated with Table 4.2 by means of both forward and backward chaining. Assume that

$$A = a$$
$$B = \text{NOT } b$$
$$C = c$$

6.9. For the rule base as developed for Table 5.6 (i.e., using VP-Expert) derive a conclusion by both forward and backward chaining when (a) eyes are bloodshot, (b) temperature is high, and (c) nose is running. Discuss your results.

6.10. Construct the rule base and select an inference procedure from this chapter to solve the following problem:

A regional motel chain wishes to construct an expert system for the selection and arrangement of furniture in a number of motels that it has just purchased. The only furniture that will be kept, from the present owners, are the beds in the rooms. Thus, the expert system must select and arrange television sets, couches, and chairs for each room. Next to the bed, the TV is the most important piece of furniture in a room. A couch is more important than a chair, but additional couches are less important than chairs. Only one item of furniture is permitted on each of the four walls—and the bed takes up one wall. A couch should be on the wall opposite the

TV, if possible. A TV should never be placed on a wall with a window. A chair should, if possible, be placed on a wall with a window.

For the purpose of discussion, assume that we need to place furniture in three rooms where the description of these rooms and their present contents are listed below:

Room	Wall 1	Wall 2	Wall 3	Wall 4
1	Bed		Window	Bed
2	Window	Bed		
3		Window	Bed	Window

Further, let us assume that there are five TVs, five couches, and five chairs available for placement in the rooms.

6.11. An alternative mode of forward chaining is achieved by following the inference process summarized below. Develop an algorithm for this approach and apply it to the aircraft identification problem of Example 6.1, and then compare these results with those obtained by the forward chaining algorithm presented in this chapter.

In this approach to forward chaining, all of the known facts are made available at the *start* of the consultation session. Thus, no queries are presented to the user once the consultation session begins. Further, the inference process considers rules according to the order in which they appear in the rule base. The search proceeds from the first rule to the last rule, which is considered a *search cycle*. If a rule can be triggered, it is immediately fired and the working memory is updated. We then move to the next (active) rule. If any rules have a false premise, they are immediately discarded and we move to the next active rule. If a rule is neither triggered nor discarded, we move to the next active rule. At the end of a cycle, we return again to the first (remaining) active rule in the rule base and repeat the search cycle. Inference is terminated whenever, for one complete cycle, no rule can be fired.

6.12. Derive, through the forward chaining procedure developed in Exercise 6.11, the conclusion for the rule base and facts of Exercise 4.9. Compare your results with those obtained in Exercise 6.4a.

6.13. Finish Example 6.2.

6.14. Solve Example 6.4 by means of
(a) Forward chaining (per the algorithm of Chap. 6 and starting with C = true)
(b) Forward chaining (per the algorithm developed in Exercise 6.11)
(c) Backward chaining

6.15. Recall the discussion of the invalid rule base in the section of Chap. 6, "On Certain Illusions about Expert System Modularity." Form, group, and order the *corrected* version of this rule base. Next assume that:

$$\text{Temperature} < 100 \text{ degrees}$$

$$\text{Severe stomach cramps} = \text{no}$$

$$\text{Throat} = \text{sore}$$

(*a*) Solve the rule base using the forward chaining algorithm of this chapter and start with temperature $<$ 100 degrees.

(*b*) Solve the rule base using the forward chaining algorithm of Exercise 6.11.

(*c*) Solve the rule base using backward chaining.

6.16. In this chapter, we have discussed one approach to mixed chaining. Another approach uses the following procedure. First, the default method of chaining is always backward chaining. Second, rules are *not* partitioned according to whether there is a priority to either forward or backward chaining. Third, whenever a rule is fired, the premises of all remaining rules are examined to see if they may be fired (i.e., we revert to forward chaining). If so, they are fired. If not, we continue backward chaining. Use this procedure to deal with the rule set of Example 6.5 and compare your results with those obtained in the text. Note that K is the final goal and that backward chaining would normally start with rule 1. However, since A and C are both known prior to the consultation session, we may immediately fire rule 8 to begin the process.

6.17. Use the mixed chaining procedure of Exercise 6.16 to deal with the knowledge base listed below. For the purpose of activating the inference process, assume that

Color of smoke is neither white nor black.

Odor is none.

Note that the conclusion to rule 5 activates the display of a file which provides warnings concerning chemical fires and actions to be taken as soon as *any* chemical fire has been noted.

R1:   *If* color of smoke is white
       Then problem is ruptured coolant system

R2:   *If* color of smoke is black
       Then problem is electrical system

R3:   *If* color of smoke is other than white or black
       and odor is similar to rotten eggs
       Then problem is chemical vessel A

R4:   *If* color of smoke is other than white or black
       and odor is none
       Then problem is chemical vessel C

R5:   *If* color of smoke is other than white or black
       Then DISPLAY (FILE 1)

6.18. Nonmonotonic reasoning is employed whenever previous knowledge may be invalidated upon the development of new facts. Discuss the impact of such reasoning on the inference procedures of this chapter.

6.19. Use EXSYS (or whatever expert systems shell you wish) to solve the problem of Exercise 6.10. If you use EXSYS, incorporate a call to an external program (a *database* representing the quantities of furniture still on hand) and the RESTART command.

6.20. Use EXSYS (or whatever expert systems shell you wish) to solve the following problems:

(*a*) Exercise 6.4*a*

(*b*) Exercise 6.4*b*

(*c*) Exercise 6.5*a*

(*d*) Exercise 6.5*b*

(*e*) Exercise 6.7

(*f*) Exercise 6.8

(*g*) Exercise 6.9

Comment on any differences noted between the results obtained by hand versus those obtained by means of the software.

CHAPTER
7
ENHANCEMENTS

CHAPTER OVERVIEW

Up until now, our focus has been on strictly deterministic, stand-alone expert systems. That is, we have assumed complete certainty in both rule validity (the conclusion, or conclusions, reached by the rule) and user response (the values supplied by the user). Further, we have addressed expert systems that, for the most part, operate essentially as independent units.

The real world, however, is characterized by uncertainty and it would seem that provisions for some consideration for this uncertainty should be attempted. In this chapter, the aspect of uncertainty in expert systems will be addressed by describing a few representative methods that may be employed. Also addressed will be one rather controversial matter, the question of whether or not the inclusion of uncertainty is really all that essential. And it should be noted that there are two distinctly different views on this topic.

Following our discussion of uncertainty, a related topic will be dealt with—the employment of the use of *unknown* in rule bases and user responses. And just how we might employ this concept in the design of specially tailored expert systems will be illustrated.

A number of times the need to address external programs (e.g., databases) from an expert system has been alluded to, and in the second half of this chapter this very important addition to the expert systems approach will be briefly reviewed. Such an enhancement, typically known as a *bridging* capability, can be a virtual necessity if one is to successfully introduce expert systems into actual, working organizations and institutions.

196

Finally, the topic of explanation and justification in expert systems will be covered. Many expert systems, and most development packages, provide such features. However, as we shall learn, they are not yet in the same league as the explanation and justification capabilities of most human experts—and it is important to recognize the actual extent of their abilities.

Together, each of the above topics reflect some very powerful enhancements or adjuncts to expert systems—and to the design of such systems. When used properly, and where one appreciates their scope and limitations, they can both improve the expert systems design and its acceptance.

SOURCES OF UNCERTAINTY

Before examining methods for dealing with uncertainty in expert systems, one should first appreciate precisely just where and why uncertainty exists. In general, there are two primary sources of uncertainty that may be encountered in an expert system:

- Uncertainty with regard to the validity of a knowledge base *rule*
- Uncertainty with regard to the validity of a *user response*

Consider first the uncertainty associated with a rule. For example, consider the heuristic rule given by "if a dog barks, then it will not bite" (from the old saying, "a barking dog won't bite"). Whoever the canine expert was who came up with this rule may either believe it is always true, or may simply believe that it is true in general. In the first case we could assign a *confidence factor* of, say, 1 to the rule, that is, it is true 100 percent of the time. In the second case, we need to assign some value less than 1. Of course, the question is, "What should that value be?" If we believe that 8 times out of 10 the rule will hold, we might then assign a confidence factor of 0.8 to the rule, and the confidence factor would thus reflect a subjective estimate of the probability of the validity of the rule. If we have no confidence whatsoever in the rule, we might assign a value of 0 as its confidence factor. It is rather unlikely that we would be so naive as to believe that any confidence factor assigned in so subjective a manner is absolutely accurate. Rather, the use of such confidence factors tends to simply indicate, *on a strictly relative basis*, our confidence (or lack of confidence) in a rule. And we simply cannot take confidence factors too literally.

The second source of uncertainty is associated with the response, or responses, of the user of the expert system; specifically the replies provided in response to generated user queries. Consider, for example, the following query to the user (in this case, a physician) of a medical diagnostic expert system:

Does the patient have severe stomach cramps?

where the expected answer is evidently either "yes" or "no." Now, if the patient is doubled over in agony, the physician would most likely answer in the affirmative.

However, what if the pain in the patient's stomach is not quite so extreme? A strictly "yes" or "no" response to such a question may be unsatisfactory. We might then ask the user to reply with a confidence factor. For example, we might (arbitrarily) use a scale of 0 to 10 where a 0 represents a judgment that there is no basis upon which to assume that there is a stomach disorder while a 10 indicates that the patient is experiencing the most intense pain imaginable. We might then rephrase our query as

`Indicate the intensity of any stomach cramps (0/10).`

If the user responds with, say, a value of 8, then we might interpret this as an indication of stomach cramps at a *very intense* level. Again, however, note that we cannot take the response too literally. Two different physicians may well provide two quite different responses, even when dealing with the same patient and data set.

Now, before we proceed further, note that there is an alternative user prompt in this situation that may be just as good, or possibly even better, than the use of (explicit) confidence factors. That is, we might present the user with the following query:

`Indicate the level of intensity of stomach cramps:`

`1. Extreme`
`2. Very intense`
`3. Moderate`
`4. Minimal`
`5. None`
`Input response: _____`

Here, rather than attempting to deal with a numeric value, the user needs only select the response that seems most appropriate. Now, we are not implying that the menu prompt is necessarily either better or more accurate than the use of confidence factors. However, in some cases it may prove more efficient and much more natural.

Returning to the thrust of our discussion, we should note that confidence factors must generally be considered to be subjective estimates of the *relative level* of confidence one should have in either a rule or user response. They are not necessarily probabilities, and are thus not subject to the rules of probability theory. This is a disturbing fact for many analysts who would like to employ a *scientifically sound* approach to the determination of uncertainty in expert systems. When one considers that expert systems are, by definition, made up of *heuristic* rules, it seems rather absurd to attempt to attach any absolute level of precision to the combination of such rules. However, to begin the discussion of various approaches to uncertainty in expert systems, we shall consider the use of Bayesian probability. And this discussion should more graphically illustrate the complexity of the problem faced.

UNCERTAINTY THROUGH BAYESIAN PROBABILITY

In order to clarify the discussion, let us focus on an expert system for the diagnosis of disk drives for personal computers. To keep things simple, let us assume that there are but two possible outcomes, or conclusions, to the consultation session. That is, the disk drive will either be found to be defective (i.e., in need of repair) or it will be considered fine. Let us further assume that we have been introduced to Mr. Mack N. Tosh, an expert in the diagnosis and repair of personal computer systems. Mack has informed us that, in his expert opinion, if a disk drive is making unusual noises, 8 times out of 10 it is defective.

In essence, what Mack has done has been to pose a production rule *plus* a confidence factor (which we shall denote as *cf*) for the validity of the rule. We might then write this rule as

Rule 1: *If* unusual noises = yes
  Then disk drive status = defective (cf = 0.8)

Such a rule has two very important implications. First, it implies that the user is able to distinguish between *normal* and *unusual* noises, a feat that may well be beyond the capabilities of the layperson. Second, and *if* confidence factors are actually probabilities, this rule implies that the probability of the alternate conclusion (i.e., the disk drive is good) is 1 − 0.8, or 20 percent. That is, if probability theory holds in such situations, there should be a companion rule such as the one given below:

Rule 2: *If* unusual noises = yes
  Then disk drive status = good (cf = 0.2)

However, and perhaps not so surprising, studies have shown that many experts will reject such an implied rule. Thus it seems evident that human experts are simply not using probability theory, at least as we know it, as the basis for their estimates of certainty.

Let us now turn to the use of empirical data in conjunction with Bayesian probabilities. Bayes' formula may be used to determine the probability of a given conclusion (C) given certain evidence, or facts (f). The formula is given as:

$$p(C|f) = [p(f|C) \cdot p(C)]/p(f) \qquad (7.1)$$

where $\qquad p(f) = p(f|C) \cdot p(C) + p(f|\sim C) \cdot p(\sim C) \qquad (7.2)$

and $\quad p(C|f) =$ probability of conclusion C given facts f
  $p(f|C) =$ probability of facts f given conclusion C
   $p(f) =$ unconditional probability of facts f
   $p(C) =$ unconditional probability of conclusion C
  $p(\sim C) =$ unconditional probability of *not* conclusion C
 $p(f|\sim C) =$ probability of facts f given *not* C

For our specific example, $p(C|f)$ is the probability of a bad disk drive given unusual noises; $p(f|C)$ is the probability of unusual noises given a bad disk drive (i.e., the drive may be defective and not produce any strange noises); $p(f)$ is the probability that *any* disk drive is noisy; $p(C)$ is the probability that *any* disk drive is defective; $p(\sim C)$ is the probability that *any* disk drive is not defective; and $p(f|\sim C)$ is the probability of unusual noises from a *good* disk drive. Why it is so difficult to employ formal probability theory for the development of confidence in production rules should now be clearer. Specifically, it is not at all an easy (or, in particular, practical) task to determine empirical (or even good, subjective) values for $p(f|C)$, $p(f)$, $p(C)$, $p(\sim C)$ and $p(f|\sim C)$.

However, for the sake of discussion, let us assume that empirical studies have taken place wherein it is noted that, when a computer has a defective disk drive, the emanation of unusual noises will occur in 90 percent of these computers (a value that might, at first, appear to lend some weight to Mack's subjective estimate provided by rule 1, earlier). Thus, we now know (if the empirical studies were performed correctly and if sufficient evidence was available and examined) that $p(f|C) = 0.90$. Let us also assume that we have enough empirical evidence to determine that disk drives are bad on 2 percent of *all* computers (i.e., $p(C) = 0.02$); and thus $p(\sim C) = 0.98$. The only information that is still required is $p(f|\sim C)$, that is, the probability of noises coming from a good disk drive. Assuming once again that we can collect data to determine this, let us presume that noises come from good disk drives about 8 percent of the time (i.e., $p(f|\sim C) = 0.08$). We can now enter all of this information into Eq. (7.2) to determine $p(f)$:

$$p(f) = 0.90 \cdot 0.02 + 0.08 \cdot 0.98 = 0.0964$$

Substituting for $p(f)$ in Eq. (7.1) we obtain

$$p(C|f) = [0.90 \cdot 0.02]/0.0964 \approx 0.187$$

Consequently, for our hypothetical example, we have determined that unusual noises indicate a bad disk drive with a probability of about 18.7 percent. Comparing this result with that originally stated by the expert (i.e., Mack), as listed in rule 1 above, we can note a very substantial discrepancy (i.e., a confidence factor for the rule of 80 percent versus 18.7 percent). It has been noted that, in actual practice, such discrepancies are both common and significant.

One might be led to believe that the probabilistic approach is likely to lead to better results, having been developed through scientifically based, quantitative methods. However, this would only be true if the probabilities used to compute $p(C|f)$ in Eqs. (7.1) and (7.2) are accurate. In many real-life situations, such accuracy simply does not exist. Even more likely, it may simply be unrealistic to gather the empirical data necessary to compute the probabilities.

In general then, the Bayesian approach is an appealing method for the determination of the certainty of production rules, but it is quite often just not practical for actual expert systems of any realistic size or complexity. We should also note that the approach described deals only with the confidence factor associated with a rule, that is, it does not explicitly consider the uncertainty associated with user responses.

"UNCERTAINTY": THE EXSYS APPROACH

Since this text includes the EXSYS expert systems shell, we will remark briefly upon the approach employed by this package for uncertainty. The demonstration version supplied with this text is based on the *standard* EXSYS package. Thus, we shall restrict our remarks to uncertainty as employed in this particular version of EXSYS.

The demonstration version permits the use of three different methods for uncertainty. These are

- The 0 to 1 system
- The 0 to 10 system
- The -100 to 100 system

The first method simply assigns a 1 to any final conclusion clause that is true, and a 0 to any one that is false. Note carefully that the only two values permitted are 0 and 1 (i.e., we do not, as in some systems, permit the fractions between 0 and 1). As such, this method is not really intended for dealing with uncertainty as the expert system treats the rule base as deterministic. Consider the following example:

Rule 0-1: *If* can fly $=$ yes
 and caped uniform $=$ yes
 and gender $=$ male
 Then Superman $=$ yes—probability $= 1$
 and Batman $=$ yes—probability $= 0$

While we personally feel uncomfortable with calling such confidence factors *probabilities,* the interpretation of rule 0-1 should be obvious. That is, if the individual sighted can fly, has a caped uniform, and is male, then we are (1) certain that he is Superman, and (2) equally certain that he is *not* Batman (since Batman cannot fly).

The second method employs *all* numbers between, and including 0 and 10. Given that we have a final conclusion clause that is reached by a number of rules, the confidence factors (or *probabilities,* as they are termed in EXSYS) are averaged. There is, however, an exception to the averaging process. If any confidence factor is either a 0 or a 10, then that value is *locked in,* and no averaging takes place.

To illustrate the 0-10 method, consider a rule base that has concluded that the user's investment strategy should be

- Bonds: with confidence values of 1, 5, and 6 (i.e., three different rules, each with a different confidence factor, have been fired and conclude that the investment strategy should be bonds)
- Blue chip stocks: with confidence values of 0, 7, 9, and 9 (i.e., four different rules have concluded blue chip stocks)

As a result, we average the values associated with bonds to derive a final confidence level of 4 for the investment in bonds. In the case of blue chip stocks, the final confidence value is 0 since the value of 0 from one rule serves to lock in that confidence factor.

The third approach uses *all* numbers between, and including, -100 and 100. Here, however, the values of -100, 0, or 100 do not lock in the confidence factor value. Further, you have a choice between simply averaging the results or combining them as either dependent or independent probabilities.

Before proceeding further, we should explain the use of negative confidence factors. EXSYS uses a range from -100 to 100. Other packages might use -10 to 10, or -1 to 1. Whatever the range, all are attempting to capture the same thing. That is, in the -100 to 100 system, we interpret

- -100 to mean that the conclusion is absolutely wrong, or false
- 0 to mean that we have no confidence in the conclusion
- 100 to mean that we are absolutely sure that the conclusion is true

The use of negative confidence values may be valuable when capturing the rules of a domain expert who reasons from negative results. For example, a physician may note that, whenever a certain diagnostic test comes back negative, we can absolutely rule out an entire category of diseases. Or, a safety engineer may use negative reasoning to exclude certain accident causes (e.g., if the expert checks for evidence of human failure, and finds that such a failure may absolutely be ruled out, then he or she may pursue purely mechanical causes).

APPROACHING UNCERTAINTY THROUGH FUZZY SETS

Both previous approaches to production-rule uncertainty focused on the determination of the level of confidence to place in a given rule. They did not, however, attempt to address the source of uncertainty as a consequence of user input. Fuzzy set theory, or fuzzy logic, has been proposed as one means for handling such a situation. Since there is neither the time nor need to provide a complete background on fuzzy sets (those who are interested in such details are directed to the references [Kickert, 1978; Zadeh, 1965; Zimmermann, 1985]), we shall simply note that the concept of fuzzy sets was developed by Lotfi Zadeh [1965] as an approach to deal with certain types of problems where a simple yes or no response is inadequate.

For example, consider the question about the height of a person. That is, suppose you are asked whether or not a given individual is *tall*. If the individual in question is an adult male and is 7 feet tall then the response is clearly affirmative. And if the subject is but 3 feet 6 inches tall, the response would clearly be negative. However, what if the person is 5 feet 10 inches in height? In this case the response is not nearly so obvious. Zadeh proposed that one use a fuzzy membership function, with values over the continuous range of 0 to 1 for such situations. A value of 1 indicates, for our example, that the person is most definitely tall (specifically, about as tall as we would ever expect a person to be). However, a *partial degree* of *tallness* could be represented by some number between 0 and 1. For example, a man who is 5 feet 10 inches in height might have a fuzzy membership value of 0.75.

In conjunction with the fuzzy membership function, a set of rules for evaluating a conclusion using fuzzy logic has been provided. Specifically,

- If premises are connected by the logical *AND* operator, we use the *minimum* of the fuzzy values associated with the premises to determine the composite value for the premises.
- If premises are connected by the logical *OR* operator, we use the *maximum* of the fuzzy values associated with the premises to determine the composite value for the premises.
- If the fuzzy value of a premise (e.g., premise i) is given as $fv(i)$ then *NOT* $fv(i) = 1 - fv(i)$.

To demonstrate, consider the following production rule:

Rule 1:   *If* disk drive noisy = yes (fv = ?)
          and disk formatting results in bad sectors = yes (fv = ?)
          Then disk drive status = defective (cf = 0.9)

Note carefully that the confidence factor associated with the rule itself is 0.9 (i.e., we might believe that the rule is correct roughly 9 times out of 10). Further, if the user provides responses concerning whether the disk drive is noisy or if formatting results in bad sectors, he or she must indicate an associated fuzzy value (fv) for each input, where this value lies between 0 and 1.

Since the disk drive of our hypothetical example is assumed to be *quite* noisy, let us further assume that the user responds to the prompt for the first premise with a fuzzy value of 0.8. However, since disk formatting only occasionally results in bad sectors (i.e., unusable portions of the disk), the value for the user's response for the second premise is assumed to be 0.3. Using the properties of fuzzy calculus as listed earlier, we would then take the minimum of 0.8 and 0.3, resulting in a composite rule premise confidence level of 0.3. Further, if we have a confidence factor of 0.9 in the validity of this rule then our confidence in the rule's conclusion (i.e., the disk drive is defective) is simply the product of this factor times the fuzzy value of the composite premise, or $0.9 \cdot 0.3 = 0.27$. That is, the value 0.27 has been obtained by multiplying the composite rule premise fuzzy value (0.3) by the rule confidence factor (0.9).

Suppose that we also have rule 2, listed below.

Rule 2:   *If* disk drive chatters when a disk is inserted = yes (*fv* = ?)
          or drive chatters when a disk is ejected = yes (*fv* = ?)
          Then disk drive status = defective (cf = 0.7)

Assuming that the fuzzy value for the first premise of rule 2 is 0.9 and that of the second is 0.2, we now take the *maximum* of these (since the two clauses are connected by the *OR* operator) to determine the composite rule premise confidence level. Under the assumption that rule 2 has a confidence factor of 0.7 the confidence in the conclusion of rule 2 is $0.9 \cdot 0.7 = 0.63$.

Note that we now have two different values of confidence in support of the same conclusion clause AV pair, that is, the conclusion that the disk drive status is defective. From rule 1, our confidence is 0.27 and from rule 2 it is 0.63. Specifically, the results of two different rules serve to support the same conclusion. So, what confidence should we now place in the final conclusion concerning the status of the disk drive? One approach is to simply *OR* the results, that is, take the maximum of the confidence levels of the supporting rules. Consequently, we would then say that our confidence that the disk drive is defective is 0.63.

To summarize, the fuzzy set approach permits the user to respond to prompts with something other than a simple yes or no answer. That is, it allows for *partial truths*. While the scientific basis for its employment in expert systems is debatable (as is the basis for most alternative approaches), it does provide a straightforward, appealing approach for the subjective inclusion of uncertainty. However, one criticism of the approach (at least in the simple form presented here) is that it fails to consider the *amount* of supporting evidence.

Consider, for example, two rules that support a specific conclusion (i.e., the same AV pair). If the confidence factor for one rule is 0.2 and the other is 0.5, then our confidence in the conclusion is 0.5. What if, however, there are *ten* rules that support this conclusion, with confidence factors of 0.2 for the first *nine* rules and a value of 0.5 for the remaining rule? Our composite confidence in the conclusion remains 0.5, and this seems counterintuitive.

CONFIDENCE FACTOR UNION METHOD

As a consequence of the result described in the previous paragraph, we may wish to employ an alternative approach to the determination of the confidence of an AV pair, given support by several rules. This method may be termed the *confidence-factor union method* as it employs an approach analogous to the determination of the union of sets. To illustrate, assume that two rules conclude the same AV pair and that one has a confidence factor of 0.2 while the other has a confidence factor of 0.5, that is,

$$cf_1 = 0.2$$
$$cf_2 = 0.5$$

where cf_i = the confidence factor of supporting rule i. Letting $C(cf)$ represent the confidence factor in the final conclusion, we have

$$C(cf) = cf_1 + cf_2 - cf_1 \cdot cf_2$$

Thus, for our problem, we may compute the final confidence factor for the AV pair as

$$C(cf) = 0.2 + 0.5 - 0.2 \cdot 0.5 = 0.6$$

This notion may be extended to any number of supporting rules by simply dealing with two rules at a time. For example, if we had three rules supporting the same conclusion with confidence factors of 0.2, 0.5, and 0.5, we combine the result obtained directly above (i.e., 0.6) with the confidence factor of the third rule, or

$$C(cf) = 0.6 + 0.5 - 0.5 \cdot 0.6 = 0.8$$

Obviously, with the confidence-factor union method, the confidence factor for a given AV pair increases with an increase in the number of rules concluding that AV pair. This may seem more intuitively appealing than simply taking the maximum value of the individual confidence factors. However, it is still a heuristic approach to the blending of confidence factors.

A WIDELY EMPLOYED APPROACH TO UNCERTAINTY

The MYCIN project (see Chap. 3) has received extensive exposure and, as a result, has served to influence subsequent expert systems software development. The manner in which MYCIN deals with uncertainty has, in particular, been widely copied. In this section, we shall present an approach to uncertainty that is based upon that used by MYCIN.

A prerequisite of the method presented here is that the use of the logical *OR* operator (i.e., in connecting premise clauses) is precluded. This is no problem as we should recall that any rule with an *OR* operator may be decomposed into two simpler rules. Thus, we only consider rules where, if multiple premise clauses exist, they are connected by the *AND* operator.

We shall use confidence factors ranging from -1 to 1, where a 1 represents a rule or response in which one has absolute certainty that it is true and a -1 indicates a rule or response that one believes to be absolutely false. A zero then indicates a lack of confidence in the rule or response. The concept of a *threshold level* will also be employed. The threshold level will be denoted as δ in this discussion. The threshold level used in MYCIN is 0.2, and we have assumed the same level in the discussion that follows (in most packages using such an approach, this level may be set by the user).

We are now ready to consider the conditions associated with the use of confidence factors in this approach. First, let us consider the computation of the composite rule *premise* confidence factor. This value is given as

$$RI_k(\text{cf}) = \min_i\{P_i(\text{cf})\} \text{ if all } P_i(\text{cf}) \geq \delta$$

or

$$\max_i\{P_i(\text{cf})\} \text{ if all } P_i(\text{cf}) \leq -\delta$$

or

$$0 \text{ if } |P_i(\text{cf})| < \delta \text{ for any } i$$

or

$$0 \text{ if any two } P_i(\text{cf}) \text{ are of opposite sign} \qquad (7.3)$$

where $RI_k(\text{cf}) =$ the *composite* rule input, or premise confidence factor of rule k

$\quad\quad P_i(\text{cf}) =$ the confidence factor for premise clause i

$\quad\quad \delta =$ the confidence factor threshold level

Next, consider the confidence factor of the *output* of any rule. This is given as

$$\text{cf}_k = RI_k(\text{cf}) \cdot [R_k(\text{cf})] \qquad (7.4)$$

where $\quad \text{cf}_k =$ the (attenuated) output confidence factor of rule k

$\quad\quad R_k(\text{cf}) =$ the confidence factor of rule k

$\quad\quad RI_k(\text{cf}) =$ the composite confidence factor of the premise of rule k

Figure 7.1 provides an illustration of the procedure. Here, we have assumed that the certainty of the rule itself is 0.8. The confidence factor of the composite input of the rule (i.e., the composite certainty across all premises) is given simply as

$$RI_k(\text{cf}) = \min_i\{P_i(\text{cf})\} = \min\{0.9, 0.7, 0.5, 0.75\} = 0.5$$

Next, to determine the confidence associated with the rule's conclusion, or output, we simply note that the confidence factor of the rule serves to attenuate the certainty associated with the rule's composite input certainty. Thus,

$$\text{cf}_k = RI_k(\text{cf}) \cdot R_k(\text{cf}) = 0.5 \cdot 0.8 = 0.4$$

We have thus far examined how to determine the confidence factor for a single rule. However, in the general case we must combine the confidence factors for the conclusions of several supporting rules in order to determine a final confidence factor for the associated AV pair. The method used to compute the final confidence factor of an AV pair, denoted as $C(\text{cf})$, as supported by several rules employs Eq. (7.5), below. Note carefully that *only rules whose confidence factors exceed the threshold level are employed in the computations*. Further, for clarity in presentation, we have assumed that just two rules support the conclusion under consideration. Extension of this formula to more than two rules should be obvious. Or, from a more pragmatic view, we may simply blend the confidence values two at a time. As should be obvious, this formula represents an extension of the confidence factor union method previously presented, that is, to encompass the use of negative confidence factors and a confidence factor threshold level.

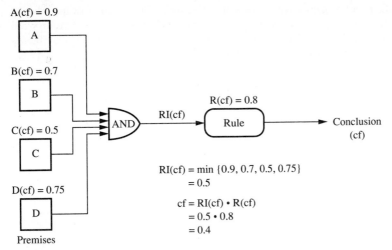

A(cf) = 0.9

B(cf) = 0.7

C(cf) = 0.5

D(cf) = 0.75

R(cf) = 0.8

RI(cf)

Conclusion (cf)

$RI(cf) = min \{0.9, 0.7, 0.5, 0.75\}$
$= 0.5$

$cf = RI(cf) \cdot R(cf)$
$= 0.5 \cdot 0.8$
$= 0.4$

Premises

FIGURE 7.1
Confidence in combined premise clauses.

$$C(cf) = cf_1 + cf_2 - cf_1 \cdot cf_2 \qquad \text{if } cf_1 \text{ and } cf_2 \geq 0$$

or $\qquad cf_1 + cf_2 + cf_1 \cdot cf_2 \qquad$ if cf_1 and $cf_2 \leq 0$

or $\qquad cf_1 + cf_2/[1 - min(|cf_1|, |cf_2|)] \quad$ if $-1 < cf_1 \cdot cf_2 < 0$

or $\qquad 1 \qquad\qquad\qquad\qquad$ if $cf_1 \cdot cf_2 = -1 \qquad (7.5)$

where $C(cf) = $ the confidence factor of the conclusion, C
$\qquad cf_1 = $ the confidence factor of supporting rule 1
$\qquad cf_2 = $ the confidence factor of supporting rule 2

and both cf_1 and cf_2 must exceed the threshold level.

> **Example 7.1: Use of Confidence Factors.** To clarify our discussion, let us apply the most recent confidence-factor approach to the knowledge base listed below, where inference will be through the forward chaining approach. Here, we are again attempting to determine if a computer disk drive is defective or not. Please note that we are using this rule base simply for the purpose of demonstration, and no real-world validity should be inferred.
>
> Rule 1: *If* drive noise is unusually noisy $[cf_{1,1}]$
> $\qquad$ *and* drive age is greater than 1 year $[cf_{1,2}]$
> $\qquad$ *Then* drive status is defective $[R_1(cf) = 0.8]$
>
> Rule 2: *If* disk insertion result is chatter $[cf_{2,1}]$
> $\qquad$ *and* screen display is distorted when disk inserted $[cf_{2,2}]$
> $\qquad$ *Then* disk drive status is defective $[R_2(cf) = 0.7]$
>
> Rule 3: *If* disk ejection result is chatter $[cf_{3,1}]$
> $\qquad$ *and* disk formatting is unreliable $[cf_{3,2}]$
> $\qquad$ *Then* disk drive status is defective $[R_3(cf) = 0.9]$

Note that the rule confidence factors $[R_k(\mathrm{cf})]$ are assumed given and that the premise confidence factors ($\mathrm{cf}_{k,i}$ = confidence factor for rule k, premise i) are to be supplied by the user upon request. Let us also assume that the confidence factors for the various premises, as ultimately obtained from the user are

$$\mathrm{cf}_{1,1} = 0.8$$

$$\mathrm{cf}_{1,2} = 0.5$$

$$\mathrm{cf}_{2,1} = -1$$

$$\mathrm{cf}_{2,2} = 0.7$$

$$\mathrm{cf}_{3,1} = 0.7$$

$$\mathrm{cf}_{3,2} = 0.3$$

First examine the propagation of uncertainty through rule 1. From Eq. (7.3), we note that the confidence factor for the premise of rule 1 is the minimum of 0.8 and 0.5, or 0.5. The confidence factor of the rule output is then given by Eq. (7.4), the product of the premise confidence factor and the rule confidence factor. This results in a value of 0.4 (i.e., $0.5 \cdot 0.8 = 0.4$). Next, for rule 2, we note that the premise confidence factors are of opposite sign and thus, from Eq. (7.3), the confidence factor of the premise of rule 2 is 0; which, for practical purposes, means that rule 2 may be discarded. From Eqs. (7.3) and (7.4), the confidence factor for the output of rule 3 is simply $0.3 \cdot 0.9 = 0.27$. As a result, we now have two rules that support our conclusion that the disk drive is defective. Since the confidence factors of the output of both rules are positive, we must employ the first condition of Eq.(7.5). Thus:

$$C(\mathrm{cf}) = 0.4 + 0.27 - 0.4 \cdot 0.27 = 0.562$$

That is, the confidence associated with the conclusion, as supported by rules 1 and 3, is 0.562. Note that had we employed the simplified version of fuzzy logic as discussed earlier, our result would have been 0.4.

Let us now try the same example but use the following premise confidence factors:

$$\mathrm{cf}_{1,1} = -0.9$$

$$\mathrm{cf}_{1,2} = -0.5$$

$$\mathrm{cf}_{2,1} = 0.1$$

$$\mathrm{cf}_{2,2} = 0.15$$

$$\mathrm{cf}_{3,1} = -0.5$$

$$\mathrm{cf}_{3,2} = -0.3$$

In this instance, the confidence factor for the output of rule 1 is $-0.5 \cdot 0.8 = -0.4$; the confidence factor for rule 2 is 0, and the confidence factor for the output of rule 3 is $-0.3 \cdot 0.9 = -0.27$. Using the second condition of Eq. (7.5) to combine these results, we obtain

$$C(\mathrm{cf}) = -0.4 - 0.27 + 0.4 \cdot 0.27 = -0.562$$

That is, the certainty associated with a defective disk drive is -0.562, or fairly strong evidence that the drive is *not* defective. And as long as we realize that the confidence factor value derived (i.e., -0.562) is simply an indication of relative certainty, and not an absolute measurement, we may use this bit of information to support our decision-making process.

WHICH APPROACH TO UNCERTAINTY IS BEST?

Since we have presented but five approaches to uncertainty (and employed only the most simple versions of these), there is hardly enough evidence upon which to draw any final conclusion as to the *optimal* approach to handling uncertainty in expert systems. Unfortunately, even had we covered every proposed approach to uncertainty, the same conclusion would still be reached. Simply put, the choice of the approach that the knowledge engineer uses to include uncertainty (*including the decision to not use any approach*) in an expert system is most often one of personal preference.

Alternative approaches thus far proposed range from those even more simplistic and naive than the ones above to those which impose a great deal of seemingly unsubstantiated complexity, and requirements for supporting (and often extensive) computation—without any truly independent evidence of significantly improved accuracy. Further, unless the user decides to develop his or her own expert systems package, he or she is usually at the mercy of the approach embedded in the particular commercial expert systems shell that is employed. As such, the main benefit of our discussion has been to provide the reader with some appreciation of the complexity and inexactness involved in the propagation of uncertainty, and of the simplifications that are typically employed so as to develop at least some indication of the relative levels of uncertainty associated with the conclusions provided by an expert system.

DO WE REALLY NEED TO CONSIDER UNCERTAINTY?

There is considerable controversy about the choice of the approach used to capture uncertainty in expert systems. As mentioned, we feel that there is no evidence that any one approach is *best*. Thus, when selecting an approach for uncertainty, we would advise the reader to *choose the approach that would appear to come the closest to actually capturing the process used by the domain expert*. Unfortunately, in much of the debate over methods for uncertainty, this particular message seems to have been lost.

On an even more controversial note, we further believe that, whether or not your software development package includes an approach (or approaches) to uncertainty, there is absolutely no need for you to feel obliged to use it. All too often, uncertainty is included in an expert system by means of putting words into the domain expert's mouth.

To clarify this last observation, consider the accomplishment of one particular knowledge acquisition effort. This effort was conducted by two graduate students, and the author was simply an observer. The domain expert under consideration deals regularly with problems of diagnosis and maintenance of petrochemical processing units. This is accomplished by taking samples from both the input and output flow lines of the processor, and then subjecting the samples to various laboratory tests. A primary objective of the procedure is to catch a potential problem before it becomes an actual one. Using the test results, coupled with his own extensive experience with the machinery, the expert must then decide on one of three courses of action: (1) do nothing, (2) change the processor's settings and take samples at shorter time intervals, or (3) shut down the processing unit for maintenance.

During one meeting, early in the acquisition effort, the domain expert was explaining how he used the results of two particular laboratory tests. He noted that, if one test was only *marginally acceptable* while the other was *somewhat on the high side*, it was *very likely* that the processing unit would have to undergo maintenance in the near future. Consequently, he would request that samples be taken more frequently from this processor. The students were bothered by the vagueness of the terminology employed (e.g., marginally acceptable, somewhat high, and very likely). They presented the domain expert with *their* formulation of *his* heuristic rule, as listed below:

Rule 1: *If* test A results = marginally acceptable ($cf_{1,1}$)
         and test B results = high ($cf_{1,2}$)
         Then sample collection rate = increased [$R_1(cf)$]

They then asked the expert the following question: "What is your confidence in this rule, that is, what percent of the time do you believe it is valid?" Their intention was to establish a confidence factor for the rule itself and to also permit the user to associate a confidence factor for each value he or she supplied during consultation. They felt that, in this way, they could transfer the domain expert's vague terminology into *hard numbers*.

However, note that the expert had never once indicated that he associated any particular numerical value, or *probability*, with the rule. Further, he never indicated that he, or any other potential user of the expert system, would be more comfortable with replying to a user prompt with a number (i.e., a confidence factor) than with a statement (e.g., marginally acceptable, acceptable, not acceptable). However, once the students *suggested* that he do so, and rather strongly implied that this was a *good thing*, he tried his very best to oblige. And this is a classic way in which knowledge engineers all too often put *their* words into the domain expert's mouth.

An alternative approach to the construction of the previous rule might be

Rule 1′:   *If* test A results = marginally acceptable
   and test B results = somewhat high
   Then sample collection rate = increased

Moreover, the user prompt for test A results might be presented using a menu, such as

```
Indicate results of laboratory test A:

1. Acceptable
2. Marginally acceptable
3. Unacceptable
4. Unknown
Input selection: _____
```

and, for test B, the menu might be:

```
Indicate results of laboratory test B:

1. High
2. Somewhat (i.e., moderately) high
3. Normal
4. Somewhat (i.e., moderately) low
5. Low
6. Unknown
Input selection: _____
```

In this latter instance, we have tried to capture just what the domain expert has stated, *rather than what we may have wanted him to say*. One must recognize that the production rule used to represent an expert's heuristic is one level of abstraction removed from the actual heuristic. Imposition of confidence factors (i.e., rather than symbols such as marginal, very good, highly likely), may simply result in yet a further level of abstraction. The more removed the rule base is from the actual knowledge base of the expert, the more unlikely it is that it will either replicate his or her expertise or, and often even more important, be accepted.

ON THE USE OF *UNKNOWN*

The two menus presented above provide the opportunity for the user to reply with the value *unknown*. You may recall, from Chap. 4, our earlier discussion of the use of this value. In addition, in certain instances, it may be appropriate to employ *unknown* as an indication of an extreme case of uncertainty.

Now, there may be two reasons why the user might select *unknown* as his or her reply to the previous two menus. First, it just may be that these particular tests have not yet been completed or, if completed, the results may have not yet been transmitted to the user. Alternatively, it may be that the user simply feels unqualified to provide any response other than *unknown*. That is, the user may lack the expertise (or, possibly, vocabulary) to respond, intelligently, to a particular prompt.

This raises an interesting issue, and one that has not really received the attention due it. Specifically, we believe that there is a certain risk to be run in using the domain expert as the sole source of knowledge—or as the sole focus of expert system performance validation. The domain expert, by definition, has substantially more expertise than the ordinary employee. And it is likely that it will be the ordinary employee who ultimately uses the expert system. If not, one of the primary justifications for an expert system has been lost.

However, things that are crystal clear to the domain expert (or knowledge engineer) may not be nearly so transparent to the ordinary user. He or she may simply not have the level of training, experience, and insight necessary to provide certain input; input with which the domain expert has no problem. Consequently, by designing the expert system (solely) about the level of proficiency of the domain expert, we may inadvertently rule out its use—and acceptance—by the ordinary user. It is thus vital, in our opinion, to solicit the opinions of the ordinary user *during* the development phase of an expert system—and most definitely before the implementation phase.

Let us refer back to our familiar aircraft identification expert system. Recall, from Chap. 5, that one rule base developed through rule induction (i.e., through the VP-Expert Induce facility) for this problem was

Rule C130′: *If* engine = prop
  and wing position = high
  and wing shape = conventional
  and tail shape = conventional
  and bulges = underwings
  Then plane = C130

Rule B747′: *If* engine = jet
  and wing position = low
  and wing shape = swept-back
  and tail shape = conventional
  and bulges = aft cockpit
  Then plane = B747

Rule C5A′: *If* engine = jet
  and wing position = high
  and wing shape = swept-back
  and tail shape = T-tail
  and bulges = none
  Then plane = C5A

Rule C141′:   *If* engine = jet
             and wing position = high
             and wing shape = swept-back
             and tail shape = T-tail
             and bulges = aft wings
             Then plane = C141

    As you may recall, we may have been somewhat surprised by this rule base. That is, it included *every* attribute provided in the example set despite the fact that only a single attribute (bulges) is required if the rule base is deterministic. On the other hand, if we are dealing with uncertainty with respect to the user input, then this rule base may seem somewhat more realistic. However, it will still require the user to provide a response to each and every attribute—and this may be extremely frustrating, even in the light of some uncertainty.

    Alternatively, we can develop the minimal rule base, using just bulges:

Rule C130:   *If* bulges are under wing
             Then plane is C130

Rule B747:   *If* bulges arc aft of cockpit
             Then plane is B747

Rule C5A: *If* bulges are none
             Then plane is C5A

Rule C141:   *If* bulges are aft of wings
             Then plane is C141

In this latter instance, we run the very real risk of not being able to identify the aircraft in the event that the user does not get a good look at the aircraft bulges. In fact, of all the attributes, bulges is certainly the one most difficult to distinguish at a distance (or at a less than optimal perspective). Thus, instead of the most recent rule base, let us consider the use of the following rule base:

Rule C130:   *If* bulges are under wing
             Then plane is C130

Rule B747:   *If* bulges are aft of cockpit
             Then plane is B747

Rule C5A: *If* bulges are none
             Then plane is C5A

Rule C141:   *If* bulges are aft of wings
             Then plane is C141

Rule A: *If* bulges are unknown
             and engine is prop
             Then plane is C130

Rule B: *If* bulges are unknown
            and wing position is low
            Then plane is B747

Rule C: *If* bulges are unknown
            and engine is jet
            and wing position is high
            Then plane is either a C5A or C141[1]

The user prompt for *bulges* in this rule base is presented below:

Indicate existence of bulges on aircraft fuselage:

1. No bulges (i.e., definitely no pronounced bulges on fuselage)
2. Bulges aft of cockpit (and on top of fuselage)
3. Bulges aft of wings (and on bottom of fuselage)
4. Bulges under wings (and on bottom of fuselage)
5. Unknown (i.e., unable to determine either the presence
   or absence of bulges)

Input selection: _____

In this menu, we have provided the user with the opportunity to input selection 5, for *unknown*. For the sake of illustration, let us assume that backward
chaining will be employed. Notice that the particular aircraft will immediately be
identified from one of the first four rules should the user be able to distinguish
the bulges. In this case, the user will only be asked about bulges. However, in
the event that the bulges cannot be distinguished, and the user replies with a
value of *unknown*, we will move to rules A through C. Further note that, if the
bulges cannot be distinguished, and if we only know that the aircraft sighted has
jet engines and high wings, then we cannot completely resolve the identification.
That is, the plane must be either a C5A or C141, but we do not know which.
This result may not be a total failure as the fact that we have been able to narrow
the type of plane down to either a C5A or C141 may still be of interest.

The potential advantages of structuring a rule base in this manner should,
hopefully, be apparent. In the best case, we should be able to reduce user prompts.
In the worst case, we should still be able to reach some meaningful conclusion.
This same approach may, in fact, be employed with rule bases that are to be accessed by users having a wide range of expertise. For example, we can structure
the rule base so that the novice user may be readily identified (he or she will answer
"unknown" to our initial set of prompts). Having identified the novice, we may

---

[1] Notice carefully that the attribute *plane* has a single value in this conclusion clause and this is
given as *either C5A or C141*. Thus, the conclusion to rule C is a single clause and we have not
violated our requirement that disjunctive clauses are not permitted in a rule's conclusion.

use a set of rules that are more adequately tailored to this individual's level of expertise. This may be illustrated in the (ungrouped and unordered) rule base provided below:

Rule 1: *If* quilt pattern from prompt is Grandma's Flower Garden
and quilt origin is prior to 1939
Then quilt value is high

Rule 2: *If* quilt pattern from prompt is Double Wedding Ring
and quilt origin is prior to 1939
Then quilt value is modest

Rule 3: *If* quilt pattern from prompt is unknown
and quilt design consists of interlocking circles
and number of intersections of each circle is 4
Then quilt pattern is Double Wedding Ring

Rule 4: *If* quilt pattern from prompt is unknown
and quilt design consists of arrangement of six sided figures
and each six sided figure is bordered by six other figures
and all figures are connected by a *path* of the same color
Then quilt pattern is Grandma's Flower Garden

Rule 5: *If* quilt pattern is Grandma's Flower Garden
and quilt origin is prior to 1939
Then quilt value is high

Rule 6: *If* quilt pattern is Double Wedding Ring
and quilt origin is prior to 1939
Then quilt value is modest

The purpose of this rule base is to determine the value of antique quilts. For someone with any previous exposure to quilts, a question about whether or not the quilt pattern is Grandma's Flower Garden would be easy. However, the novice collector may well have absolutely no idea of how to reply to this question. The terminology may be totally unfamiliar. However, such a user is given the opportunity to reply with "unknown," which results in the implementation of rules 3 to 6. These rules, in turn, allow the novice to identify the particular quilt pattern through more detailed questioning—at a level more appropriate with his or her background.

Notice carefully, in the rule base, the difference between the attributes: *quilt pattern* and *quilt pattern from prompt*. The former is found through the inferencing process (i.e., of either rules 3 or 4). The latter is determined by the user, in response to the following prompt:

```
Indicate type of quilt under consideration:

1. Grandma's Flower Garden
2. Double Wedding Ring
```

```
3. Dresden Plate
4. Maple Leaf
5. Lone Star
6. None of the above
7. Unknown
Input selection: _____
```

Finally, note the distinction between selections 6 and 7. A reply of "none of the above" indicates that the user is familiar enough with quilt design to note that the quilt he or she is viewing is not listed (e.g., it might be a pattern such as Sun Burst, or Baby Blocks). However, the choice of selection 7 indicates that the user simply has no idea about the quilt pattern in question.

BRIDGES IN EXPERT SYSTEMS

There is an old saying, "no man is an island," where the implication is that we all (men and women) rely on other humans for emotional as well as physical support. Early expert systems shells tended to ignore this sage advice and functioned very much as islands unto themselves. Consider, for example, the following quote from the magazine, *High Technology*, in their March 1985 issue [Kinnucan, 1985]:

> A shortcoming of most available shells is that they lack traditional computer capabilities, such as database management and sophisticated mathematical computations that many applications require. For example, few of the systems have an interface to an external database management system. "Essentially, the current systems are closed boxes that you can't get out of," says Henry Seiler, president of Level 5 Research.

However, as with so much else that has happened in this field, the buyers and users of expert systems shells quite simply, and quite forcefully, demanded that provisions be made for communication with other software programs and existing databases.

We term the ability to communicate between other programs and databases as the *bridge* between expert systems and the rest of the world, and consider this facility to be a virtual necessity in the majority of (successful) real-world implementations. To demonstrate the importance of such bridges, consider the following actual example. The author and some of his colleagues were asked to develop an expert system for the scheduling of processing activities at a large (U.S.-based and owned) oil refinery [Deal, forthcoming]. All scheduling was being accomplished, on a strictly heuristic basis, by a few human experts and, consequently, their knowledge base was acquired, refined, and cloned into the prototype version of the expert system initially developed. A commercial expert systems shell was employed in the effort, and this shell had the capability of communicating with a variety of external programs as well as with many commercial, personal computer resident databases. However, in this particular case we were unable to convince the sponsors of the importance of such an interface, and they restricted the effort to the development of what was termed a *bare bones* expert systems scheduler in order to save on development costs.

However, the primary factors that served to drive the expert systems scheduler rule set were those associated with data pertaining to

- The amount of raw material (e.g., crude oil, natural gas, and intermediate products) in storage in the refinery tanks or scheduled for delivery to the refinery
- The amount of finished, certified products in storage in the final product tanks
- The availability and technical specifications of the processing units
- Information about customer orders (e.g., amounts and desired delivery dates)

All of the above data were being kept by the company, but were dispersed among several databases and several different computers, located both inside and external to the refinery. The "owners" (i.e., the various divisions within the firm) of each portion of the database seemed, in our opinion, much more interested in amassing and controlling data than in disseminating it to those who actually might use the information for some productive purpose. Through some fairly heroic, and quite unnecessary endeavors, the human schedulers were able to access these different databases and use the information in their scheduling decisions.

It seemed obvious to us that a major benefit of an expert systems scheduler would be to tie all of the external (and partially redundant) databases together, and to use this unified database to support the expert system that would then automatically make the process scheduling decisions. That is, if immediate access was provided to the data, it would be rare to even require a response from the user. To initially provide such interfaces would, of course, require more development time (and, as mentioned, funds) than that necessary to develop the expert systems scheduler alone. The decision was made by the sponsor to input all supporting data *manually*. The expert systems development effort was completed in a year and the final result was an expert systems scheduler that could consistently produce as good, if not better schedules than the human schedulers, but one that was handicapped by the necessity to continually update all supporting data by hand. While the firm had indeed saved on development cost by eliminating direct access to their databases, those who had to use the expert system were frustrated by the need to manually input an extensive amount of data each time the system was employed. As a consequence, the expert system has thus far been rejected for actual implementation in the organization.

TYPES OF BRIDGES

The above example serves to indicate the importance, within the expert systems development package, of a bridge to databases. The lack of such a bridge was, in fact, the sole reason for the ultimate lack of acceptance of the subject expert system, despite its otherwise excellent performance. In our opinion, the ability to access external databases is the single most important bridge in expert systems implementation. Without this ability, one is typically forced to (unnecessarily) replicate existing databases within the expert system, a requirement that can well serve to negate the advantages expected from the expert system. Or, as in the case just described, one must manually input extensive amounts of data.

Fortunately, most commercial expert systems shells, even those at the lowest end of the price spectrum, now incorporate some means of access to external databases. However, the vast majority of existing shells (particularly those at the lower end of the price range) also run primarily on personal computers. Such shells usually access only those databases (and spreadsheets) that run on such systems (e.g., in the DOS environment). This means that, if the organization in which the expert system is to be implemented employs such software to presently store their data, the shell can both acquire data from existing, external files—*as well as revise these files* if so desired.

However, in many instances, expert systems are being constructed on personal computers while the existing databases reside on mainframes. While access between such systems is possible, the provision for this access may, and usually will require far more time and expense than would seem worthwhile. One alternative approach that has been employed has been to download (either electronically or even manually) the mainframe resident databases to personal-computer resident databases. While this is also not a particularly pleasant or simple undertaking, the approach may be, in the long run, cost effective and may even result in improved, overall productivity (i.e., through the switch to a primarily personal computer-based system). The major obstruction to such downloading is typically the complaints of the organization's computer systems department which foresees, quite properly, the eminent decline of its role and power in the organization. Alternatively, one may upload the expert system, as developed on a personal computer, to a mini- or mainframe environment. And some shells do run on both environments.

There are, of course, other external programs besides those dealing with databases, and most commercial shells provide some degree of bridging to these programs. For example, some particularly useful external programs for the support of expert systems include

- Graphics programs
- Simulation programs
- Statistical packages
- Optimization software (e.g., linear programming, goal programming)
- Frame-based knowledge representation

Consider, for example, an expert system for the classification of building architectures. Whenever the user is asked to provide a response, he or she may also be provided the opportunity to actually see computer-displayed graphical examples in support of the query. For example, if one is asked about the types of columns in the building (e.g., Gothic, Corinthian, etc.), the user can ask for help, and the expert system might then access a graphics program containing supporting architectural illustrations. Similarly, an expert system for the identification of ships could access a graphics program that contains various views (e.g., side, head-on, and aerial views) of the classes of ships that are to be recognized.

An example of just where one might wish to include access to a graphics routine has recently been presented. Recall our earlier discussion (on the use of *unknown*) of the expert system for the determination of the value of antique quilts. There, for the novice collector, we had to very precisely describe each quilt pattern so that the novice user might identify the quilt type. However, as the old saying goes, "a picture is worth a thousand words." Just think of how much easier it would be to present the novice with a screen display of the various quilt patterns, and to permit him or her to make a selection in this manner.

Yet another example of a graphics bridge is demonstrated in a recent, prototype expert system developed to support the handling and installation of piping in oil refineries. The purpose of this system is to control a robot that picks up, maneuvers, and places pipes into the pipe racks that run throughout the typical refinery. The pipes themselves are used to transmit various products from one point to another in the refinery. Removal, repair, rerouting, and installation of new pipes are ongoing efforts in the typical refinery and the expense associated with such maintenance is far from negligible.

The system developed may be described as follows. A human operator, at some remote location, can view both the actual materials handling operation (using a television camera) and a simulated, graphics representation of the particular materials handling movements that are planned for each piece of pipe. If desired, the human operator may override the expert systems procedure and implement, instead, his or her own maneuver (as first tested by the graphics simulator).

The materials handling system described above also serves as an example of a bridge to a simulation package, in this case a graphics-based simulator. Another example of a system that bridges to a simulation package is one for the scheduling of testing at a certification laboratory. In this problem, samples (each of which may require one or more tests, from among a wide possibility of test types) are received at a laboratory. Based on the tests required by the sample, the availability of personnel and equipment, and the priority assigned to the sample, a supervisor would previously catalog the tests and then manually develop, on paper, a schedule for the times and sequences of the samples through the laboratory workstations (or, if necessary, the transmittal of the samples to an outside laboratory). The expert system has cloned the knowledge base of the supervisor to produce a test schedule similar to that which would have been developed by the human operator. However, the system also develops several alternative schedules, and evaluates each of these through a computer simulation. The schedule that provides the best results is then selected and implemented. In this case, a commercial expert systems shell (written in C) interfaces with an external simulation program (written in BASIC) specifically for the laboratory under consideration. The program actually also bridges to an external graphics routine, used to display the Gantt charts for each schedule under consideration as well as a color-coded routing of each sample through the laboratory.

Bridges to statistical support packages are also quite common and extremely useful. Specifically, rather than having to construct one's own statistics software, the bridge permits calls to existing software whenever necessary. One such bridge was constructed in support of an expert system used in the control of a manufacturing process. In this process, the diameter, thickness, and various other physical properties of metal tubing are constantly monitored by a system of sensors. The sensor output is delivered to an A/D (analog-to-digital) converter and the resulting data are stored in a database. Periodically, this database is accessed by a statistics package that computes, among other things, the means and variances of the data. The results are then transmitted to the expert system. Using these results, the expert system either allows the processing to continue, or sends out signals which adjust the processing, or (in extreme cases) shuts down the processing line for major repair. While some of the decision making is analytically based, a considerable amount of judgment is required, and this is captured in the expert systems portion of the overall process controller.

In some instances, a portion of the problem addressed may be amenable to solution using optimization procedures (e.g., linear programming, goal programming, dynamic programming, etc.). If this is the case, bridging will permit calls to external, commercial optimization software. An example of this type of bridge appears in an expert system that is currently in development. The purpose of this particular system is to aim a high energy particle beam at a target. The beam initially passes through a *beaming* device that serves to direct the beam through a set of magnets. Adjustments to the beam are available through corresponding adjustments of the energy delivered to each magnet along the beam path. The assignment of these energy levels, given a specific target to be illuminated (or destroyed), may be determined using the solution of a nonlinear, goal programming model [Ignizio, 1982, 1985]. However, a portion of the problem that is not nearly so analytical is that associated with the analysis and interpretation of the signals sent back from the sensors located along the beam path. And this diagnostic problem has been represented as an expert system. The expert system then determines, from the sensor data, whether or not the beam path needs adjustment, and by how much. This information is presented to the nonlinear goal programming package which serves to determine the precise amount of adjustment to each magnet along the beam path. As such, we see a continual interchange between the expert systems diagnostic program and the nonlinear goal programming software.

The final bridge to be discussed is the bridge to a frame-based knowledge representation program. For a number of reasons, this text is focused on production rule-based expert systems. However, some analysts find the combination of rule-based systems and frame-based systems to be attractive. Consequently, a number of frame-based representation programs have been developed as adjuncts to existing, commercial, rule-based expert systems. As a result, one may choose to represent a portion of the knowledge (in particular, certain facts) in the frame-based adjunct, and another portion in the production-rule format. Calls between one format and the other are accomplished through the bridging capability.

Before leaving the topic of bridges, it should be noted that, in addition to expert system shells, there are also commercial expert systems *environments*. Such environments (e.g., KEE, ART, Nexpert Object), which may require an expensive, special purpose computer, offer the user the capability to employ graphics, simulation, frame-based representation, and other supporting tools without leaving the environment. That is, these capabilities are *built into* the environment. While this results in a powerful development package, it serves to vastly increase the complexity of the software—and learning how to (properly) use such packages is no trivial endeavor. On the other hand, by means of bridging from rule-based expert systems shells, we may divide the development effort according to the interests, expertise, and experience of the participants. Thus, someone familiar with databases will deal with that element, someone experienced in the use of graphics routines may focus on that area, and the knowledge engineer is free to concentrate exclusively, or almost exclusively, on the development of the rule base.

EXPLANATION AND JUSTIFICATION

The ability to explain (i.e., why a particular question has been asked of the user) and justify (i.e., just how a certain conclusion was reached) are features that are incorporated in some, but certainly not all (e.g., DENDRAL, considered by many to be the very first expert system, does not have such an ability) expert systems and expert systems development packages. Further, it should be recognized that while some knowledge engineers are strongly in favor of such features, others are considerably less enthusiastic. In light of such diverse feelings, we have chosen to include our discussion of these capabilities within this chapter, that is, as an enhancement to expert systems.

Before proceeding further, let us address one other point. All too often, one is given the impression that expert systems have the ability to explain and justify much on the same level as does a human expert. This is not really an accurate representation. Actually, explanation and justification are accomplished within an expert system by means of a fairly rudimentary *trace and match* routine. And this should be made clear in the discussion fo follow.

We shall describe explanation and justification in expert systems by means of a simple example. Recall the rule base associated with Table 6.23 of the previous chapter, as repeated below:

Rule B/Y: *If* B is true
             Then Y

Rule A/Y: *If* A is true
             Then Y

Rule C/B/T: *If* C is true
             Then B is true

Rule C/B/F: *If* C is false
              Then B is false

Rule A/X: *If* A is true
              Then X

Rule C/D: *If* C is true
              Then D is true

Given that C is true and A is false, and employing backward chaining to determine the goal Y, the final table for a consultation session with this rule base is given in Table 7.1.

Examining the cells under the Working Memory table, we may note that a new column has been added. This column serves to indicate just how each element of the working memory was determined, and why. For example, in this consultation, the user will be asked about C (i.e., whether it is true or false) as noted in the first row of the Working Memory table. Should the user ask "why?" (i.e., why am I being asked about the value of C?), we may simply examine the corresponding row of the Goal table. Here, we note that C is associated with rule C/B/T. Thus, when the user asks about C, he might simply be shown rule C/B/T plus some associated text. The text itself is easily generated, for each rule in response to an inquiry. For example, the display presented to the user might be

```
The value of C is being asked because B is
determined by C:

Rule C/B/T:   If C is true
              Then B is true
```

And if the user wants to know how Y was concluded, we simply examine the cell in the Working Memory table next to the conclusion Y. That is, Y was concluded from rule B/Y. Thus, the presentation shows the user might look something like that shown below:

```
We know Y because B is true:

Rule B/Y:   If B is true
            Then Y
```

TABLE 7.1
Final table—with inference trace

Rule/Premise table			
Rule number	**Rule status**	**Premise clause number**	**Premise clause status**
(1)	A, ~~TD~~, FD	(1)–1	~~FR~~, TU
(2)	A, ~~TD~~, FD	(2)–1	~~FR~~, TU
(3)	A, D	(3)–1	~~FR~~, FA
(4)	A	(4)–1	FR
(5)	A, TD	(5)–1	~~FR~~, TU

Goal table	Working Memory table	
~~C (rule C/B/T)~~	C = true	From user
~~B (rule B/Y)~~	B = true	From rule C/B/T
~~Y~~	Y	From rule B/Y

And if the user next wants to know how B was determined, we may simply examine the table to note that it was found as a consequence of the firing of rule C/B/T—and present the explanation to the user.

Rather obviously, our explanation and justification capabilities have been achieved by simply matching the facts developed in the working memory with their sources, be these user responses or the firing of other rules. And beyond this, there is little else that may be explained or justified.

SUMMARY

Uncertainty exists in two forms in expert systems. First, there is almost always some uncertainty about the validity of a rule. Second, there is often uncertainty about the expert systems user response to a given query. Although numerous approaches for dealing with uncertainty have been proposed in the literature, we have dealt with but a small sample of these methods in this section, and each in a simplified format.

Before considering the inclusion of uncertainty in an expert system, we strongly advise the reader to

- Determine whether or not such inclusion will actually and measurably enhance the expert system—and, if not, do not feel obliged to incorporate uncertainty
- Determine whether or not the method for uncertainty, as employed by the software, is actually consistent with the method that is employed by the domain expert

- Make sure that the inclusion of uncertainty reflects the needs of the domain expert, and not just those of the knowledge engineer
- Do not forget the user

Further, before mindlessly incorporating uncertainty into user prompts (i.e., by means of requiring that the user input a numerical confidence factor), consider the use of menu-driven prompts, wherein the user may select from among a number of familiar terms.

We also discussed the concept and usefulness of bridges in expert systems. Of particular importance is the ability of any (useful) expert system to directly access supporting databases.

The final section of the chapter dealt with the explanation and justification capabilities of expert systems, and attempted to illustrate just how such features are generated. In some instances, such capabilities have led to increased confidence in the employment of expert systems, and a resultant improvement in the likelihood of their acceptance. While this can certainly be an important result, the knowledge engineer must still realize that present-day explanation/justification capabilities in expert systems pale in comparison with that of an articulate, thoughtful, human expert.

In general, unless we construct our own expert systems software, our choices with regard to the manner in which uncertainty is handled, the bridging capability, and even the explanation/justification features, are limited to that employed by the specific software package. Fortunately, these capabilities are being expanded and enhanced as commercial expert systems packages evolve.

EXERCISES

7.1. In this chapter, we have discussed two fundamentally different schemes for dealing with uncertainty in user responses. The first employs a numerical scale [e.g., indicate the intensity of stomach cramps (0 to 10)]. While the second employed a menu (e.g., extreme, very intense,). Present a statement of the advantages and disadvantages of both approaches.

7.2. Discuss the advantages and disadvantages of the Bayesian probability approach to uncertainty, as presented in this chapter.

7.3. A criticism of *simple* methods for computing uncertainty (i.e., such as the 0 to 10 and -100 to 100 methods in EXSYS) is that they are *unscientific* and thus lack credibility. Can you develop a sound argument in support of such methods? If so, list that argument.

7.4. In choosing between various methods for handling uncertainty (including the choice of not using any formal method), what factors should be considered?

7.5. A domain expert notes that, when an oily residue appears on a bearing, the likelihood of need for maintenance is minimal. Further, when white particles appear on the bearing, the need for maintenance is also minimal. However, if *both* an oily residue and white particles appear on the bearing, the need for maintenance is high. List these rules and discuss the selection of the most appropriate uncertainty scheme from among those described in this chapter.

7.6. Use the MYCIN approach (i.e., in the section, "A Widely Employed Approach to Uncertainty") to determine the uncertainty of the final conclusion for the inference network depicted below:

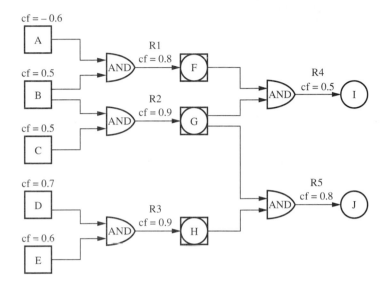

7.7. Repeat Exercise 7.6 if the cf for assertion A is 0.4.

7.8. Discuss those instances in which the use of the *unknown* response may be appropriate.

7.9. Refer to Exercise 4.23. How might uncertainty be employed in such a rule base? Would the use of uncertainty (i.e., confidence factors) likely be compatible with the approach used by the domain expert?

7.10. The use of *unknown* as a value for an attribute may lead to some problems, depending on the inference process employed. Consider the rule listed below, which is assumed to be the first rule of a rule base that is used to select candidates to be interviewed for a job.

    Rl: *If* applicant's job record = unknown
            and applicant's references = unknown
            and applicant's educational record = unknown
            Then decision is insufficient data

Can you conceive of two different results, depending on the inference process, when backward chaining is employed and the goal is *decision*?

7.11. Assume that our rule confidence factors are restricted to nonnegative values and that we wish to employ the confidence-factor union method for the case when *three* rules support a given AV pair. Develop the associated formula for the confidence factor for the AV pair.

7.12. Consider the rule base listed below. What must the minimum value of the confidence factor for the value of A be in order for the goal to be true when the MYCIN-type approach is used and the threshold level is set to 0.2?

R1: *If* B = b
Then C = c (cf = 0.7)

R2: *If* C = c
Then goal = yes (cf = 0.6)

R3: *If* A = a (cf = ?)
Then B = b (cf = 0.8)

7.13. Describe the difference between a menu listing of *none of the above* and one of *unknown*. Do you think a user will appreciate this distinction? If not, what might be done to help the user?

7.14 The following rule base lists those rules used to determine if an airplane needs to make an emergency landing. Given that

$$cf_{1,1} = 0.7$$
$$cf_{1,2} = 0.6$$
$$cf_{2,1} = 0.3$$
$$cf_{4,1} = 0.8$$
$$cf_{4,2} = 0.5$$
$$cf_{4,3} = 0.4$$

find the confidence factor for the final conclusion using:
 (*a*) The fuzzy set approach of this chapter
 (*b*) The confidence-factor union method
 (*c*) The MYCIN-based approach
 (*d*) The 0 to 10 EXSYS approach

R1: *If* hydraulic fluid quantity is low
and rudder movement is sluggish
Then there is a hydraulic failure (cf$_1$ = 0.9)

R2: *If* there is a fuel leak
Then emergency landing is necessary (cf$_2$ = 0.85)

R3: *If* there is an electrical failure
and there is a hydraulic failure
Then emergency landing is necessary (cf$_3$ = 0.95)

R4: *If* cockpit lights are dim
or there is a flight instrument malfunction
or flaps move slower than normal
Then there is an electrical failure (cf$_4$ = 0.9)

7.15. An expert system in a hypothetical nuclear power plant is used to determine whether or not the reactor should be shut down due to a level 2 radiation leak. Ten radiation measurements are taken at a distance of 0.5 miles from the reactor core. One of the rules in the expert system uses the average of these ten measurements and is stated as:

If average of radiation measurements > 0.18
Then there is a level 2 radiation leak

Plant personnel have extensively researched this type of failure and have a substantial database in support of their analyses. In particular, they have calculated that the probability the average radiation measurements are greater than 0.18 when there is *not* a level 2 radiation leak is 0.12. The unconditional probability of a level 2 leak is 0.003. Further, the probability of a radiation leak greater than 0.18 when a level 2 leak is present is 0.91. Using Bayesian analysis, calculate the confidence factor for the conclusion to the above rule.

7.16. Lubricant performance in a high speed turbomachine is monitored using a variety of sensors. From previous tests and experience with similar machines, a fair amount of data are available with regard to the behavior of the machine under various lubricating conditions. Associated with the machine is a sensor which measures the operating temperature. Whenever this temperature reaches or exceeds a certain level, a *high-temperature* alarm is activated. Given that this alarm sounds, what is the probability of lubricant damage if

- Probability of high temperature due to a damaged lubricant is 0.75
- Probability of high temperature occurring at *any* time is 0.12
- Probability of never experiencing a high temperature during operation is 0.88
- Probability of high temperature even when the lubricant is undamaged is 0.03

7.17. Two knowledge engineers are engaged in a heated discussion of uncertainty in expert systems. Tom, an advocate of a *scientifically*-based approach, argues that whatever the method selected, it absolutely must satisfy the laws of probability. Harry, an advocate of *purity in cloning*, notes that the single criterion must be that of reflecting, as precisely as possible, the manner by which the domain expert reaches a conclusion—even if the laws of probability appear to be violated. Whose side, if any, would you take? Why?

7.18. A firm wants you to build an expert system. From your preliminary investigation, you believe that such a system, if it is to be accepted by the user, must have an efficient bridge to the firm's database. The firm, however, rejects the inclusion of a bridge because of the extra time and cost required to develop and validate such a facility. Develop a brief presentation that might convince the firm of the necessity of the bridging facility.

7.19. If, after giving your presentation in Exercise 7.18, the firm still insists on proceeding without a database bridge, what might you do?

7.20. Employ EXSYS to solve any arbitrary rule base where the shell's bridging capabilities are demonstrated with a call to an external program by means of
  (*a*) A call at the start of the consultation
  (*b*) A call from an attribute in a premise clause
  (*c*) A call from an attribute in a conclusion clause

CHAPTER 8

VALIDATION

CHAPTER OVERVIEW

As we mentioned in Chapter 1, one perspective of expert systems holds that the methodology is, in essence, simply an alternative to conventional computer programming. As such, one approach that has been proposed for the validation of expert systems is to perform a process somewhat analogous to that used to validate (i.e., debug) conventional computer programs. While we do not share this opinion, it is true that there are certain rather universal guidelines that are applicable to either the validation of computer programs or rule bases. These generally involve notions from structured programming and relate to the organization, logic, and clarity of statements—whether they be procedural statements or heuristic rules. In particular, such guidelines serve to directly or indirectly address the notions of consistency and completeness within rule bases. As such, a number of recent papers have addressed rule-base validation through such concepts (or enhancements of such concepts) and several of these have served to influence the material in one section of this chapter, the discussion of which will be our second phase of expert systems validation.

However, we personally believe that most discussions of expert systems validation have been too narrowly focused. Further, we feel that the use of the term *debugging* only tends to lend credence to the perception that we are dealing with computer programming. Our feeling is that validation of an expert system must address the following three issues:

- Phase one: the justification for the employment of an expert system
- Phase two: the validation of the consistency and completeness of the expert system's rule base
- Phase three: the verification of the overall performance of the expert system

Let us begin with an examination of the importance of phase one.

ANOTHER STORY

Not long ago, I attended a meeting hosted by a government agency. The purpose of the meeting was to solicit advice (and, ultimately, formal proposals) on how one might develop a meaningful, practical, and effective means for the validation of expert systems. It seems that this agency was literally being overrun by expert systems that had been developed either through its funding, or through funding from other government agencies. However, while such systems were *evidently* able to produce answers, no one was quite sure just how much confidence (or suspicion) to place in the results.

A number of those attending the meeting were a bit shaken by the fact that, after already spending millions of dollars in expert systems development, it was only now that the rather fundamental aspect of validation was being considered for (serious) funding.[1] However, we all sat there obediently listening to one of the agency's key personnel expound, for over 2 hours, on his personal views concerning just how the validation of expert systems should be conducted. And, to be fair, he did raise one or two points of interest. Basically, however, his premise was that one should validate an expert system in much the same way that one validates the performance of, say, a refrigerator or television set. That is, *upon delivery*, the expert system should be required to undergo a series of tests that serve to confirm that it is indeed working according to specifications. Thus, the three fundamental questions to be answered were, in his opinion,

- What types of tests should be considered?
- In what manner should the tests be conducted?
- On what basis should the results be judged?

At the conclusion of the monologue, I asked if I might make one observation. Specifically, I stated that, while I certainly agreed that the subject of validation

---

[1] Even more troubling is the fact that a number of such systems have already been placed into operation, without having subjected the system to (evidently) any rational, or at least thorough, validation. The tragic shooting down of the Iranian commercial airliner in July of 1988 was alleged, in some quarters, to be a result of the AI battle-management system used by the U.S. Navy; a system that could not effectively discriminate between a large, slow airliner and a small, fast F-14 fighter plane. As such, the importance of the validation of such systems should be apparent.

was one of major importance, I had noted the absence of any acknowledgment, on his part, of the importance of *first* making sure that one should actually use an expert systems approach in the first place, for example, as opposed to the use of discriminant analysis, cluster analysis, linear programming, permutation search, and so on. In particular, I felt that he had neglected to discuss the far more critical, *global* issues concerning the use of expert systems.

His reply was, and I quote: "I'm not familiar with the methods you have mentioned (i.e., discriminant analysis, cluster analysis, linear programming, and permutation search).[2] If they are so powerful, why aren't they receiving the amount of attention that expert systems and AI are?[3] Besides, the purpose of this organization is to fund the development of AI and expert systems; we really don't have any interest in other methods." Case closed.

I would hope that the views of this particular individual represent a minority perspective within that agency. In my opinion, and as I have stressed throughout this text, *before* one decides to use (or purchase, or develop) an expert systems approach (or, for that matter, *any* solution methodology) he or she should first make certain that such an approach is practical, appropriate, and cost effective — and more so than any readily available alternative.

Unfortunately, the vast majority of the literature in AI and expert systems fails to even mention this fundamental notion. Thus, the implication (whether intended or not) is that, given a problem, *any problem*, we immediately proceed to construct an expert system to solve it. The issue of alternative approaches is, for the most part, never even raised. As such, it is probably not all that surprising that individuals educated and trained in such a manner exhibit such a lack of familiarity and appreciation of other approaches to decision analysis.

Our assertion is that the very first step in the validation of an expert system, be it of a completed application or of a proposal for the development of such a system, is to provide an informed, objective, and comprehensive justification for the use of the expert systems approach. Then (and only then), if the expert systems approach can be defended, we may proceed to consider the other, less global aspects of validation.

YET ANOTHER STORY: DMES AND AALPS

For those who remain unimpressed with the importance of first ascertaining the validity of the very use of expert systems, consider the following real-life example. In the late 1970s and early 1980s, a fairly extensive effort took place in the specific area of the development of practical solution methods for the loading of military

---

[2] Evidently, there is only a single tool in this individual's decision analysis tool kit. And this is evidence of what we have termed the *one-trick pony* syndrome of far too many advocates of expert systems.

[3] A good question. Does the reader have an answer?

cargo aircraft. And such efforts continue, in fact, today. As typical of many endeavors within DoD (the Department of Defense), these efforts were neither coordinated nor overseen by a single agency. As a result, efforts large and small, and often redundant, were undertaken by several branches of the military as well as by various contractors.

One of these efforts, designated as the Automated Air Load Planning System (AALPS), was begun in 1978 under contract to the U.S. Army and DARPA (Defense Advanced Research Projects Agency). It is our understanding that, as of 1988, funding on the order of approximately $10 million had been expended toward this work.

Another effort, titled the Deployable Mobility Execution System (DMES), was initiated by the Air Force (specifically by AFLMC, the Air Force Logistics Management Center at Gunter Air Force Base, Alabama) in 1981. This effort was undertaken, internally, by two AFLMC personnel, Douglas D. Cochard and Kirk A. Yost. At the completion of DMES,[4] in 1983, about $100,000 in total had been expended (including all salaries, travel, and expenses) on its development and implementation. That is, the DMES effort was completed at a cost of approximately one percent of that of AALPS.

In 1985, Cochard and Yost published, in the open literature, an article detailing the development, operation, and implementation of DMES [Cochard and Yost, 1985]. And they described DMES as, *simply*, a *heuristic method* for air cargo loading—which was implemented, in PASCAL, on a microcomputer.

In 1987, an article on AALPS appeared in the open literature [Anderson and Ortiz, 1987]. The authors of this particular paper stated that AALPS was a *knowledge-based expert system*. Originally, it had been developed in PASCAL but, more recently, had been converted into PROLOG (an intriguing decision in itself although certainly, to some, a more acceptable AI language). Oddly, no mention was made in the AALPS article of either the existence of DMES or of the earlier article by Cochard and Yost. *Yet both papers stated the claim that their method had been **the one** used to implement the actual loading of cargo aircraft for the Grenada invasion of October 1983*. To quote from the paper on DMES:

> During the Grenada rescue operation in October 1983, the system [DMES] saved over $2.5 million in flying-hour costs and provided timely planning.

> Lieutenant General Jack Mackmull, the commander of the 18th Airborne Corps, stated: "Without DMES, we could not have carried out our mission in a timely and efficient manner."

And, quoting from the AALPS paper:

---

[4] Evidently, the effort continues, at a low profile, under the designation of CALM [telephone conversation with Captain Kirk Yost, 1988].

In addition, AALPS supported the deployment of the XVIII Airborne [i.e., the 18th Airborne Corps] to Granada [sic].

Thus, those who have read both of these papers (and we would most strongly advise the reader to do so) cannot help but be, at the very least, puzzled. That is, was the simple, cheap ($100,000) *heuristic method* the one used to so successfully plan the aircraft loading for Grenada, or was it the expensive (multimillion dollar) *expert system* that was used? Why did the commander of the 18th Airborne Corps give a testimonial on behalf of DMES if it was indeed AALPS that was employed? Or, could possibly both methods have been employed?

Another obvious question that comes to mind is, just what are the actual, meaningful similarities and (in particular) differences between DMES and AALPS? Also, why is one a *simple heuristic* while the other is a *knowledge-based expert system*? After carefully reading the two articles, documentation and reports on both systems, and after discussing the matter with those who actually participated in one or the other effort, we have concluded that, as best we can determine, the significant similarities between DMES and AALPS are that

- Both systems are intended for cargo (and military personnel) loading of military aircraft (i.e., C130, C141, and C5A)
- Both systems were originally developed in Pascal[5]
- Both systems incorporate, for the most part, an almost identical set of heuristic rules (or, as termed in AALPS, knowledge base), as acquired from interviews with expert cargo loadmasters (evidently, in many cases, the same individuals)
- Both claim to be *the* loading scheme actually implemented in support of the Grenada invasion

And, the only evident major differences would seem to be

- The cost and duration of the two efforts
- The assertion that one (DMES) is a heuristic method while the other (AALPS) is a knowledge-based expert system

As we mentioned in Chap. 3, the DMES and AALPS articles have raised a number of questions, both within and outside the AI community. In particular, those who are critics of expert systems use these two papers to buttress their argument that expert systems is nothing more than heuristic programming in a slightly different format.

Should one then draw the conclusion that AALPS is not an expert system?

---

[5] DMES, now known as CALM, has been converted into C. AALPS has been converted to PROLOG, and is being run on SUN workstations.

Or that DMES really is? Our view is that, if there truly is such a thing as an expert system, then DMES (and, thus, AALPS) is certainly such an entity. DMES was developed through the acquisition of the heuristic rule set of experts, and this knowledge was then coded into a form that mimics the performance of the human cargo loadmasters. And it makes no difference what language was used (i.e., PASCAL, C, PROLOG, or any other), the result still satisfies the definition and purpose of an expert system.

PHASE ONE:
EXPERT SYSTEMS JUSTIFICATION

So, how does all this fit into our discussion of the validity of an expert system? To answer this, let us assume that you have been provided the opportunity to either fund the development, or to purchase a final version, of AALPS, or any other *expert systems* package. Further, to make things even more interesting, let us assume that we are talking about the investment of a significant amount of money. Our assertion is that, *before* you even begin to consider such a development (or purchase), and certainly well before you engage in any evaluation of the completed package, you should first ask yourself the following five questions:

1. Does the package address a problem that is really of concern, and whose solution (or enhanced solution, or faster solution) would be of significant benefit?
2. Do you have someone in your organization (or, access to an independent, outside consultant) who is knowledgeable in the fields of expert systems *and* decision analysis, and who can provide you with an informed, objective evaluation of the package? If not, be particularly wary. If so, seek his or her assistance with regard to the remainder of the questions.
3. Is the problem being addressed one that *can* and *should* be approached by expert systems? In particular, do alternative solution methods exist that can do a better job, in less time, with less expenditure of funds (i.e., over the entire life of the implemented system), and with less training?
4. Has this particular application been addressed by others and, if so, could their results be used?
5. Is the only significant difference between this package and others (i.e., that perform essentially the same task) simply in the area of embellishments (e.g., eye-catching graphics, multicolored displays)? If so, are these embellishments really necessary, or merely *nice* to have?

Next, you should ask the potential developers, or providers, of the package the following five questions:

6. Do they have actual experience in decision analysis in general, and are they aware of the approaches (e.g., the various forms of mathematical program-

ming, applied statistics, cluster analysis, discriminant analysis, heuristic programming) typically employed in that area?

7. Can they explain, in layperson's terms, just exactly what an expert system is, and what its *tangible* benefits (i.e., as compared to alternative approaches) are?

8. Can they provide a rational explanation as to just why they employed (or are proposing) expert systems rather than some alternative methodology?

9. Did they pursue a literature survey in search of other, related efforts? If not, why not? If so, what were the results (and ask for a copy of the survey)?

10. Finally, can they provide a detailed estimate concerning:

- The expected benefits of implementation of the package
- The expected cost of development and implementation
- Any special requirements (e.g., the purchase of a LISP machine)
- The amount of training required (and cost and duration of such training)
- The amount of time before the system will be *up and running*

Question 1 would certainly seem to be rational. It is simply not logical to spend any significant amount of time and funds toward the pursuit of the solution of a problem that is not really important. And, even if the problem is of importance, why waste time on the development of any tool that will not provide some measurable difference in the quality and timing of the solutions derived? Consequently, if we cannot answer "yes" to question 1, we should drop the matter.

The second question is simply: "Is there someone who can make a reasonably intelligent, objective evaluation of the package?" If not, you are going to have to depend on hunches and emotions—not a particularly reassuring or scientific way to make a decision that involves a substantial amount of cash up front, as well as a continued investment in time and training. Our strong recommendation is to find someone that is knowledgeable in *both* expert systems and the general field of decision analysis.

Question 3 deals with the rather fundamental issue of *appropriateness*. That is, is an expert system the *only* way to solve the problem in a practical and efficient manner? If not, can you objectively justify the use of expert systems over that of the alternatives? You might do well, at this point, to try to honestly evaluate your personal motivation for considering an expert system. Was it just because you *felt* that expert systems (or AI) was a hot topic? And that you needed to show some involvement and/or interest in such a hot topic? If so, that is not a very sound basis on which to make a rational decision.

The fourth question is certainly one that we hope that the sponsors of the AALPS effort considered. It would seem that, had they been aware of DMES, had they investigated the methodology in depth, and had they made an objective comparison of the two methods, they might have at least wondered why one system cost so much more than the other. In general, if we are aware of alternative, existing approaches to the same problem, it is ill-advised to not at least make some rudimentary comparison.

Question 5 is directly related to the previous question. For example, if the sponsors of AALPS were aware of DMES, and still decided to pursue the AALPS development, was that decision based on the fact that AALPS offered additional features—*features that were worth the substantial additional cost*. Unfortunately, in all too many cases, we have noted that a customer purchases one software system over the other simply because of features that are nice, but not necessary. Elaborate color graphics and animation are features that come most readily to mind. It is a fact, and software developers are well aware of it, that many systems are sold almost exclusively on the basis of the graphics output and screen displays. In some cases, far more time and effort are devoted to this area than to the actual algorithms or knowledge bases. As a result, you can obtain truly magnificent displays of truly wretched results. The tendency to *gold plate* such packages seems particularly prevalent in the military sector (or, perhaps, it simply receives more publicity within that sector).

Let us now move to question 6, the first question asked directly of the software developer. Again, this would seem to be a commonsense sort of inquiry. And again, we can list any number of expert systems development efforts that were undertaken by individuals who did not have any experience whatsoever in the general area of decision analysis. Even worse, they did not have any interest in that area or appreciation of its relationship with expert systems. Yet, expert systems is a tool, and just one tool, for decision analysis. The analogy that comes to mind is an old joke, of which we will only tell part. It seems that a man was given a screwdriver as a gift. It was, in fact, the first and only tool he had ever possessed. He then spent the next few weeks tightening every screw in his entire house. When that was done, he was still not satisfied. He then went to the hardware store and bought a file. After learning how to use the file, he filed a groove in the head of each and every nail in the house, and then used his wonderful screwdriver to *tighten* each of these.

Silly story? You bet it is. And it is just as silly to let someone develop a procedure to solve your problem if the *only* tool they have is that of expert systems.

Now, consider question 7. There is an old, apt saying that if a person really and truly understands something, he or she should be able to explain it in simple, precise terms. If they don't understand the topic, any attempt at explanation will be gibberish. If they are trying to impress you, the explanation will generally be lengthy and (overly) complex. If they can't explain the method, they will be equally hard pressed to explain the results, if any. If they try too hard to impress you, then we would certainly be wary of the integrity of the results.

Question 8 can be a real conversation stopper. All too often, it is the very last question that the developers of an expert system could ever conceive of being asked. And if they have no knowledge of decision analysis in general, they are going to be particularly hard pressed to explain why they selected expert systems over some other approach. Not long ago, we had the opportunity to ask this question of a group of individuals who had developed an expert system to solve a problem of project selection and funding. We had already been provided a description of the problem, along with a fair amount of documentation about the procedure the particular firm had been employing (i.e., prior to the introduction of the expert systems approach). It was clear that the problem could be solved, for an *optimal* solution, by means of integer programming [Kaufmann and Henry-Labordère, 1977; Murty, 1976; Taha, 1975].

The response to our question, after a period of stunned silence, was that they had been asked to demonstrate the versatility of expert systems in the organization and, once this was confirmed, an internal group on expert systems was to be formed. The problem they had selected was one that had always been performed (within their firm) on a strictly heuristic basis, and they had access to the individual that performed the ranking and selection of projects. Thus, it was clear (to them) that expert systems was *the* way to proceed.

We asked if they were familiar with mathematical programming, or with the well-known capital budgeting model and its various solution methodologies. They were not. Still, they were adamant that they had achieved their purpose. As they put it, they were able to replicate the decisions that had been arrived at by the firm's expert. As such, they were achieving results of equal validity, and probably on a more consistent basis. The fact that there were alternative methods that could not only solve their problem, but that could arrive at better results only momentarily silenced them. "So what," they said, "our method is free of complex mathematical statements—and can be understood and appreciated by our management." And, of course, they were right. However, is that reason enough to accept a procedure when an alternative exists that could enhance the cash flow of the firm by literally hundreds of thousands of dollars each year?

We now are at question 9, the request for the results of a literature survey. This too can be a stunner. Most often, it would seem, such surveys are not even considered; and that probably explains why so many methods are reinvented, and why so many lessons must be consequently relearned. However, even when a survey is available, one cannot be assured that it is (even reasonably) complete. One immediate example is that of the AALPS paper [Anderson and Ortiz, 1987]. That is, even though the article on DMES appeared almost two years earlier, in the open literature, on precisely the same topic, on exactly the same application, absolutely no mention of DMES appears in the AALPS references.

If a key reference is overlooked, one must question the thoroughness of the developers. And if such a reference is intentionally omitted, one has to be alarmed.

Finally, we come to the last question. An expert systems effort should be held to the *same set of standards* as any other effort undertaken by an organization. Specifically, one should expect to be provided with a full, defensible accounting of the effort involved, along with an objective, reasonable estimate of the benefits to be derived. Simply having an expert system for the sake of having an expert system is a factor that is going to mean less and less as time goes by, and as the glamor associated with such *exotic, high technology* ultimately diminishes.

PHASE TWO: RULE-BASE VALIDATION

Once it has been established that the use of the expert systems approach is indeed justified, and once we have developed a portion of the rule base (i.e., a prototype expert system), we may proceed to phase two of the validation process. Specifically, we may now direct our attention toward the systematic examination of the logic and integrity of the rule base.

Some existing expert systems shells include facilities for some partial checks on knowledge-base rule development. This is typically accomplished within the rule editor and user interface of the shell. For example, the EXSYS shell checks each rule as it is entered for possible conflicts that may exist between that rule and any existing rules in the knowledge base. A number of shells incorporate features that permit the check of syntax, for example, an examination of each rule entered for spelling errors or unbalanced parentheses.[6] However, for the most part, existing shells do not yet provide a particularly comprehensive or thorough approach to the validation of the rule set.

TEIRESIAS [Davis, 1976] was, evidently, the first software system developed to assist the knowledge engineer in the validation of expert systems rule bases. Later, various rule checking packages were developed in support of such expert systems and expert systems development packages as ONCOCIN, TIMM, and KES.

Recently, however, Nguyen, Perkins, Laffey, and Pecora [1985, 1987] have reported on the development of a far more comprehensive validation program named CHECK. CHECK works with knowledge bases involving both forward and backward chaining and, in particular, with LES (Lockheed Expert System), an expert systems development package. The system addresses problems of both inconsistency and completeness. Much of the material that immediately follows is, in fact, based on the facilities available through CHECK[7] and has been influenced by the 1987 paper by Nguyen et al.

Prerequisites and Assumptions

Nguyen et al. state: "Knowledge-base problems can only be detected if the rule syntax is restrictive enough to allow one to examine two rules and determine whether situations exist in which both can succeed and whether the results of applying the two rules are the same, conflicting, or unrelated." If an unrestricted syntax is employed, the methods to be described cannot, for the most part, be implemented.

It is also assumed that the rules have been grouped, according to subject. In the event of backward, or goal-driven rules, the set of rules for each goal may thus

---

[6] Checks that, in general, are not necessary if one uses a menu-driven expert systems shell.

[7] The actual algorithms for the performance of the checking process appear in Nguyen et al., 1985.

be checked independently. Actually, this is not a strict requirement—but, as we have already discussed in Chap. 4, it certainly makes rule checking considerably simpler. Further, and again while not a strict requirement, rule checking is made immensely easier if only conjunctive clauses are employed in the premise and if but a single clause appears in the conclusion. In essence, by following the guidelines presented earlier, in Chap. 4, all of these conditions will have been satisfied.

Finally, we shall *initially* assume that all the rules are *backward* chaining rules and that the rule set is *deterministic* (i.e., no confidence factors). Later will be described how both forward chaining rules and confidence factors may be considered.

Checks for Consistency

There are essentially five types of inconsistency that may be identified. These are

- Redundant rules
- Conflicting rules
- Subsumed rules
- Unnecessary premise clauses
- Circular rule chains

and each of these will be addressed in turn. However, let us first introduce some simplifying notation. Specifically:

$\{P(k)\}$ are the set of premise clauses for rule k
$\{C(k)\}$ are the set of conclusion clauses for rule k
$p(k)$ is one premise clause from the set $\{P(k)\}$
$c(k)$ is one conclusion clause from the set $\{C(k)\}$
$\{S\} = \{T\}$ implies that the set S is equal to the set T

REDUNDANT RULES. Consider the two rules given below. Notice that rule 1 is made redundant by rule 2. That is, both rules are triggered by the same set of premise clauses while the conclusions of one (rule 1) are a subset of the other (rule 2).

Rule 1:   *If* A = X   Rule 2:   *If* B = Y
           and B = Y   *and* A = X
           Then C = Z   *Then* C = Z
                                      and D = W

Specifically, rule r is redundant with respect to rule s if

$$\{P(r)\} = \{P(s)\}$$

and $\{C(r)\}$ is included in $\{C(s)\}$.

Consequently, rule r (i.e., rule 1 in the example) may be removed from the rule set.

CONFLICTING RULES. Consider the two rules listed below. These rules are said to be in conflict since, for the same set of premise clauses, different (*and contradicting*) conclusions are reached. Specifically, in the first case C = Z while in the second C = W.

Rule 1:   *If* A = X   Rule 2:   *If* A = X
          and B = Y   *and* B = Y
          Then C = Z   *Then* C = W

That is, rule r is in conflict with rule s if

$$\{P(r)\} = \{P(s)\}$$

and $\{C(r)\}$ contradicts $\{C(s)\}$.

If two rules are in conflict, this *may* mean that one or the other (or both) has been erroneously formed. However, as you may recall from our earlier discussion on multivalued attributes, there are cases in which two rules can be in conflict and both can still be correct. Consider, for example, the two rules below:

Rule 1:   *If* red spots are observed
          and fever is noted
          Then the patient is allergic to medicine A

Rule 2:   *If* red spots are observed
          and fever is noted
          Then the patient is allergic to medicine B

That is, the patient can exhibit precisely the same symptoms for two different allergies. Such a result *most likely* indicates that we may not have included enough symptoms to discriminate between the two allergic reactions. However, it just may be that the patient is allergic to both medicines — and thus multiple values for the conclusion attribute may be appropriate. Knowledge engineers, using their judgment, must determine which cause is more likely and revise, if necessary, the rule base accordingly.

SUBSUMED RULES. One rule is subsumed by another if both have identical conclusions, but the first contains additional (and thus unnecessary) premise clauses. In the rules below, rule 1 is subsumed by rule 2.

Rule 1:   *If* A = X   Rule 2:   *If* A = X
          and B = Y   *Then* C = Z
          Then C = Z

More formally, rule r is subsumed by rule s if

$$\{C(r)\} = \{C(s)\}$$

and $\{P(s)\}$ is a subset of $\{P(r)\}$ and rule r

may be removed from the knowledge base.

UNNECESSARY PREMISE (IF) CLAUSES. Consider the rules shown below. Note that the same conclusion is reached regardless of the value of the attribute B. Thus, the premise clause containing B is unnecessary, and may be removed to form a more efficient rule. Specifically, rules 1 and 2 should be replaced with a single rule: IF A = X THEN C = Z.

Rule 1:   *If* A = X   Rule 2:   *If* A = X
           and B = Y   *and* B = NOT Y
           Then C = Z   *Then* C = Z

Formally, rules r and s have an unnecessary premise clause if:

$$\{C(r)\} = \{C(s)\}$$

and some $p(r)$ is in conflict with some $p(s)$

and all other premise clauses are equivalent.

CIRCULAR RULES. A set of rules is said to be circular if the chaining of these rules results in a loop, or cycle. This may be illustrated by rules 1 through 3, below. In these rules, we have assumed a single goal, designated here simply as *decision*. Thus, *using backward chaining*, we shall first move to rule 2, the only rule having the goal in its conclusion. The first premise clause of rule 2 appears as the conclusion clause of rule 1. Thus we chain from rule 2 to rule 1. Since the premise of rule 1 is the conclusion of rule 3, we move to rule 3. And, from rule 3, we chain *back* to rule 2—completing one cycle.

R1:   *If* A = X   R2:   *If* B = Y   R3:   *If* decision = yes
     Then B = Y   *and* C = Z   *Then* A = X
                  Then decision = yes

    An effective checking procedure for circular rules is accomplished by use of a *rule dependency graph*. The rule dependency graph associated with the three rules directly above is depicted in Fig. 8.1. The guidelines for the establishment of a rule dependency graph are given as

• Establish a node for each rule and each goal clause
• An edge connects two nodes if either
    one or more premise clauses in one rule matches one or more conclusion
      clauses of another (the edge is directed from the first rule to the second)
    or a goal clause matches one or more conclusions of a rule (the edge is
      directed from the goal clause to the rule)

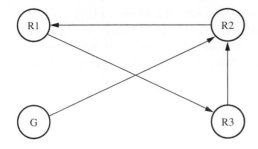

FIGURE 8.1
Rule dependency graph.

Following these guidelines, node R1 (rule 1) is connected to R3 (rule 3), node R2 (rule 2) is connected to node R1, node R3 is connected to node R2, and node G (goal) is connected to R2. *Note carefully the direction of the arrows.* As may be seen, there is a loop formed by nodes R1, R2, and R3, which serves to indicate the existence of circular rules. The loop itself is given as R2–R1–R3–R2. Consequently, whenever a loop appears in the rule dependency graph (and this includes a loop from a node directly back to itself, or a self-circular loop), the existence of a circular rule set has been determined.

The existence of circular rules indicates the *possibility* (depending on the specific capabilities of the chaining procedure used) of an infinite loop at run time. As such, they should rather obviously be avoided.

Before we leave this topic, let us consider another set of rules, as given below. If we plot the rule dependency graph for these three rules, the topology of the graph will be identical to that of Fig. 8.1, *except for the direction of the arrowhead on the edge from node 2 to node 3.* Thus, there is no loop for these rules and they are not circular. The reader is advised to construct the associated graph and to verify this conclusion.

R1:	*If* A = X	R2:	*If* B = Y	R3:	*If* C = Z
	Then B = Y		*Then* decision = yes		*Then* A = X
					and B = Y

Checks for Completeness

There are also five types of incompleteness that may be detected. These are

- Unreferenced attribute values
- Illegal attribute values
- Unachievable intermediate conclusions
- Unachievable (final) conclusions, or goals
- Unachievable premises

and we shall address the checks for each of these conditions.

UNREFERENCED ATTRIBUTE VALUES. Consider a knowledge base for investment advice wherein one of the attributes is that of *interest rates*. Further, let us assume that the three possible values for interest rates are *high, medium,* and *low*. Recall, from Chap. 4, that these three values are termed the legal values of the attribute, *interest rates*. Now, if one or more of these values never appear in any rule's premise clauses, we have a situation of unreferenced attribute values. The two rules below serve to indicate just such a case. Specifically, there is no rule to deal with the interest rate value of *medium*. As such, the knowledge base will either never reach a conclusion for this value or it will reach a wrong conclusion (i.e., if the user responds with a value of medium for an interest rate query).

R1: *If* interest rates are high   R2: *If* interest rates are low
     Then invest in bonds   *Then* invest in stocks

Consequently, a rule validation system should be able to alert the knowledge engineer whenever an attribute value does not appear in any premise clause of the rule set. This can be achieved quite easily by forming a matrix of attribute values (as rows) and premise clauses (as columns). Asterisks (*) may then be placed at the intersection of any row and column associated with a reference to the attribute value. Thus, whenever an entire row in such a matrix is empty, that corresponding attribute value is unreferenced. Consequently, either a rule is missing, or a premise clause is missing, or the attribute value should be removed from the set of legal values.

ILLEGAL ATTRIBUTE VALUES. Consider again a knowledge base for investment advice where the two attributes are *interest rates* and *inflation*. Further, let us again assume that the legal values of interest rates are *low, medium,* and *high* while the legal values for inflation are *low* and *high*. Next, consider the rule cited below. The values of *very high*, for interest, and *hi*, for inflation, are not in the set of legal values, and the rule needs to be corrected. Such illegal values may be detected by matching all rule attribute values against the set of legal values. If no match exists, an illegal attribute value has been identified.

Rule: *If* interest rates are very high
       or inflation is hi
       Then invest in gold stocks

UNACHIEVABLE INTERMEDIATE CONCLUSIONS. Consider the rule listed below. An intermediate conclusion is unachievable whenever the intermediate conclusion does not appear in the premise of any rule. Again, we emphasize that we are dealing with an intermediate conclusion and not a final goal.

Rule: *If* A = X
       Then R = W

Thus, the intermediate conclusion (i.e., R = W) to this rule would be considered unachievable if the clause R = W does not appear in the premise of any other rule. Such a rule is either extraneous (and may be removed) or may have been incorrectly formulated. Another example of an unachievable intermediate conclusion has previously been presented in Example 6.4 where the (intermediate) conclusion clause of rule 5 in that example is unachievable (i.e., through backward chaining alone).

UNACHIEVABLE GOALS. A *goal* (i.e., a *final* conclusion) is termed *unachievable* whenever

- There is no query for the premise of the goal
- *And* the premise cannot be deduced from any other rule (i.e., it is not the conclusion of some other rule)

Consider the rule set listed below. If this is the complete rule set, *and* there is no query associated with the attribute *E*, the goal (in rule 3) can never be achieved since the value of *E* cannot be determined from any other rule.

R1: *If* A = U R2: *If* C = W R3: *If* E = Q
     Then C = W   *Then* D = X   *Then* goal = yes

UNACHIEVABLE PREMISES. A rule *premise* is termed *unachievable* whenever

- There is no query for the premise
- *And* the premise cannot be deduced from any other rule (i.e., it is not the conclusion of some other rule)

and, in fact, the premise of rule 3, as given directly above, is unachievable if there is no user query associated with it. Note also that, when we follow the guidelines that we listed in Chap. 4, unachievable goals and premises are less likely to occur.

RULE DEPENDENCY MATRIX. We may utilize a rule dependency matrix [Ignizio, 1987b; Kim, 1989] to isolate the last three types of incompleteness. Such a matrix may be most easily explained by a simple example. We shall use the three rules listed below to illustrate the procedure. For the sake of illustration, we shall assume that the only user prompt for this rule base is that one associated with attribute C (in rule 2). The associated rule dependency table is presented in Table 8.1. *The* columns *of this matrix are associated with rule* conclusions, *while the* rows *are indicative of rule* premises. Further,

- An asterisk (*) appears at the intersection of a row and column if the premise associated with the row appears in the conclusion associated with the column.
- A pound sign (#) appears at the intersection of a row and column if the conclusion associated with the column appears in the premise associated with the row.

TABLE 8.1
Rule dependency matrix

Premises	Conclusions			
	Rule 1	**Rule 2**	**Rule 3**	**Query**
Rule 1		*#		
Rule 2				*
Rule 3			G	

- An asterisk (*) appears in the column labeled "Query" if the premise for that row has an associated query (i.e., a user prompt).
- The letter "G" should appear at the intersection of any row and column whose rule includes a goal in its conclusion.

R1: *If A = X*   R2: *If C = W*   R3: *If E = Q*
  Then B = Y   *Then A = X*   *Then* goal = yes

The rules for the detection of incompleteness from such a rule-dependency matrix may be summarized as follows:

- If there are no asterisks in any rule row, the associated premise is unachievable
- If there are no pound signs in any rule column, the associated intermediate conclusion is unachievable.
- If there is a "G" in any column and there are no asterisks in the corresponding row, then the associated goal is unachievable.

Using these rules, and upon examination of the rule dependency matrix of Table 8.1, we may quickly conclude that

- The premise of rule 3 is unachievable.
- The intermediate conclusion of rule 1 is unachievable.
- The goal (i.e., of rule 3) is unachievable.

Validation of Forward Chaining Rules

A set of forward chaining, or data-driven rules may be checked in a manner quite similar to that described for backward chaining rules. In fact, the only difference is with respect to a check for unachievable intermediate conclusions, unachievable goals, unachievable premises, and circular rules. In a set of forward chaining rules, the detection of these four cases is simply not applicable. For example, in the three rules listed directly above, the conclusion of rule 1 is unachievable only if backward chaining is employed. However, if forward chaining is used it may well happen that a value for the premise of rule 1 (i.e., for the attribute *A*) is part of the data; and thus the conclusion to rule 1 may be reached.

The Impact of Confidence Factors

Confidence factors serve to complicate the validation procedure. The CHECK program of Nguyen et al. deals with confidence factors that follow, in general, the format used in MYCIN. Specifically, a value of +1.0 means that the rule, or user response, is definitely true; a value of −1.0 means that it is definitely false; and a value of 0.0 implies that the truth of the rule, or response is unknown. Actually, like MYCIN, a threshold value of 0.2 is employed. The reader may wish, at this time, to refer back to the discussion in Chap. 7 for complete details concerning the employment of such confidence factors. Let us now proceed to a summary of the impact, on the debugging process, of the use of such confidence factors.

REDUNDANT RULES UNDER UNCERTAINTY. The mechanics of the check for redundancy is the same, whether or not confidence factors are used. Further, redundant rules that remain in a knowledge base can lead to serious problems when confidence factors are employed. This is because each redundant rule may be fired, resulting in multiple countings of the same information which, in turn, erroneously increases the final weight given to the associated conclusions.

CONFLICTING RULES UNDER UNCERTAINTY. One may also check for conflicting rules by the same process as under certainty. However, quite often there is really nothing wrong with the inclusion of such rules *when confidence factors are employed*. For example, consider the rule set given below.

R1: *If* intake pressure > 200   R2: *If* intake pressure > 200
      and temperature = rising   *and* temperature = rising
      Then replacement decision   *Then* replacement decision
         = flange (cf = 0.75)   = entire unit (cf = 0.5)

Here, the same premises lead to two different replacement action suggestions, each with a different confidence factor value. Evidently, the likelihood of a bad flange is higher than that of having to replace the entire unit (i.e., for whatever piece of machinery that is under consideration). Although the checking procedure would indicate that these two rules are in conflict, the knowledge engineer may either override this result (i.e., leave the rules as is) or combine the two rules (i.e., add the conclusion clause of rule 1 to that of rule 2, through the AND operator).

SUBSUMED RULES UNDER UNCERTAINTY. Subsumed rules, like conflicting rules, may be appropriate when confidence factors are used. Consequently, we may not wish to remove rules that are subsumed. Specifically, a knowledge engineer may wish to develop rules in such a manner that the more restrictive lend added weight to the conclusions reached by the less restrictive. Consider, for example, the two rules listed below. Rule 1 is technically subsumed by rule 2. However, by leaving both in the knowledge base, the weight lent to the conclusion, $C = Z$, is increased if both are fired. That is, the fact that $B = Y$ evidently lends extra weight to the conclusion of rule 1.

Rule 1: *If* A = X
        and B = Y
        Then C = Z (cf = 0.7)

Rule 2: *If* A = X
        Then C = Z (cf = 0.6)

UNNECESSARY PREMISE CLAUSES UNDER UNCERTAINTY. Unnecessary premise clauses may actually be necessary if confidence factors are employed. The reasoning is similar to that noted above for subsumed rules. For example, rules 1 and 2 below have an *unnecessary premise clause* (the second clause, as associated with attribute B) *only if either of the rules are deterministic or if the rules have the same confidence factor.* However, if the confidence factors are different, the unnecessary clauses are actually necessary. Specifically, the certainty of the conclusion (i.e., $C = Z$) is considered different according to the values of the second premise clause.

Rule 1: *If* A = X
        and B = Y
        Then C = Z (cf)

Rule 2: *If* A = X
        and B = NOT Y
        Then C = Z (cf)

CIRCULAR RULES UNDER UNCERTAINTY. Circular rules may still be detected in the same way when using confidence factors. However, it should be noted that a cycle may be "broken" whenever confidence factors are employed. That is, if the confidence factor of any component of the loop falls below the threshold value (i.e., 0.2), the loop is broken.

UNREFERENCED ATTRIBUTE VALUES AND ILLEGAL ATTRIBUTE VALUES UNDER UNCERTAINTY. One may check for unreferenced and illegal attribute values in precisely the same manner as under certainty. And precisely the same conclusions may be drawn.

UNACHIEVABLE INTERMEDIATE CONCLUSIONS UNDER UNCERTAINTY. The detection of unachievable intermediate conclusions becomes a more complex task when confidence factors are used. This is because an intermediate conclusion may become unachievable even though its premise clause matches a conclusion clause in some other rule. This may be best illustrated by example. Note that, according to the guidelines cited earlier, the conclusion of rule 2 (i.e., $R = Z$, which is assumed to be an intermediate conclusion), below, is considered achievable. However, note also what happens if the (attenuated) confidence factor associated with the conclusion of rule 1 is less than the threshold value. In this event, and if the only way to determine the premise of rule 2 (i.e., $D = W$) is through the conclusion of rule 1, then the conclusion of rule 2 is unachievable. For example, let us assume that the value of attribute A is supplied by the user and the associated confidence factor for this value is 0.6. Now, if the confidence factor for rule 1 is 0.3, the attenuated confidence factor for the conclusion that $D = W$ is $0.3 \cdot 0.6 = 0.18$, and this is below the threshold value. Thus, rule 1 will not be either triggered or fired—and the intermediate conclusion of rule 2 cannot be achieved.

Rule 1: *If* $A = X$ $(cf_{1,1})$ Rule 2: *If* $D = W$ $(cf_{2,1})$
   Then $D = W$ (cf_1) *Then* $R = Z$ (cf_2)

UNACHIEVABLE PREMISES AND GOALS UNDER UNCERTAINTY. The detection of unachievable premises and goals, although certainly not an impossible task, is also complicated by the use of confidence factors. For example, an unachievable goal may occur even though its premise appears as the (intermediate) conclusion of another rule. Again, this result is because the confidence factor of the other rule is less than the threshold level. Consider, for example, the rule set listed below and where we assume no query is associated with the premise of rule 2. Now, if the only rule in which $B = Y$ appears as a conclusion is rule 1, *and* if the attenuated confidence factor of rule 1 is less than the threshold level, the goal in rule 2 will be unachievable. The same problem would exist if a chain of rules lead to the conclusion $B = Y$, and the resultant confidence factor for the chain was less than the threshold value.

Rule 1: *If* $A = X$ Rule 2: *If* $B = Y$
   Then $B = Y$ (cf) *Then* goal $=$ yes (cf)

The detection of an unachievable premise is complicated in a similar manner. Specifically, if the attenuated confidence factor for rule 1, in the above rule set, was less than the threshold level, the premise of rule 2 would be unachievable (if it had no associated user query).

PHASE THREE: PERFORMANCE VERIFICATION

The final aspect of expert systems validation is that of the verification of the performance, or output, of the expert system. That is, just *how well* does it accomplish its intended role in actual practice. Performance verification of any system, and particularly that of a complex system (as in the case of most real-world expert systems) is a broad topic that has been effectively addressed in a number of other texts [Cleland and King, 1968; Sage, 1977; Shinners, 1967]. Consequently, we shall only attempt to briefly summarize some of the key issues of performance verification as they relate to expert systems in particular.

The primary issues that must be dealt with in this phase are

- Precisely what is the intended role of the expert system?
- What specific measures of performance should be used to evaluate the conclusions reached by the expert system (i.e., how do we measure just how well the expert system is performing its intended role)?
- How should these measures of performance be collected?
- On what basis should the resultant measurements be evaluated?

As mentioned, one oft-cited proposal for the validation of a completed expert system is, in essence, to compare the performance of the system against that of the human expert, or experts, from whom the knowledge base was acquired.[8] As noted earlier, we have several problems with such a simplistic view, and we shall deal with this topic in more detail later. However, even if the comparison of the performance of an expert system with the human expert is to be conducted, this is not a trivial task. That is, one does not simply run a number of consultation sessions by both the human expert and the expert system and then (somehow) compare results. Rather, we still must deal with each of the four issues raised above. Consequently, let us now address each of these issues in turn.

Intended Role of the Expert System

The intended role of an expert system *should,* rather obviously, have been determined well before we even considered the development and/or implementation of the approach. Specifically, an expert system will provide expert advice concerning decisions to be made within a specific domain. As such, we should have laid down the precise objectives, goals, and requirements of the system under consideration, and these should appear as part of the expert system's documentation.

For example, if we are considering the development of an expert system for, say, battle management for naval forces, then we have identified the general domain of expertise. The role of such an expert system is thus to provide advice, *on a timely basis,* upon which decisions may be made in the heat of battle. Such advice should correspond to, ideally, that which would be provided by the best military minds *if* they

- Had access to all of the data developed during the battle
- Had time to digest and base their decisions on these data

---

[8] Of course, this is only appropriate if the expert system was, indeed, developed through the acquisition of a knowledge base as cloned from human experts. In the event that historical data are used for rule induction, an analogous approach would be to compare the decisions reached by the induced rule base with those recorded in the database.

On this basis, we should then attempt to spell out precisely what the expert system is expected to accomplish. In addition, we should cite such items as

- List of threats (i.e., strategy, tactics, and weapons) likely to be faced
- Range of data that will be available to the system
- Specific type of user input that it may expect
- Specific type of advice that is to be provided to counter the threat
- Actions that the expert system must perform either automatically, or in conjunction with a human operator
- Extent and scope of the expert system's responsibility
- Specific limitations of the expert system's responsibility
- Safeguards to be provided to protect both the integrity of the expert system (e.g., passwords) as well as some indication of when the bounds of the knowledge base have been exceeded (i.e., the ability to know what it does not know)

Once all of this has been accomplished, the development and validation of any resultant expert system may be accomplished in a more orderly, objective, and systematic fashion.

Measures of Performance

If the intended role of the expert system is well documented, the determination of the measures by which its performance will be judged is a considerably easier task. Typically, the introduction of an expert system will have an impact in three primary areas: productivity, organization, and personnel. Each of these, in turn, may be broken down into more specific subtopics. A hypothetical breakdown for each of these is listed below:

- Impact on productivity
    - Solution accuracy
    - Improvement over existing procedures
    - Computational aspects (e.g., computation time, storage requirements)
    - Ease of use, updating, and revision
    - Clarity and timeliness of output
    - Appropriateness of software/hardware selection
- Impact on the organization
    - (Life cycle) cost of software and hardware
    - Cost savings of system (e.g., reduction in personnel)
    - Profit potential (e.g., via better, faster results)
    - Unification of databases and improved decision making
    - Reorganization, staffing requirements
- Impact on personnel
    - Morale (short and long term)

Training/education requirements
Job enhancement/degradation

Associated with each of the above subtopics should be some measure by which the achievement of that particular subtopic may be evaluated. Some of these will be relatively easy. For example, reorganization and staffing requirements can usually be estimated with some degree of accuracy, and the associated measures are rather obvious. Other measures will be somewhat more difficult. For example, an estimate of the life-cycle cost for the implementation of an expert system, while essential, will require a fair amount of effort and expertise. Other subtopics will be quite difficult to measure. One example is that of the effect of the introduction of the expert system on the morale of the personnel within the organization. This is not something that can be reduced to dollars and cents, and is certainly far more subjective than some of the other subtopics. However, in most cases, what we are seeking is the identification of *trends,* or *relative* measures. That is, will morale rise or fall? Is the system's ease of use excellent, average, or poor? In any case, whether or not a factor may be precisely measured, it should at least be listed *and* considered if it serves to play a role in the acceptance of the introduction of the expert system.

Data Collection

Once we have identified the various (pertinent) performance measures, and determined the manner by which they are to be measured, we then need to establish a plan for the collection of these data. Obviously, the measures cover a wide spectrum and thus a systematic, coordinated effort is essential. However, this aspect of the verification process is really no different from that which is (or should be) employed in the verification of any other complex system (e.g., the introduction of a management information system, a new computer system, etc.).

One way in which we may collect data is to obtain access to the past records of the organization, as they pertain to the decision domain that the expert system is to operate within. The expert system may then be run using these data as the input and the expert system's conclusions will then form the resultant test database. Another is to establish a set of trial cases (e.g., as suggested by the human expert from whom the knowledge base was cloned) and use these as input to the expert system. A third is to run the expert system in parallel with the existing approach to compare the results provided by both techniques.

Data Evaluation

One typical, although far from perfect mode of performance evaluation is through the use of the so-called cost-benefit, or cost-effectiveness methodologies. The premise of such an approach is that one is able, somehow, to combine all of the performance measures into a single composite measure (typically, a *utility* measure [Cohon, 1978]). Then, we simply divide the composite measure by the system's cost to derive a number that allegedly represents the value of the system. The military sector, in particular, has focused on this approach as is noted by references to judging a system on its "bang per buck." The critics of such an approach either claim or note that

- Its fundamental premise (i.e., that one may find a *single, credible* utility measure) is hopelessly idealistic
- Many measures of performance are noncommensurable (e.g., how does one place a credible dollar value on the loss of a human life, and how can the loss of life be realistically combined with, say, labor turnover?)
- The combination of a number of different measures of performance, each subject to its own inherent errors, results in an aggregate measure that may well be just an accumulation of errors

Another, more recent, and generally more powerful approach is through the use of various methods of multicriteria decision analysis. Some of these also provide a single, aggregate measure. Others, however, deal with the entire, explicit array of performance measures [Cohon, 1978; Ignizio, 1982, 1985]. To do justice to this topic would require an extensive discussion, and this is not the intent of this text. However, the references provide the details of such techniques and may be consulted by the reader interested in actually applying such approaches.

A COMMENT ON THE REPLICATION OF ADVICE

Whenever one thinks about the verification of the performance of an expert system, one typically first (if not exclusively) thinks about just how the advice given by the expert system will compare with that given by a human expert; in particular the human expert from whom the knowledge base was acquired. Thus, the implication is that, if the expert system replicates the advice that would be provided by the human expert, the performance of the system has been verified.

However, note the various measures of performance that have been listed above. Solution *accuracy* is *just one* of these measures. If we focus exclusively on this measure, we may well neglect others that play just as important a role in performance. For example, what good is an expert system that duplicates the advice of a human expert but takes an excessive amount of time to reach its conclusions? Or requires an excessive amount of money to implement? Or has a disastrous impact on the morale of the organization's personnel? Rather obviously, the replication of advice is important, but it is only one part of the equation.

A CASE STUDY: VALIDATION OF DMES

While the developers of DMES (i.e., the Deployable Mobility Execution System for the loading of military aircraft) did not label their tool as an expert system, it would certainly qualify as one under the definition employed in either this text or most other works on expert systems. As such, we shall use DMES as an example of just how one might validate an expert system. While the approach used in the validation of DMES may not satisfy all of the guidelines that have been outlined earlier, most of the important issues were dealt with in a satisfactory manner. Further, when one considers the limited DMES budget, it is difficult to find too much fault in the process that was employed.

First, consider phase one of the validation process. The problem faced by DMES was certainly a vital one, and one that could benefit by either faster or improved decision-making capabilities. Second, it is a problem of combinatorics and thus only heuristic programming or expert systems would seem to be a viable solution procedure. Third, the existence of true expertise could be confirmed (i.e., the expertise of certain of the cargo loadmasters). As such, one would certainly seem to be able to justify the employment of an expert system. Further, the role of DMES was made clear: to provide an enhanced method for the loading of certain military aircraft.

We cannot comment on the second phase, the validation of the rule base, as this effort was not documented. Further, the DMES rule base was not separated from the inference procedure and thus the specific approach that we have proposed here would not be directly applicable.

Finally, consider phase three. While not specified explicitly, some of the measures of performance by which DMES was evaluated include

- A replication of the load schemes that would have been developed by the cargo loadmasters
- A reduction in the time necessary to develop a cargo loading scheme
- A decrease in the number of aircraft necessary to achieve a given mission
- A compatibility with the methods and documentation normally used by the human loadmaster (i.e., a consideration of the impact of the introduction of the tool on both personnel and the organization)

After testing the software on a number of artificial cases, the developers of DMES conducted a relatively extensive comparison of results in actual (noncombat) operations. Specifically, for a number of training operations, the loads achieved (on paper) by DMES were compared to those actually achieved by the human loadmasters. Once satisfied that the system was able to load the aircraft as efficiently (or more so) as the human expert, and much faster, it was deemed ready to operate on its own—as it did with such success in support of the Grenada rescue operations.

SUMMARY

The validation of an expert system is a fundamental part of the development process. In the past, validation of an expert system, when it has been done, has too often been accomplished as an afterthought. We believe, however, that one must begin to plan for the validation of an expert system just as soon as the decision has been made to develop such a tool.

In this chapter, we have divided validation into three phases. The first phase is that of the justification of the use of an expert system for the problem at hand. An expert system that cannot be defended cannot, we believe, be validated.

The second phase of validation is concerned with the logic and integrity of the rule base. If the guidelines for rule-base development as presented in Chap. 4 are followed, this phase is made considerably easier. Validation of rule bases consist, primarily, of various checks for consistency and completeness. Since a rule is an autonomous element, one can begin such validation as soon as the first few rules have been generated, and should continue these checks until the complete rule base has been developed.

The final phase of validation is focused on the verification of the performance of the expert system. That is, does it do what it is supposed to do—and does it do this efficiently and effectively? If so, the development effort will most likely be a success. If not, the developers most likely have no one to blame but themselves.

EXERCISES

8.1. Just how might the perception that "expert systems is (just) an alternative to conventional programming" affect phase one of the validation process?

8.2. Critics of expert systems (i.e., those who believe that expert systems is merely another name for heuristic programming) have used the DMES [Cochard and Yost, 1985] and AALPS [Anderson and Ortiz, 1987] articles to buttress their argument. Prepare a response which serves to show that there are, indeed, certain significant differences between the two methodologies.

8.3. Can *you* answer the seventh question in the list provided in the section of this chapter titled: "Phase One: Expert Systems Justification?" Try your answer on someone who is unfamiliar with either expert systems or artificial intelligence. Was he or she satisfied with your explanation?

8.4. Critique the rule bases presented below. Assume that backward chaining will be employed.

(*a*) (Let goal = disease)

     R1:   *If* eyes = bloodshot
             Then test A = yes

     R2:   *If* test A = yes
             and test B = no
             Then disease = flu

     R3:   *If* disease = flu
             Then eyes = bloodshot

(*b*) (Let goal = animal type)

> R1: *If* hair color = red
> *and* marsupial = yes
> *Then* animal type = red wombat
>
> R2: *If* hair color = red
> *and* habitat = Tasmanian burrow
> *Then* animal type = red dingo

(c) (Let goal = goal, and assume a user prompt for attribute *A* only)

> R1: *If* A = X
> *and* B = Y
> *Then* D = V
>
> R2: *If* F = W
> *Then* goal = no
>
> R3: *If* D = V
> *Then* B = Z

(d) R1: *If* hair color = red
> *and* marsupial = yes
> *Then* animal type = red wombat
>
> R2: *If* hair color = red
> *and* marsupial = yes
> *and* habitat = burrow
> *Then* animal type = red wombat
>
> R3: *If* hair color = red
> *and* marsupial = yes
> *Then* animal type = red wombat
> *and* location = Tasmania

(e) R1: *If* hair color = red
> *and* marsupial = yes
> *and* habitat = zoo
> *Then* animal type = red wombat
>
> R2: *If* hair color = red
> *and* marsupial = yes
> *and* habitat = Tasmania
> *Then* animal type = red wombat

8.5. What is the difference between an unreferenced and an illegal attribute value?

8.6. Check the rule base of Exercise 4.8*a* for completeness—assuming that the legal values for the attributes are

Order backlog = (large, average, minimal, none)
Labor supply = (excess, adequate, nil)

Historical data = (available, not available)

Customer = (important, not important, unknown)

8.7. An expert system has been designed for the military for use in decision making when one's forces are under nuclear attack. How should the performance of such an expert system be verified?

8.8. A discount sporting goods company believes that it might be losing potential new customers due to the lack of expertise, on the part of its sales force, in taking orders over the telephone. Might this be a good application of expert systems? If so, how would you suggest that the resultant expert system's performance be validated?

8.9. In support of the rule validation process for an expert system that you are developing, your client suggests that you use his top computer programmer. This programmer has assured your client that he can substantially reduce the size and complexity of your rule base. What is your response?

8.10. Critique the following rule base. Assume that backward chaining is to be used and that the final goal attribute is *status*.

R1: *If* temperature is high
Then pressure is high

R2: *If* pressure is high
and fluid level is high
Then status is dangerous

R3: *If* chiller is on
Then temperature is high

R4: *If* status is dangerous
Then chiller is on

8.11. Critique the following rule base:

R1: *If* your grade point average is high
Then your image is that of being smart

R2: *If* your grade point average is high
Then your image is that of being a nerd

R3: *If* your grade point average is high
Then a lot of other people are likely to be jealous

8.12. Critique the following rule base:

R1: *If* process is injection molding
and state of the material is paste
Then material is ceramic

R2: *If* process is injection molding
and state of the material is paste

and injection pressure is intermediate
Then material is ceramic

8.13. Critique the rule base of Exercise 8.12 for the case where confidence factors are employed for both user input and rules.

8.14. Check the following rule base for circular rules.

R1: *If* A = B
   Then goal (2) = yes

R2: *If* C = D
   Then goal (1) = yes
   and A = B
   and G = H

R3: *If* goal (2) = yes
   and E = F
   and G = H
   Then K = J

R4: *If* K = J
   Then goal (1) = yes
   and C = D

R5: *If* Y = X
   Then C = D
   and E = F

8.15. Generate a rule base that permits you to distinguish between various types of computers (i.e., microcomputers, minicomputers, mainframes, and supercomputers). Use the data provided in the table below in constructing your rule base.

Type	Cost (dollars)	Speed (MIPS)
Microcomputer	100–5000	≤ 1
Minicomputer	5000–75,000	0.1 to 3
Mainframe	30,000–5,000,000	0.5 to 30
Supercomputer	> 3,000,000	>20

8.16. Referring to the problem of Exercise 8.15, what are your thoughts about the appropriateness of using an expert system for this situation?

CHAPTER
9

HYBRID
EXPERT
SYSTEMS

DEFINITION

Some of the definitions of the word *hybrid* include (1) anything of mixed origin, (2) the offspring of two animals or plants of different races, varieties, or species, or (3) a composite substance. The development of hybrids has played a particularly significant role in agriculture. Here, hybrid plants (e.g., drought resistant corn and wheat) have been developed that exhibit characteristics that are often far better than those of either of their parents.

Another hybrid that is of considerable recent interest is that of composite materials. Davidson [1988] states that "composite materials are considered among the most exciting developments in modern technology." One such hybrid is that known as carbon-carbon, a composite material being used in the top-secret Stealth Bomber effort. However, as Davidson notes, composites have actually been with us a long time. The Bible cites one of the earliest examples: clay mixed with straw, from which primitive dwellings of exceptional strength were built. In the case of certain hybrids, the whole truly seems to be far greater than the sum of its parts.

In this text, we use the term hybrid to designate a tool for decision making that has been formed through the combination of two, or more, different classes of decision-analysis tools. More specifically, when we refer to a *hybrid expert system*, it is to be inferred that we are discussing a decision-analysis tool formed through the judicious combination of an expert system and some other methodology. As just one example, a decision tool formed through the combination of an expert system and a linear programming algorithm would be classified here, as a hybrid expert system.

Within the AI community, there is an alternative definition of hybrid expert systems. Specifically, this term is sometimes reserved for expert systems that use more than one form of knowledge representation (e.g., such as both rules and frames). However, this is not the definition that we shall employ, nor the type of hybrid expert system that we shall describe here.

THE IMPORTANCE OF
HYBRID EXPERT SYSTEMS

As we have attempted to stress throughout this text, not all problems can or should be solved by means of an expert system. Further, even among those whose solution by expert systems is appropriate, the results found through such an approach may, in some instances, be only marginally acceptable. Or perhaps they may be acceptable, but nowhere near that which actually could be obtained through the expenditure of a reasonable level of additional time and effort.

Analysts who have dealt with heuristic programming have, for some time, noted that it is often possible to obtain substantially improved results by means of simply combining two or more heuristic programs [Ignizio and Harnett, 1974; Ignizio, 1980; Langston, 1987]. In such situations, one typically feeds the output of one heuristic program into the input of another. Thus, the solution developed by one method becomes the initial solution for the next, where, hopefully, an even better result will ultimately be obtained. The result of such a combination may also be considered to be a hybrid solution approach.

There is a perception, among some advocates of expert systems, that the be all and end all of expert systems is to mimic the performance of the human expert from whom the knowledge base was cloned. As such, these individuals are, in our opinion, far too easily satisfied with an expert system that exhibits performance (in the domain of interest) on the same level as the human expert. As discussed, such a comparison is often presented as *the* way in which one should validate an expert system.

However, as we discussed in Chap. 5, not all (alleged) domain experts, really exhibit any true expertise (at least, to any significant level). In fact, a fair number of the alleged domain experts that we have encountered are doing little more than producing *results*, where such results may be nothing more than, at best, a feasible solution (i.e., a solution that may be implemented without the violation of any constraints). Actually, this should not be surprising; in fact, it would be far more surprising to discover that human experts could actually find *good* solutions for a large portion of the types of problems now commonly dealt with using expert systems. The reason for this is that expert systems are often being used to deal with problems exhibiting combinatorial explosiveness (i.e., problems that are computationally complex, or NP-complete [Karp, 1975;

Kaufmann and Henry-Labordère, 1977; Murty, 1976]). If a given problem has a near infinite number of possible conclusions, it seems rather presumptuous to think that a human being can readily identify the single, best conclusion—or even a relatively good conclusion.

Consider, for example, the well-known traveling salesperson problem [Taha, 1975; Kaufmann and Henry-Labordère, 1977; Lin and Kernighan, 1973; Murty, 1976]. In this problem we are given N cities and a listing of the distances between each city. We are then to route a salesperson, starting at one city (e.g., the home office site), to each of the N cities and then finally back to the initial city. Further, no cities (except for the initial city) are to be visited more than once on a given tour. And this is to be done to minimize the total length of the salesperson's tour.

Now, if we have, say, just four cities and the road network as shown in Fig. 9.1, there are but a total of six possible routings (and half of these are simply reversals of the others). Consequently, through nothing more than brute force enumeration, it is easy to determine that the optimal routing is: A-B-D-C-A (or, A-C-D-B-A, the reverse of the tour), for a total distance of 24 units. However, the number of possible routings in a traveling-salesperson problem is given by $N!$ if all cities are interconnected. This indicates that the number of routings increases exponentially with the number of cities. Thus, with only 100 cities, the total number of routings is 100!, and the solution to this problem could not be found in the reader's lifetime, even on the most powerful computer, via exhaustive enumeration.

When humans are given a combinatorially explosive problem to solve, they are most often completely oblivious to the true extent of the inherent difficulty of the task. Most typically, they resort to some sort of heuristic procedure to ultimately develop *a* solution to the problem—a solution that is invariably nowhere near the optimal. However, they (and, often, their employers) generally feel quite satisfied simply to have found *any* solution. Consequently, in the real world, the designation of *expert* is a title that can be all too easily obtained.

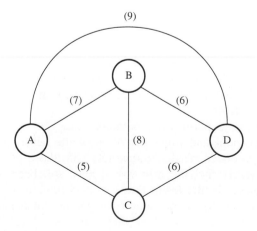

FIGURE 9.1
Example road network.

To provide an actual illustration of such misplaced confidence, let us cite a situation that we encountered some years ago. A small manufacturing company in the southeastern United States asked us to investigate the possibility of automating their dispatching procedure (i.e., the routing of shipments from the firm to the retailers used by the firm). It seems that they had one person in charge of all dispatching. He had worked for the firm for almost 40 years (over 30 years as their chief dispatcher), and was about to retire. The owners of the firm considered the man a true genius in dispatching (he was most certainly "getting the job done"), and viewed what he did with unconcealed awe. They were understandably concerned about the forthcoming loss of this expertise.

Our initial reaction was that the fastest as well as most promising course of action would be to interview the dispatcher, determine his heuristic rule set, and then develop a (computer-based) heuristic programming procedure (or, using today's terminology, an expert system) that would handle his duties. In the past, we had encountered a number of similar problems and such an approach had worked quite well. However, in our discussions with the dispatcher, and through the construction of a model of the problem, it became clear that this particular dispatching problem could be represented as a relatively straightforward distribution network and, in fact, was but a special case of a well-known class of linear programming problem. The model could be solved, for an *optimal* solution, by an existing network analysis algorithm [Murty, 1976].

Unfortunately, due to the size of the problem involved, the network analysis algorithm took some considerable time to arrive at the solution. However, it is well known that such computation times may be reduced if the process is started with a *good* solution. Consequently, we developed a heuristic program that served to replicate the rules used by the dispatcher. The solution developed by this procedure was then used by the network analysis algorithm as the initial start point in its search process. In this way, the optimal solution could be derived in a reasonable time. And such a combination may be considered a hybrid solution process.

Once the technique was working properly, we compared our results with those of the dispatcher (using past company records). On the average, the network analysis algorithm resulted in more than a *30 percent savings* in shipping costs per year.

Now, had we restricted our effort to simply cloning the expertise of the dispatcher, we would have done little more than support the perpetuation of his rather mediocre level of performance. And this *should not* be what expert systems is all about. Unfortunately, this appears to be precisely what is happening in a number of expert systems development efforts.

There is, however, an interesting (and perplexing) paradox here. Had we restricted our effort to just the development and implementation of the heuristic program—and had such a system produced results comparable with those of the human (which it did), it is most likely that the firm would have still been completely satisfied. In fact, that is precisely the result that they had originally sought. Further, we would have received precisely the same fee regardless of the approach.

Returning to the concept of hybrid expert systems, it may be noted that the real power of such approaches is that they may often be used to find better (often, far better) solutions to problems in which the performance of the human expert leaves something to be desired [Ignizio et al., 1987; Ng, 1989]. Further, they may accomplish this improvement in a cost-effective manner. In this chapter, the reader will be presented with just a few such hybrid expert systems. However, through these examples, the overall philosophy and general approach to the development of hybrid expert systems should be apparent.

A HYBRID EXPERT SYSTEM FOR RESOURCE-CONSTRAINED SCHEDULING

A problem that seems to exist in virtually all organizations, albeit in a variety of guises, is that known as the *resource-constrained scheduling problem* [Spitzer, 1969; Wiest, 1966; Yang and Ignizio, 1987,1989]. This class of problem is known to be NP-complete and, as such, the only rational approaches to its solution appear to be through either heuristic programming or expert systems, or some clever combination of the two. The characteristics of such a problem may be most readily presented by a simple example.

> **Example 9.1: a single resource example.** In this example, the problem to be addressed will be solved for an initial solution using an expert system—and that solution then improved upon by means of a simple heuristic program. Let us now consider the first phase of the solution process.
>
> *The development of an initial solution using an expert system.* Consider a problem where a number of tasks are to be performed on several processors (e.g., machines, workstations). Each task has a specific, and possibly distinct processor routing. For example, task A may have to be processed via a routing from processor 3 to processor 1 to processor 4 while task B may have to be processed via a routing from processor 3 to processor 4. In addition, let us assume that only one task may be performed at a time on a processor and, once a task is started, its processing cannot be interrupted.
>
> Thus, given M tasks and N processors (along with the routing of each task and the processing time required for each task to processor assignment), our job is to determine the *best* schedule for these tasks. Table 9.1 serves to summarize the routings and processing times for our illustrative example. Figure 9.2 depicts the task routing schematic. Notice, from Table 9.1, that task A requires 2 units of time on processor 1 (the first number in the cell associated with task A, processor 1) and that processor is the second one on its route (as noted by the 2, as shown in parentheses in the cell). If the cell entry is blank, no processing takes place for that particular task/processor combination. Thus, we may note that task A proceeds from processor 3 (where it takes 4 units of processing time) to processor 1 (where it takes 2 units of time) to processor 4 (where it requires 4 units of time). For purpose of discussion, let us assume that each unit of time is equal to one day.
>
> Consideration of such a problem should immediately raise the question about what measure, or measures, to use in order to determine whether one schedule is

TABLE 9.1
Task processing
and routing summary

Task	Processor			
	1	2	3	4
A	2(2)		4(1)	4(3)
B			6(1)	2(2)
C	4(1)	6(3)	2(2)	

better than another. One common measure of scheduling effectiveness is that known as *makespan*. The *makespan* of a schedule, in turn, is simply the *total elapsed time between the start of the first job and the completion of the last job* in the schedule. Other measures of scheduling effectiveness often employed include the minimization of the number of tardy jobs (i.e., number of jobs that are late), the minimization of idle times on processors, and the maximization of throughput (i.e., number of jobs completed per unit time).

Consequently, to this point, we might think of the scheduling problem as one in which we are to assign a start time for each task, on each processor that is a part of its route, to minimize the schedule makespan while satisfying all problem constraints. Thus, for our example of Table 9.1, one schedule is shown in the Gantt

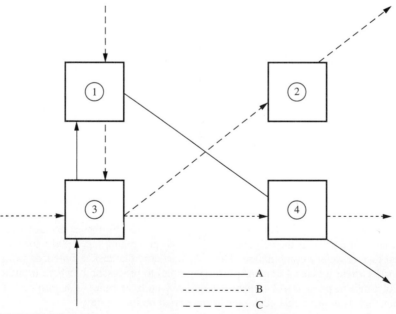

FIGURE 9.2
Task routing schematic.

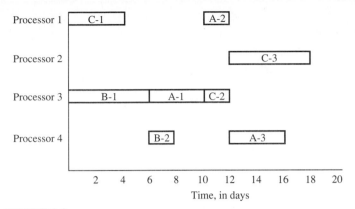

FIGURE 9.3
A feasible schedule in the absence of resource constraints.

chart of Fig. 9.3. A Gantt chart serves to graphically summarize any given schedule. For example, we may note that processor 1 services task C (specifically C-1, the first operation of task C) for four days, that is, from day zero through the end of day 3. It is then idle until the start of day 10, when it starts to process the second operation of task A. This takes two days at which time the processor is once again idle. Notice that the makespan for the schedule shown is 18 days (i.e., the time at which the third operation of task C is completed). [1] Also observe that, in the development of this schedule, absolutely no consideration has been given to the amount and level of any resource (or resources) consumed by each task.

The *resource constrained* scheduling problem has one additional complication. Specifically, each scheduling activity (e.g., processing operation, routing operation, processor setup) may consume one or (generally) more scarce resources. As such, any schedule developed must exhibit a *resource consumption profile* that does not cause the system's *resource availability profile* to take on any negative values. In other words, we cannot schedule an activity if there are not enough resources to support it.

To keep things simple, let us assume that our problem involves only a single resource that is consumed by processing operations. Units of this resource are assumed to be periodically received. Specifically, we receive 30 units of the resource every 8 days. Consequently, at time zero we are provided with 30 units of the resource. Eight days later (i.e., at the end of day 7, or the start of day 8), we receive another 30 units, and so on. Thus, the amount of the resource that we have at any time is simply the difference between that initially available, plus any amount delivered, minus that consumed. For example, at the start of day 8, we will have 30 units (from time zero) plus 30 more units (as delivered at the start of day 8) minus whatever number of units were consumed during the previous 8 days.

---

[1] We are not implying that this is an optimal schedule. Rather, it is simply a feasible schedule in that all of the tasks are performed in their required order—in the absence of any resource constraints.

TABLE 9.2
Task processing resource consumption rates

	Processor			
Task	1	2	3	4
A	1		1	4
B			4	1
C	1	2	2	

Let us then assume a resource consumption rate for each of the tasks given in Table 9.1. These data are summarized in Table 9.2. For example, note that task A, on processor 1, consumes the resource at a rate of 1 unit per day. Note also that the schedule developed earlier in Fig. 9.3 is most definitely not a feasible schedule. That is, it would require consumption of more resources than are actually available. Notice, for example, that by the end of day 6 (the start of day 7), we have already consumed all units of available resources.

Now, let us assume that a human scheduler exists in this situation and that the procedure he uses to schedule tasks is based upon the use of a heuristically derived *rank ordering* of the set of tasks (specifically, an iterative ranking *of only those tasks whose predecessor activities have been completed*). This rank ordering is accomplished, in turn, according to the consideration of the following attributes:

- Processing time requirements
- Earliest feasible start time
- Resource consumption rate

Initially, and rather obviously, the only tasks that may be considered for scheduling are task A/processor 3, task B/processor 3, and task C/processor 1. The way in which our expert rank orders these tasks is listed below:

1. B-1/3[6,0,4]
2. A-1/3[4,0,1]
3. C-1/1[4,0,1]

Where the notation used is defined as

    Task-sequence/processor [PT, EST, RCR]

and

    PT = processing time requirements
    EST = earliest (*feasible*) start time
    RCR = resource consumption rate

Note in particular that the earliest start time must be feasible in the sense of there being sufficient resources to support the task during its processing. In our example, initially, the top-ranked task is that of the first step of job B, as performed on processor 3. This task takes 6 days on processor 3, may be started at time zero, and consumes the scarce resource at a rate of 4 units per day. Since there are 30 units of the scarce resource, and since the first operation of task B will consume but 24 units of these, it is feasible to schedule task B at this time.

Now, let us examine why the three tasks have been ordered as shown above. Our expert believes that processing time is the most important factor, and that the task with the longest processing time should be scheduled first. If there are ties for this factor, he then breaks them in favor of the task with the earliest feasible start time. If a tie still exists, he breaks that tie in favor of the task with the lowest resource consumption rate. And, if we still have a tie, he simply picks a task at random from those which are tied. Now that we understand his rule set (we don't have to agree with it, only understand it), we should be able to construct a corresponding knowledge base.

In developing the knowledge base, it should be apparent that we are dealing with a type of construction problem. That is, we wish to construct the components of a schedule. Further, as has been discussed, this class of problem is usually best approached through forward chaining. With this in mind, we might develop the following set of production rules to represent the scheduling technique of our expert.

Rule 1: *If* start = yes
      Then rank order all jobs
      and set start = no
      and set partition 1 = yes
      and set partition 2 = no

Rule 2: *If* partition 1 = yes
      and jobs left = 0
      Then STOP

Rule 3: *If* partition 1 = yes
      and jobs left > 0
      Then clear partition 2 rules
      and schedule next job
      and remove scheduled job from listing
      and rank order all remaining jobs
      and set partition 1 = no
      and set partition 2 = yes

Rule 4: *If* partition 2 = yes
      Then clear partition 1 rules
      and set partition 2 = no
      and set partition 1 = yes

Note that we are using dummy variables designated *partition 1* and *partition 2* (a concept useful for an iterative process, and which was previously described in

Example 6.2 of Chap. 6). At most, only one of these partitions may be active at any time.[2] Initially, we will supply our expert system with the data from Tables 9.1 and 9.2 and also set

> Start = yes
> Partition 1 = no
> Partition 2 = no

Thus, we will begin the procedure at rule 1. Rule 1 will obviously be triggered and fired. This will result in a rank ordering of all tasks whose predecessors have been completed (since this may be handled through a straightforward, number crunching algorithm, we might call an external algorithm to perform the rank ordering or, alternately, an internal production rule set could be developed), and cause partition 1 to be set active.

The forward chaining operation will then move to rules 2 and 3. Rule 2 will only be triggered and fired when all jobs have been scheduled. Until then, we will (repeatedly) trigger and fire rule 3. Each time rule 3 is fired, the result will be

- Clearing of the rules in partition 2
- Scheduling of the next job
- Removal of the scheduled job from the list of active jobs
- Rank ordering of all remaining jobs
- Setting of partition 1 = no
- Setting of partition 2 = yes

That is, we schedule the next job and then we deactivate partition 1 and activate partition 2. This results in a move to rule 4. Rule 4 is used simply to allow for the procedure to iterate, that is, to continue to schedule all remaining jobs.

If we utilize this simple rule set, along with the rank ordering procedure described above, we will develop a feasible solution to the resource-constrained scheduling problem. The solution found for our example is presented in the Gantt chart of Fig. 9.4. Notice that the makespan for this schedule is 20 days, as compared to 18 for the schedule of Fig. 9.3. Of course, the schedule in Fig. 9.4 is feasible in light of the added resource constraint, while the previous schedule is not.

To further clarify how this latest schedule was derived, we have presented the listings of the rank orderings at each iteration of the scheduling process. This is shown in Table 9.3.

Examining Table 9.3, notice that, at each iteration, the task that was determined to be scheduled is shown in boldface. For example, at the first iteration, we assign the first operation of task B to processor 3. Also, from the data in the

---

[2] We may also notice, as in the case of Example 6.2, there is relatively little *expertise* captured within this rule base. Again, our purpose is to illustrate an approach rather than to develop a more realistic, and complex, rule base.

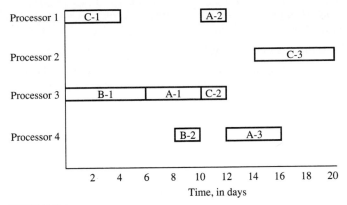

FIGURE 9.4
A feasible schedule under limited resources.

brackets (i.e., 6,0,4), we may note that this task takes 6 days on processor 3, starts at time zero, and consumes 4 units of the resource per day (i.e., for a total of 24 units consumed over the 6 days). The readers are advised to go through the process of rank ordering for this problem, and to compare their results with those shown in Table 9.3.

Enhancement of the expert system solution. The schedule shown in Fig. 9.4 is that obtained through our simple expert system. Let us next see how we might improve this solution. To accomplish this, we shall use an *extremely simplified version* of the Yang/Ignizio heuristic programming algorithm for resource-constrained scheduling [Yang and Ignizio, 1987, 1989]. This algorithm requires that one first has a feasible starting solution. We may thus use the schedule obtained above to satisfy this requirement. We then seek to improve this initial schedule. This is accomplished, when possible, by following the general procedure listed below. More specific details

TABLE 9.3
List of rank orderings used in schedule development

Iteration	Rank order of tasks
1	**B-1/3**[6,0,4]; A-1/3[4,0,1]; C-1/1[4,0,1]
2	**C-1/1**[4,0,1]; A-1/3[4,6,1]; B-2/4[2,6,1]
3	**A-1/3**[4,6,1]; B-2/4[2,6,1]; C-2/3[2,7,2]
4	**B-2/4**[2,8,1]; A-2/1[2,10,1]; C-2/3[2,10,2]
5	**A-2/1**[2,10,1]; C-2/3[2,10,2]
6	**A-3/4**[4,12,4]; C-2/3[2,10,2]
7	**C-2/3**[2,10,2]
8	**C-3/2**[6,14,2]

with regard to the actual implementation of this procedure may be found in the references.

1. List the task that is the last to be finished in the present best schedule.
2. Attempt to interchange an operation of the task identified in step 1 with some task (or block of tasks) that was completed well prior to the completion of this task.
3. If the interchange of step 2 is possible, make this interchange and then see if it is then possible to readjust some other tasks. Specifically, attempt to move some other task (or tasks) to an earlier start time.
4. Continue with steps 2 and 3 until no further improvement is possible.

If we employ this heuristic algorithm, we will first move task C (on processor 3) to time four—while moving task B (on processor 3) to time six. The move of task B on processor 3 will then make it necessary to move task B on processor 4 to time 12. These moves then permit us to make the following adjustments:

- Task A on processor 3 is moved to time zero
- Task A on processor 1 is moved to time four
- Task C on processor 2 is moved to time six
- Task A on processor 4 is moved to time 14

No further adjustments may be accomplished through this approach and so our new, improved (and still feasible) schedule is as shown in Fig. 9.5. Notice that we have reduced the makespan from 20 units to 18 units, an improvement of 10 percent. While a 10 percent reduction in makespan is not insignificant, in larger, more realistic problems, the improvement would generally be much greater. In fact, in real-world problems (and using the complete form of the algorithm) we have obtained improvements ranging from about 10 percent to more than 50 percent, as compared to the schedules developed by the alleged human experts.

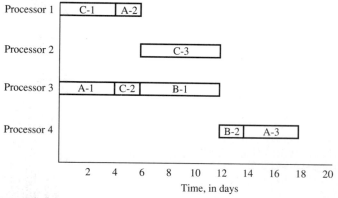

FIGURE 9.5
The improved schedule.

AN EVALUATION OF
THE SCHEDULING PROCESS

The simple hybrid expert system of Example 9.1 serves to indicate, to some small degree, the potential of the hybrid approach. The first phase of the approach is a simple expert system, that is, an expert system in the sense that it represents the approach used by a human scheduler. The second phase of the approach is an even simpler heuristic programming algorithm which serves to provide an enhanced result by using the solution derived by the expert system as a point of departure.

The observant reader may, however, note that we could have, just as easily, constructed a *single* production rule set for the entire solution process—or, for that matter, we could have constructed a *single* heuristic programming procedure to implement the entire process. And it is not always clear which approach is the *best*. Again, the problem lies in the difficulty of drawing any meaningful distinction between expert systems and many, if not most, heuristic programs. However, when one carefully examines the complete approach for scheduling as has been described above, it should be obvious that the majority of the effort is involved with numbers and procedures, rather than with symbols and expertise.

A SPECIFIC EXAMPLE
WITHIN THE MILITARY SECTOR

In Example 9.1, we presented a simple, illustrative example. Let us next consider an actual example of the resource-constrained scheduling problem and the use of hybrid expert systems.

In the late 1970s, we were asked to help solve a problem in the military sector. The particular problem under consideration was that of determining a way in which the *combat readiness* of our military forces might be enhanced [Ignizio, 1978]. Combat readiness, in turn, was defined as the number of days of training that would be required to prepare a battalion for actual deployment to a combat site. For example, a *30-day readiness* status meant that the battalion could be made available for deployment in no more than 30 days. Rather obviously, the military would like to have all of its battalions ready for nearly instantaneous deployment. Even more obviously, such a goal is impossible to satisfy.

The military recognized that, to attempt to satisfy the combat readiness goal, one must determine (for each battalion) the prescribed number and frequency of each military exercise that supposedly serves to prepare the unit for combat. This portion of the problem was formulated as a goal programming model [Ignizio, 1978, 1982, 1985]. The primary objectives of the problem were to maximize the number of combat-ready battalions and to minimize the time required to achieve the desired level of readiness. Furthermore, this was to be accomplished in light of limited resources (e.g., ammunition, test ranges, instructors, etc.). As a consequence, the output of the goal programming model indicated, for each battalion, the number of times that each exercise should be performed (over the planning horizon).

However, the goal programming result represents only a part of the solution. That is, once we know *how many* exercises to perform, the next question to be answered is, "*When* should these exercises be scheduled?" And it was the scheduling aspect of the problem that was, by far, the most difficult.

In our previous example, we had to determine when to schedule activities in order to minimize the schedule makespan while satisfying constraints on

- Limited resources (i.e., the resource consumption profile of the schedule cannot cause the resource availability to go negative)
- Precedence (i.e., a given task cannot be started until its predecessors are completed)
- Noninterruption (i.e., once a task is started, it cannot be interrupted)

The military readiness scheduling problem has all of these characteristics while, in addition, imposing the need to consider *coordinated* exercises. That is, certain exercises can only be performed in conjunction, or cooperation with one or more other battalions.

Because the problem faced was obviously NP-complete, it was determined that the appropriate mode of solution should be through a heuristic program, an expert system, or a hybrid expert system.[3] Further, since the schedules being developed by the human schedulers were considered suspect on our part, we decided to tentatively focus on a hybrid expert system. In such a system, the initial solution would be provided by an expert system based upon the knowledge base of the human schedulers, and this solution would then be used as a starting point for a heuristic programming method.

The result of our effort was the development of what we believe to be a very efficient and effective hybrid method. Specifically, it was observed that the hybrid expert system was able to consistently develop schedules that reduced schedule makespan by more than 50 percent over those developed by the human schedulers. Further, the time required to produce such schedules was only a fraction of that required by the human counterparts. It should be noted that the fact that we were able to reduce schedule makespan by such an amount is indicative of the rather mediocre performance of the human schedulers—which should not be surprising in the light of the complexity of the problem faced. This observation led us to virtually drop the expert system phase of the solution process. That is, we replaced the expert system by a less complex scheme for simply generating feasible schedules—and then fed these results into the second phase (i.e., a heuristic program) of the solution procedure. The overall performance of the system remained

---

[3] To illustrate just somewhat the complexity of the problem, with just 15 battalions, 50 exercises per battalion, 50 resources, and a 360 day planning horizon, the mathematical model for this problem involves more than 13.5 million variables (specifically, zero-one variables). And a real problem would be considerably larger.

virtually the same while we were able to even further decrease the time required to develop the schedules. Details with respect to this approach may be found in the references [Yang and Ignizio, 1987, 1989].

We may observe that, had we simply cloned the knowledge base of the human schedulers, the sponsoring agency would most likely have been completely satisfied. That is, we would have a system that replicates the decisions made by the human schedulers, and which does so in far less time and on a more consistent basis. However, through an effort requiring relatively little in the way of additional time and resources, we were able to develop a system that was a vast improvement over those results provided by the expert system alone. As such, we view such results to be supportive of our belief that the truly competent knowledge engineer simply must be aware of alternative approaches and, in particular, of previous developments in heuristic programming—whatever their source.

A HYBRID EXPERT SYSTEM FOR TRANSDUCER DEPLOYMENT

A *transducer* is a physical device that transmits energy, or power, from one system to another. For example, in the field of electromagnetics, an antenna is a transducer, and transmits its energy to another antenna that serves to collect the received energy, and then to transmit it on to yet another receiver (e.g., a radio or television set). In acoustics, we have analogous transducers. These can be ceramic elements that convert vibrations into acoustic energy, and radiate this energy through the air or water. One example of an acoustic transducer is that used in sonar systems. Another would be the so-called ultrasound devices now in wide use in the medical profession.

Generally, the transmission of power is accomplished by an ensemble of transducers, known as an *array*. Each element in the array is thus akin to a single instrument in an orchestra. If all elements are in tune, the result is a pleasing sound, the sound that was intended to be produced. However, if one or more elements are out of tune, the result can vary from mildly unpleasant to revolting, for example, the *music* made by a *heavy metal* band.

The focus of our specific example is the design and, in particular, deployment of acoustic transducers for sonar arrays (i.e., as used on ships, submarines, and torpedoes). The trend in arrays, either electromagnetic or acoustic, is to use larger and larger arrays (i.e., more elements), and to arrange these arrays to conform to the shape of the object on which they are deployed (e.g., the surface of a ship or submarine). As such, the design of acoustic arrays, never a simple task, is becoming increasingly more complex and costly.

In the early days of acoustic (or electromagnetic) arrays, the design process was most often accomplished through a *trial-and-error* approach. More recently, (most) array designers have taken advantage of the use of mathematical optimization (e.g., linear programming, nonlinear programming, and goal programming [Ignizio, 1981]). Through such procedures, the array designer may, in a systematic fashion, readily determine such design parameters as

- The amplitude of the signal to be delivered to each transducer
- The precise position of the transducers (i.e., the coordinates of each device on the mounting surface)
- The phase of the signal to be delivered to each transducer

In particular, one may form a mathematical model wherein the design objectives (e.g., maximize power delivered, minimize susceptibility of the array to jamming) are to be optimized subject to the satisfaction of system constraints (e.g., limited number of elements, limited deployment positions, maximum allowable sidelobe levels). Typically, one wishes to produce a *beam* of acoustic energy that takes on a particular shape. The solution to the mathematical model then tells the designer where to place each transducer, as well as the strength and phase of the energy to be delivered to that element. However, the results are only valid under the unrealistic assumption of an error-free model. More specifically, if each and every transducer element is not precisely identical (i.e., in terms of operating characteristics), the results achieved in practice may (and generally will) differ substantially from those predicted on paper.

The most obvious way to alleviate such a problem would seem to be to place tighter restrictions on the manufacturing process of the transducers. That is, tighten the manufacturing tolerances. However, manufacturing costs rise *exponentially* as a function of design tolerances. Consequently, it may simply not be economically (or physically) feasible to attain the tolerances necessary to obtain the necessary reduction in the variances between transducers.

Another approach that has been employed is to physically arrange the imperfect transducers in such a way as to attempt to cancel out the effects of the differences. That is, if an unwanted signal is of about the same amplitude, and roughly 180 degrees out of phase with another unwanted signal, they will tend to cancel one another out. With small arrays, this may represent a possibility. For example, with a four-element array, we need only examine 4!, or 24 different arrangements of the transducers—and select the arrangement providing the best results. However, with, say, a 1000-element array we would have to examine 1,000! different arrangements—an impossible requirement.[4]

However, arrays obviously have been, and are being designed and built, and so we were determined to find out just how array design engineers managed to deal with this situation. We discovered that a number of procedures were being employed. However, the best of these, in our opinion, was based on the use of what are known as Kendig error terms [Kendig, 1971]. Details on this concept are available in the references. Here, we simply note that such terms may, *in theory*, be computed for any array deployment scheme. However, in actual practice the number of computations necessary is combinatorially explosive. As such, the more clever of the array design engineers have developed and employed heuristic pruning methods so as to reduce the number of terms computed.

---

[4] Further, each *examination*, it should be noted, is by no means a trivial task.

There are two things of particular interest about this pruning process. First, it seemed to do a fairly decent job. Second, the procedure is very much analogous to that used in the DENDRAL expert system (see Chap. 3 for a brief discussion of DENDRAL), although it has never received anywhere near the amount of attention given to DENDRAL.[5]

A decision was thus made to develop a hybrid expert system. The first phase of the hybrid was to be an expert system based upon the array design engineers' rule set (i.e., the set of pruning heuristics used to deal with the Kendig error terms). The solution thus developed was to then be used as a starting point for a numerical search method known as permutation search. A brief summary of permutation search is provided in the appendix at the end of this chapter. A complete description of the resulting hybrid expert system may be found in the references [Ignizio et al., 1987]. We need only mention here that the results were truly dramatic. Not only did the hybrid expert system permit us to develop arrays whose performance almost matched their theoretical limit, it enabled us to actually (and substantially) relax the manufacturing tolerances on the transducers. As a result, one may design and fabricate better arrays, at lower costs.

Table 9.4 serves to summarize the results that were obtained, on actual hardware, for an 8×8 acoustic array (an array with eight rows and columns, for a total of 64 transducers) that has been, for some time, of intense interest to the Navy. The *paper solution* to this design showed that it was theoretically possible to design such an array with a main beam 20 degrees in width (as measured at the -3 dB levels) and with no sidelobe levels higher than -40 dB below the main beam. However, in several years of experimentation (i.e., the generation of numerous random deployments), the best array developed had required the use of array elements with extremely tight, and costly manufacturing tolerances (plus or minus 1 percent in amplitude and plus or minus 0.6 degrees in phase), and the sidelobes of this array were still only -31 dB below the main beam.[6]

We employed the hybrid expert system approach to solve precisely the same problem. However, we first dramatically *relaxed* the manufacturing tolerances of the transducers. Specifically, the tolerances were set to plus or minus 9 percent in amplitude and plus or minus 11 degrees in phase. This was done primarily because of the extensive costs associated with attempting to hold to the tighter tolerances. Even with this relaxation, we were able to obtain substantially *improved* performance. As may be seen from Table 9.4, the expert system portion alone resulted in sidelobes of -35 dB (a 4 dB improvement) while the hybrid expert system reduced sidelobes to -39 dB (with -40 dB being the theoretical limit).

---

[5] Possibly because the method has most typically been implemented in FORTRAN, or some other "non-AI" language.

[6] A dB (decibel) is given as 10 times the log (base 10) of the ratio of the power of the sidelobe to the power of the main beam. Thus, for example, a -3 dB level means that the power is one-half that of the main beam.

TABLE 9.4
Results on an 8 × 8 array

Deployment method used	Highest sidelobe level (dB below main beam)
Random assignment	−31 dB
Expert system (alone)	−35 dB
Hybrid expert system	−39 dB

A NEURAL NETWORK-BASED HYBRID EXPERT SYSTEM

Neural networks share a great deal in common with expert systems, including the following:

- Their primary use (thus far) has been in the solution of problems of classification.
- Realistic assessments of their potential have been distorted by intense hype and overly optimistic predictions.
- Entrepreneurs, of widely varying degrees of competence—and ethics—have been attracted to the methodology.
- If employed properly and wisely, the approach can be effectively applied to a number of important, real-world problems.

One approach involving neural networks that appears to hold considerable promise is the use of the methodology in conjunction with an expert system.

Since this is not a text on neural networks, we do not intend to even attempt to describe the methodology of neural networks in any detail. Instead, we shall simply discuss how one may construct a particularly simple hybrid system—composed of a neural network and an expert system. To do this, we will briefly describe the most elementary neural network—a single-layer perceptron.

The Single-Layer Perceptron

A representation of the single-layer perceptron is presented in Fig. 9.6. This network consists of only a single node, or *processing element* (i.e., a PE) with a number of continuous valued input signals and a single output. The output signal will be either at a level of $+1$ or -1, depending upon the specific set of input signals.

Each input signal corresponds to a single attribute. For example, if we are attempting to determine if an airplane is a fighter or bomber, one attribute might be the top speed of the plane while another might be its maximum payload. Given a hypothetical plane, the values of each attribute (i.e., top speed and maximum payload) would then be input into the processing element.

The purpose of the processing-element output is to assign the object under investigation to one class or another. For example, class I might be fighter aircraft

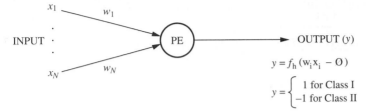

FIGURE 9.6
Single-layer perceptron.

while class II is associated with bombers. Or, class I could be loan applicants that have been classified as good risks while class II represents those that are considered to be bad risks. If the output signal is $+1$, then the input is class I. If the output signal level is -1, the input is recognized as class II.

The formula that serves to determine the output of the PE is given as

$$y = f_h \left(\sum w_i x_i - \phi \right) \tag{9.1}$$

where x_i = the signal level at input i (i.e., for attribute i)
w_i = the weight associated with input i
ϕ = a *threshold level*
f_h = the so-called hard-limiting function

The hard-limiting function may be described as follows. If the weighted sum of the input signals exceeds the threshold level, the PE is said to fire, that is, its output level becomes $+1$ (i.e., class I). Otherwise, the output level remains at -1 (i.e., class II).

The only adjustment in most neural nets are those of the input weights. Initially, these weights are typically set to some small (randomly generated) values. The network is then trained by presenting it with examples and then critiquing the network's output. Training involves the readjustment of the input weights to develop the correct response for the training set. The training rules for the single-layer perceptron are listed below [Lippman, 1987].

Training Rules: Single-Layer Perceptron[7]

1. *Initialization.* Let $t = 0$. Set all $w_i(t)$ and ϕ to small, random values. Note that $w_i(t)$ is the weight on input i at time t and ϕ is the output threshold.
2. *Training input.* Present a new, continuous valued input $(x_1, x_2, \ldots, x_N)$ plus the desired output, $d(t)$.

---

[7] Adapted, with permission of the IEEE, from Lippman, R. P., "An Introduction to Computing with Neural Nets," *IEEE ASSP Magazine*, April 1987, pp. 4–22. © IEEE, 1984.

3. *Calculate weighted output*[8]:

$$y(t) = f_h \cdot \left(\sum_{i=1}^{N} w_i(t) \cdot x_i(t) - \phi \right)$$

4. *Modify interconnection weights.* Change weights as follows:

$$w_i(t + 1) = w_i(t) + \delta \cdot [d(t) - y(t)] \cdot x_i(t) \qquad \text{for all } i$$

$$d(t) = +1 \text{ if input is from class I}$$

and

$$-1 \text{ if input is from class II}$$

where

$$\delta = \text{a positive gain factor } (< 1)$$

$$d(t) = \text{the desired, correct output for the training input}$$

5. *Repeat.* Return to step 2 and continue with training inputs.

Given a suitable number of examples (which, unfortunately, can be considerable), the weights on the inputs to the network will gradually converge to those necessary to discriminate between class I and class II. Perfect discrimination will occur if the examples may be separated by a linear discriminant function (i.e., a straight line or a hyperplane). More complex networks (e.g., multilayer perceptrons) may be used to classify examples that are not so conveniently separated.

An Application

The combination of a neural network and an expert system would seem to be appropriate where the problem faced involves two particular phases: classification followed by some other activity. First, an object must be classified. Once the classification has been made, then something must be done with, to, or for the object.

For example, the hybrid system may be one used by the military to seek out and destroy enemy tanks. Data from sensors (e.g., radar, infrared, television) would be transmitted to the system wherein the neural network would first be used to determine if the associated target is actually an enemy tank. If it is, this determination is transmitted to the expert system where a production rule set is employed to conduct the required military exercise (e.g., a determination of which weapon system, in which mode of operation, to deploy against the tank).

SUMMARY

In this chapter, we have presented just a few examples of hybrid expert systems. Such systems are typically characterized as a two-phase process. Most frequently,

---

[8] Above the threshold, $y(t)$ is to be $+1$; below it, it is to be -1.

in the first phase, an initial solution is obtained by means of an expert system (i.e., as based upon a knowledge base acquired from a human expert or experts). In the second phase, the output of the expert system is used as a starting solution for, most typically, a heuristic programming approach. Together, the combination is often able to produce results that *exceed* those produced by a human expert.

One of the lessons of this chapter, and a point stressed throughout this text, is that not all human experts really exhibit any significant degree of true expertise. Further, there are some classes of problems that may simply be too complex to be efficiently dealt with by humans, regardless of their experience and insight. Consequently, if all we do is to clone the decision-making process of such pseudoexperts, we do little more than perpetuate mediocrity. The hybrid expert systems approach offers one way to alleviate this situation.

A particularly promising form of the hybrid concept is that which employs a neural network followed by an expert system. Typically, the neural network is used for classification and the expert system takes some action, based on the specific classification.

EXERCISES

9.1. On what types of problems, classification or construction, would you guess that the employment of a hybrid expert system would be most attractive?

9.2. Let us assume that you have constructed and validated an expert system. The domain expert, to the best of your knowledge, appears to be an expert in the area under investigation and the expert system develops results comparable to those derived by the domain expert. Should you, in this case, consider a hybrid expert system? Why or why not?

9.3. Read the DMES article [Cochard and Yost, 1985]. Do you believe that the approach used is, fundamentally, a hybrid "expert system"? Why or why not?

9.4. For the scheduling problem subject to resource constraints of Example 9.1, develop an "expert system" for solution based on the following philosophy. If there are any unscheduled tasks whose predecessors have been completed, schedule the task whose processing time is the shortest. If there are ties for the shortest processing time, then break these in favor of the task with the highest resource consumption rate. If ties still exist, break these in favor of the task with the earliest feasible start time. If we still have ties, then arbitrarily pick a task.

9.5. Solve the problem of Example 9.1 by means of the expert system developed in Exercise 9.4.

9.6. Both the expert system of Example 9.1 or that developed in Exercise 9.4 involve, for the most part, mathematical operations and comparisons. Expert systems, however, are most appropriate for problems involving symbols. Can you think of any heuristic rules, involving symbols, that might be appropriate for a more realistic version of this scheduling problem?

9.7. Another type of scheduling problem is known as the M-machine, unconstrained flow shop. Here, $N(n = 1, 2, \ldots, N)$ jobs are to be processed by $M(m = 1, 2, \ldots, M)$ machines where each job must be processed on each machine and the sequence of jobs must be the same on all machines. That is, if there are five jobs and the processing sequence is $\{4, 1, 3, 2, 5\}$ on any one machine, it must be the same on all the remaining machines. Further, once a job is in process it must not be interrupted.

Let us assume that our scheduling "expert" uses the following heuristic: sum the processing times of each job and order the processing sequence in descending order of these sums (i.e., the first job should have the largest sum of processing times). Use this rule to determine the processing times for the four jobs listed in the table below. What is the makespan of this schedule?

Job	Time on machine A	Time on machine B
1	2	3
2	4	2
3	5	2
4	2	6

9.8. One way to determine a schedule makespan is to construct the associated Gantt chart (as done in this chapter for Example 9.1). However, this can be a tedious process. A considerably easier way is to employ a *makespan table* [Ignizio and Gupta, 1975]. The makespan table associated with the schedule { 2, 1, 4, 3} for Exercise 9.7 is shown below. Notice that the makespan for this particular schedule is shown in boldface in the bottom-right cell (i.e., 17 units).

Sequence position (p)	Job (n)	$t(p,1)$ / $T(p,1)$	$t(p,2)$ / $T(p,2)$
1	2	4 / 4	2 / 6
2	1	2 / 6	3 / 9
3	4	2 / 8	6 / 15
4	3	5 / 13	2 / **17**

Further, note that

$t(p, m)$ = the processing time, on machine m, of job p in processing sequence P [e.g., $t(3, 2) = 6$]

$T(p, m)$ = the completion time, on machine m of job p in processing sequence P [e.g., $T(3, 2) = 15$]

The table may then be established according to the rules listed below:

(a) Develop a table with N rows and M columns (not counting the two columns for position and job, i.e., the first two columns). Use the first two columns of the table to depict the job-processing sequence under consideration (e.g., see table above).

(*b*) For each row associated with each job n, enter the process times of that job in the *upper* half of the cells for that row.

(*c*) $T(1, 1) = t(1, 1)$

(*d*) For row $p = 1$ and $m > 1$:

$$T(p, m) = T(p, m - 1) + t(p, m)$$

(*e*) For column $m = 1$ and $p > 1$:

$$T(p, m) = T(p - 1, m) + t(p, m)$$

(*f*) For all other elements in the table (i.e., $p \neq 1$ and $m \neq 1$):

$$T(p, m) = t(p, m) + \max\{T(p, m - 1), T(p - 1, m)\}$$

Using this approach, compute the makespans (for the data of Exercise 9.7) for the following schedules

(*a*) $\{4, 3, 2, 1\}$

(*b*) $\{3, 4, 2, 1\}$

(*c*) $\{4, 2, 3, 1\}$

9.9. Use the *expert system* schedule developed for Exercise 9.7 as the starting solution for permutation search (see the appendix to this chapter) and derive, if you can, an improved solution.

9.10. Use an initial schedule of $\{1, 2, 3, 4\}$ for the problem of Exercise 9.7 and then employ permutation search to find, if you can, an improved solution.

9.11. Outline a hybrid expert system for the location of warehouses. Assume that there are M possible warehouse sites and that there are $K(K < M)$ warehouses to be constructed. The cost of each warehouse, wherever it is built, is strictly a function of the number of customers (retail centers) served. The primary objective in this problem is to reduce the sum of the total distance between the customers and the warehouses that serve these customers (assume that a customer must use the nearest warehouse).

9.12. Use the single-layer perception training rules to develop a perception that can distinguish between airplanes and birds. For attributes, simply use wings, feathers, and engines.

APPENDIX TO CHAPTER 9:
AN OVERVIEW OF PERMUTATION SEARCH

Permutation search is a remarkably simple, as well as an often remarkably effective means of dealing with a wide variety of problems in combinatorics. Nicholson [1971] provides a fairly in-depth description of the method, particularly in its employment as a solution technique for scheduling problems. In the array deployment problem of this chapter, we employed a particularly simple version of permutation search, and it is this version that we describe herein.

Given, say, four objects, one permutation of these objects may be listed as

$$[P^0] = [1, 2, 3, 4]$$

where $[P^0]$ is the initial permutation of objects 1, 2, 3, and 4. That is, if we are dealing with a schedule, we would process job 1, followed by job 2, followed by job 3, and finally job 4. We may define a "neighborhood" of a permutation as the sets that contain all those permutations formed by making a one-for-one adjacent exchange of the objects of the original permutation. As such, there will always be $N - 1$ permutations within a given neighborhood, where N is the total number of objects under consideration. Thus, the neighborhood of our previous permutation is limited to

$$[P^1] = [2, 1, 3, 4]$$

$$[P^2] = [1, 3, 2, 4]$$

$$[P^3] = [1, 2, 4, 3]$$

The steps of (this version of) the permutation search algorithm may then be summarized as follows:

Step 1: Obtain an initial permutation and term this the base permutation. Designate this as permutation $[P^0]$.

Step 2: Investigate, *one at a time*, the permutations in the neighborhood of the present base solution. As soon as such a permutation provides for an improved solution, make this the new base solution and repeat step 2. If all permutations in the neighborhood are examined and none provide an improved solution (over the present base solution), then stop the search procedure.

Quite often, step 2 is modified to permit a restart of the search at some new, randomly selected, start point. Let us, however, demonstrate the basic process on a simple example.

Consider an array (or, for that matter, any other problem in combinatorics wherein one is able to evaluate a given solution) consisting of but four transducers and four empty slots, into which these transducers are to be positioned. For any positioning, it is possible to compute the resulting array beam pattern (e.g., the sidelobe levels with respect to the main beam). Let us further assume that we have a starting solution wherein transducer 1 is placed into slot 1, transducer 2 in slot 2, transducer 3 in slot 3, and transducer 4 in slot 4. This may be represented as:

$$[P^0] = [1, 2, 3, 4]$$

Further, the sidelobe level for this arrangement is computed and assumed to be -27 dB. Our next step would be to examine the first permutation in the neighborhood of $[P^0]$, that is, $[P^1] = [2, 1, 3, 4]$, and compute its sidelobe levels. If they are better (i.e., lower), this becomes the new base solution. If not, we would examine the next permutation in the neighborhood. And if no improvement can be found after examining all of the permutations in the neighborhood of a base solution, we terminate the process (or start again with some randomly selected initial permutation). It should not be difficult to see how precisely the same process might be applied to the solution of a scheduling problem.

CHAPTER

10

ALTERNATIVE DEVELOPMENT MODES

CHAPTER OVERVIEW

As we discussed in Chap. 4, there are a number of alternative approaches to the representation of expert knowledge. And we have already presented our reasons for focusing, so exclusively, on the use of rule-based knowledge representation and rule-based expert systems within this text. However, it is most certainly not our intent to lead the reader to believe that rule-based expert systems are the only means by which one may, or should, approach the development of expert systems.

Rule-based expert systems are, we believe, an excellent way to start one's education and training as a knowledge engineer—and, happily, they serve as a means to quite adequately approach the majority of expert systems development efforts. However, alternatives do exist that permit certain attractive extensions, or increase one's control over the inference process. And a knowledge engineer who considers himself or herself to be truly skilled in expert systems development should, at the very least, be aware of such alternatives. Consequently, in this chapter we shall attempt to provide the reader with a brief overview of the various modes of expert systems development—including alternatives to rule-based expert systems.

OUR CHOICES: AN OVERVIEW

In essence, there are four choices that must be made prior to and during the expert systems development effort. As has been emphasized throughout this text, the very first choice is that of the selection of the methodology to be employed, that is, whether to employ expert systems or some alternative approach. If it is decided that expert systems is indeed the best way in which to deal with the situation at hand, then we must make the following choices:

- The mode of knowledge representation within the model
- The software development package
- The hardware upon which the expert system is to be implemented

The first choice in our list is concerned with whether we use a rule-based mode of representation, OAV triplets, semantic networks, frames, logic programming, or some combination of these. However, an OAV triplet is a special class of semantic network and, for most purposes, a semantic network may be considered to be a special class of frame. Consequently, our choice of representation is mainly from among production rules, frames, logic programming, or some combination of these methods.

Further, the choice of knowledge representation has a direct impact upon the choice of the software development package that might be employed. That is, if we believe that production rules are appropriate and adequate, we can use a rule-based expert systems shell. Or, if we so choose, we may decide to construct our own expert systems shell—from scratch—by means of some computer programming language (e.g., C, FORTRAN, LISP, or PROLOG). If we feel that frames are necessary, we might then either use a rule-based expert systems shell with a bridge to frames (and many shells now offer such enhancements) or else employ an expert systems "environment." Expert systems environments are sometimes termed *multiparadigm* packages as they include capabilities for more than one mode of knowledge representation. Most environments, however, emphasize the use of frames. Finally, if we feel the need for representation through logic, we would likely employ a software package employing logic programming—such as PROLOG.

Actually, our choice of the software will most likely have to be made in conjunction with our choice of hardware. While some packages may be run on a variety of general-purpose computers, others may require special processing support. For example, should we decide to build our expert system from scratch using, for example, the LISP programming language, then we might wish to select a micro-, mini-, or mainframe computer with LISP capabilities—possibly one that has a special *LISP chip* that serves to expedite processing speed. Or, we may even wish to select from among a number of dedicated (and very expensive) LISP computers. Alternatively, should we wish to employ an expert systems environment, we may also be forced to select a special-purpose computer.

Another factor that serves to complicate the matter of our choices of software and hardware is that of the continuing evolution of such support. At one time there was a very clear distinction between expert systems shells (e.g., cheap, simple, not very powerful) and expert systems environments (e.g., expensive, complex, and powerful). However, and as we shall discuss, these distinctions are, for the most part, no longer nearly so definitive. In fact, it is becoming more and more difficult to tell such packages apart. A similar evolution has occurred in the area of hardware. Not very long ago, the use of LISP would have required a dedicated, expensive computer. Today, LISP may be employed on a personal computer. We shall explore these topics in somewhat more detail in the sections to follow.

DEVELOPMENT LANGUAGES

Before the advent of the proliferation of commercial expert systems development packages, the construction of one's own expert systems software package was often a matter of necessity. Today, however, the only legitimate reason we can conceive of for such an endeavor is the need for a degree of control over the inferencing process that is above and beyond that available through commercial expert systems software packages. Even still, if all we wish to do is to apply expert systems to a particular task, we find it difficult to believe that any additional flexibility achieved through such a heroic effort will offset the expenditure of time and funds that will be consumed by the endeavor. Despite this, there are still some brave souls who insist on constructing their own packages—and we shall discuss the choices of development languages in support of such an effort in this section.

To a certain portion of the AI community, consideration of development languages for expert systems is restricted to the so-called AI languages. These AI languages, in turn, include LISP and PROLOG in particular, as well as a number of hybrid combinations of the two (e.g., POPLOG, LOGLISP, FROG). In addition, and more recently, we see an interest in so-called object-oriented languages (e.g., SMALLTALK). From the reading of much of the literature in AI, one might draw the conclusion that this is the extent of interest in development languages.

However, when one begins to actually examine the current set of expert systems tools, particularly those that have gained or are gaining widespread commercial acceptance, a very different story emerges. Quite simply, *very efficient expert systems are being developed by means of virtually any computer programming language*; and it would appear that most recent developments utilize *non-AI languages* such as FORTRAN, BASIC, APL, PL/1, PASCAL, and (in particular) C. In fact, in several instances, expert systems have been developed using only the internal command languages available in commercial database software. As such, if one intended to provide an in-depth review of all pertinent expert systems development languages, such a review would have to cover virtually all available programming languages. Since this is not our intent, we shall restrict our overview to just the four following general *categories* of programming languages:

- List processing languages
- Logic programming languages
- Object-oriented languages
- Conventional programming languages

LIST PROCESSING LANGUAGES. The parent of all present list processing languages, and by far the most well known of the so-called AI languages (at least in the United States), is the LISP language. The name LISP, in fact, is simply an abbreviation for LISt Processing. LISP is also, in terms of computer programming languages, a very old language, having been originally developed by John McCarthy in 1959.

LISP actually handles only two specific entities, atoms and lists. In LISP, a list is a sequence of elements, where each element is either an atom (i.e., a single object) or another list. Thus, LISP can deal with *lists of lists*, and, as such, is naturally suited to recursive operations—a particularly convenient facility in constructing efficient inference engines. Further, any LISP program may be used as data for another LISP program. This means that LISP data and program instructions are in exactly the same format. As such, declarative knowledge (i.e., information about the properties of an object) may be integrated with procedural knowledge (i.e., information about what actions to perform). This capability has also resulted in the development, over the years, of numerous *built-in* LISP functions that are extremely useful for support, in particular, of AI research.

LISP is also both interactive and interpretive. While the interpretive feature (i.e., the ability to run programs as they are being written) is an attractive aid in program development, the lack of computational speed in the interpretive mode led to considerable early criticism of LISP. However, LISP compilers are now available and have, to some degree, alleviated this problem. Despite this, LISP programs are still notorious for their lack of speed (and their consumption of memory), particularly when dealing with extensive numerical calculations.[1]

List functions (known as primitives) may be used to perform certain operations on lists. Consider, for example, the following list:

'(A B C)

The list, always enclosed in parentheses, in this case contains three elements (atoms): A, B, and C. Preceding the list is a single quote which serves to indicate that we are dealing with a list and not a function. One list function is CAR, which serves to return the first element of a list. Thus,

(CAR '(A B C)) returns A.

Another function is CDR, which returns the remainder of the list after the first element. Thus,

(CDR '(A B C)) returns '(B C).

---

[1] In all fairness, it must be stressed that LISP was never intended for *number crunching*. Rather, it is a tool for working efficiently with symbols.

We may also use list functions for truth tests. For example, we may use the EQUAL function to determine if two lists (or atoms) are equal. Thus,

(EQUAL 9 9) returns T,

which simply indicates that it is true that 9 equals 9. On the other hand,

(EQUAL '(A B) '(A C)) returns NIL,

which means that the list (A B) is not the same as the list (A C).

We may also instantiate variables through list functions. For example, the function SETQ may be used to set the atom A equal to the atom 5, as shown below:

(SETQ A 5)

COND is yet another useful list function that permits us to test conditions, and to then take some action if the test is true. We thus may use the COND function in conjunction with the SETQ function to form production rules. Consider, for example, the following production rule:

Rule: *If* A = 5
      Then B = Red

This may be stated, via LISP, as

(COND ((EQUAL A 5) (SETQ B RED)))

Although we have but scratched the surface of LISP functions, it should now be at least marginally clearer as to just how one may develop an expert system using LISP.

LISP does not exist in a single version; rather, it is available in a variety of dialects. This diversity has led to dialects that can differ substantially from one another, which can lead to considerable frustration if one wishes to use or implement an application developed in some other dialect, or on another machine.

Further, until relatively recently, the use of LISP required an expensive, special-purpose computer: a so-called LISP machine. With the cost of such machines in the range of $50,000 to $100,000 (or more) each, the initial investment in the choice of LISP can be substantial. More recently, versions of LISP that run on personal computers have been made available, at a far more reasonable cost. However, the computational performance of such implementations has still left a lot to be desired. Even more recently, special-purpose LISP chips have been made available for incorporation into personal computers, resulting in some improvement in performance. These hardware developments will be discussed in more detail when we discuss expert systems hardware.

While the implementation of LISP on personal computers can serve to substantially reduce the initial investment, hardware is only a part of the cost associated with LISP—and usually only a relatively small part. Specifically, training costs associated with the introduction of LISP are often substantial, and LISP programmers generally demand higher salaries.

LOGIC PROGRAMMING LANGUAGES. By far, the most popular of the logic programming languages is PROLOG. PROLOG was developed in France at the University of Marseille by Colmerauer and Roussel in 1973 and has been selected by the Japanese as the official AI programming language for their well-publicized Fifth Generation Project. The name itself stands for PROgramming in LOGic. The fact that PROLOG was developed in France, and has been blessed by the Japanese, may serve to explain, to some degree, the often intense animosity between the advocates of LISP and those of PROLOG. Bruce Buchanan, a professor of computer science at Stanford University, responded in this way to the question about why PROLOG is so widely used in Europe and Japan [Mishkoff, 1985]:

> In Europe, they have not had the tools we've [the U.S. AI community] had for the last 20 years. They've had to work with smaller machines; LISP requires much more memory than PROLOG. The Japanese have a somewhat different reason. LISP was so identified with AI research in the United States over the past 25 years that, as a matter of national pride in their flagship Fifth Generation Project, they had to find some other vehicle. I don't think that is a good scientific reason at all.

Personally, we find such a statement to be a bit difficult to accept. In light of the fact that the Japanese have never been at all hesitant to accept and use a good idea, whatever its source, it seems odd that they would avoid LISP simply because it is "so identified with AI research in the United States." However, statements such as these seem all too common on both sides of the argument.

Leaving the real and imagined deficiencies of PROLOG to be recited by the LISP advocates, the three most widely cited advantages of the PROLOG language are:

- It requires considerably less memory than LISP, and may be run on smaller, less expensive machines
- The language is simple to learn and implement, and is certainly much simpler than LISP
- The language is declarative

The first advantage is only pertinent if one restricts himself or herself to a choice between just LISP and PROLOG. The second is true in general, and particularly true when comparing LISP and PROLOG. But it is the third factor that is of most interest. To appreciate the third advantage, one must understand the fundamental difference between procedural languages and declarative languages. Procedural languages require the programmer to specify the precise, step-by-step set of instructions to be carried out in the solution of a problem.[2] Declarative languages, on the other hand, only require that one provide the program with the facts and relationships (e.g., data and knowledge base) that exist for a problem; the solution to the problem is then accomplished by PROLOG's own internal inference engine. And this is a truly radical difference for those schooled in the use of more conventional, procedural languages. We may note, however, that such a separation is essentially what is accomplished in any well-designed expert system in which there is a separation of the knowledge base and inference engine.

The PROLOG inference engine employs backward chaining and backtracking without memory. Search is conducted depth first, without restriction, and all the rules are scanned. However, it is also possible to use PROLOG in the forward chaining mode, and a variety of recent embellishments provide for considerable additional flexibility.

The development of an expert system in PROLOG is, quite possibly, easier than with any other language. As an indication, consider the following production rule:

Rule: *If* inflation is high
        Then invest in gold

The same rule, in PROLOG, might be written as

gold_investment (yes):-write ('INFLATION'), read (INFLATION),
INFLATION = high.

Here, the conclusion (invest in gold) is written first and the premise (inflation is high) follows. The symbol ":-" is used for IF. The statement may be read as invest in gold if inflation is high. The complete PROLOG statement may be interpreted as follows. The phrase *write ('INFLATION')* is used to simply cause INFLATION to be printed on the user's terminal, as a prompt for the user to enter the value for inflation. The phrase *read (INFLATION)* will result in the value entered by the user to be assigned to INFLATION. The next phrase, *INFLATION = high*, is true if the user entered *high* in response to the prompt. In essence then, all one has to do is to place the production rules into the PROLOG format, specify a goal, respond to any user prompts, and then wait as PROLOG solves the problem.

Initially, PROLOG was implemented on mainframes. Like LISP, although not yet to such an extent or diversity, a number of dialects have evolved. However, one dialect in particular has received a fair level of acceptance. This is TURBO PROLOG (a product of Borland International) [Townsend, 1986]. TURBO PRO-

---

[2] To illustrate, the reader need only consult any text on procedural languages such as BASIC, FORTRAN, or C.

LOG runs on personal computers and has a surprising degree of computational efficiency. It also sells for only about $100. As such, the package has already been used in a wide variety of applications, including the construction of expert systems and expert systems shells. However, and as just one indication of the continued lack of interest in PROLOG within the United States, Borland International has noted that it does not intend to enhance TURBO PROLOG beyond its current release—which would certainly indicate a lack of adequate market.

OBJECT-ORIENTED LANGUAGES. Recently, there has been considerable interest shown in so-called object-oriented languages. The much heralded HYPER-CARD (for Macintosh computers) may be considered a type of object-oriented language (although its developers consider it to be far more than just that). Evidently, however, the oldest recognized object-oriented language is SMALLTALK. And we shall use this language to demonstrate the (basic) properties typical of an object-oriented language. More specifically, we shall briefly describe the characteristics of SMALLTALK-76.

In SMALLTALK-76, or any object-oriented programming language, the fundamental entity is the *object*. In turn, each object is characterized by its *state* and its *actions*. The state of an object is simply the attributes and values of the object. Actions are then the operations that the object is capable of executing.

Each object is also a member of a *class*. For example, a class might be that of all the houses that are for sale in the town of Carmel, California. This class, in turn, may be a member of a *superclass*, that is, *all* of the houses that are for sale in northern California's Monterey peninsula. Each class is characterized by its *fields* and *methods*. Further, each class inherits the fields and methods of its superclass. A field contains one of the states of the object while a method contains one of the actions of the object. To demonstrate, consider the following two objects:

- Object 1: All houses for sale in Carmel, California in the price range of $300,000 to $400,000.[3]
- Object 2: All houses for sale in Carmel, California in the price range of $400,000 to $500,000.

These two objects may then be represented as shown in Table 10.1.

From Table 10.1, we note that the attributes associated with each object (i.e., price range of houses) includes (1) asking price, (2) location, (3) required down payment, and (4) salesperson to be contacted. These four items are thus the fields for the state of the two objects.

---

[3] This section of the text is being written while I am living in a rented cottage in Carmel by the Sea, better known as the town in which Clint Eastwood was once mayor. And we are sorry to say that there just aren't any houses (or even empty lots) in this price range.

TABLE 10.1
Carmel real estate — object listing

	Object 1 cheap houses	Object 2 moderately priced houses
States	Asking price Location Down payment Responsible salesperson	Asking price Location Down payment Responsible salesperson
Actions	Display Mortgage calculations	Display Mortgage calculations

We may also perform certain operations, as determined by the methods associated with the object's actions. In particular, we may (1) display a picture of the house or, (2) calculate the required monthly mortgage payments for whatever choices of mortgage that may be available. However, before we perform any operation, a message must be transmitted to the object, stating the name of the action and the arguments to use. For example, if we wish to display a picture (i.e., graphical representation of the exterior views and perhaps the floor plan) of the house, we transmit a message listing this action by name along with the house identification code. For example, the message format may be DISPLAY (HOUSE #33217).

The similarity between SMALLTALK (or object-oriented programming in general) and frames, as discussed in Chap. 4 and as used in many expert systems environments, should be apparent. And if, at this point, you are wondering whether there is any meaningful difference between frames and objects, you are most definitely not alone. To quote Shafer [Shafer, 1989]:

> If objects and frames hadn't started from different historical points, they would be the same thing. For all practical purposes, they are anyway. There are some arguable conceptual differences between them, but these are so subtle as to be practically useless to those of us whose jobs include building expert systems. An object has properties; a frame has demon slots. I suspect that in the final analysis these two concepts merge. As it is, any distinctions made between them are distinctions without a real difference.

CONVENTIONAL PROGRAMMING LANGUAGES. Let us once again note that an expert system may be developed via *any* programming language. Moreover, a substantial number of existing commercial expert systems tools employ allegedly *non-AI languages*. Further, a number of tools originally developed through LISP or PROLOG have recently been reprogrammed using conventional languages.

The objections, on the part of some of those in the AI community, to the use of conventional languages seem to be that such languages do not incorporate certain *built-in* functions that are useful in constructing inference engines (e.g., pattern matching, recursion) or dealing with knowledge bases. We have already quoted Buchanan with regard to his opinions as to LISP versus PROLOG. In that same interview [Mishkoff, 1985], he also stated the following concerning the use of conventional languages versus LISP:

> I had a similar experience with a person who worked for me. He said that he could create AI systems in FORTRAN and he did. But one of the first things he did was implement list processing and recursion in FORTRAN; he needed the tools that LISP already provides. . . . People who have been programming in FORTRAN for 25 years don't think that there is anything new that they can't do. And, of course, the answer is that they're right. Of course you can do any of this in other languages—it just may take twice as long.

While we know of no independent study that compares the times necessary to build an expert system in LISP as opposed to FORTRAN (or any other conventional language), it is indeed true that LISP has *built-in* functions that are not available in other languages (*as well as vice versa*). However, and in defense of those "people who have been programming in FORTRAN for 25 years," even if one has to develop such facilities, few programmers would be so shortsighted as to have to develop them more than once. Furthermore, the programmer employing conventional languages would also never have to go through the extensive amount of training and practice that is required if one is to become reasonably proficient in the use of LISP.

At the present time, the trend in expert systems development is certainly away from AI languages and toward conventional ones (C in particular). This does not mean that conventional programming is the *best way*, but rather that it is *presently* viewed (particularly by the most important ingredient in the formula, the customer) as the most cost-effective approach to the development of expert systems.[4] One example of this is depicted in the following comparison. Specifically, various AI languages were compared against C on a problem involving the redesign of a 24-transistor cell [Van Horn, 1986]. The results were

- OPS5 (using an interactive version of LISP): 8 hours
- LISP (with reduced interactive capabilities): 45 minutes
- PROLOG: 10 minutes
- C: 1 minute

While newer, faster versions of these languages are now available, the relative differences in computational efficiency have not appreciably changed.

---

[4] However, if one is interested in the development of tools within the more general area of artificial intelligence (i.e., other than expert systems), then LISP, as well as PROLOG, certainly have some quite substantial advantages.

Advantages and Disadvantages of Languages

The primary advantage of building your own expert systems software package is that of control. Simply put, you have control over the entire consultation/inferencing process, as well as the ability to modify, revise, or enhance any portion of the package that you so wish. One can also derive a certain degree of personal satisfaction from such an effort. And it is certainly an excellent way to put one's education into action. Further, it will most definitely enhance one's appreciation of the inference process.[5]

However, if one wishes to construct a really efficient package, with the features and support comparable to those readily available through most commercial rule-based shells, such an effort is a truly major endeavor. And to us it would seem to make little sense to do so unless your goal is build a shell that is to be marketed in competition with those now available.

RULE-BASED EXPERT SYSTEMS SHELLS

Should we not wish to go through the time and effort necessary to construct our own expert systems development package, we will likely focus on a selection from among either the many commercial rule-based expert systems shells that are available, or from among a somewhat more limited number of expert systems environments. In this section, we shall focus on the choice of a rule-based expert systems shell.

The first expert systems shells were viewed with a certain degree of contempt by a fair portion of the AI community. These early shells were typically

- Constrained to strictly a rule-based mode of knowledge representation
- Constrained to very small rule bases (in some cases, on the order of tens of rules)
- Constrained to a single inference mode (most usually, backward chaining)
- Lacking in such features as bridges to external programs

Based on these limitations, such shells were often designated, sarcastically, as toys—and unworthy of consideration. In addition, these early shells were

- Developed, most often, by means of *conventional* computer programming languages

---

[5] We have, in fact, assigned the development of an expert systems shell (a very basic, elementary form of such a shell) as a project in support of our courses in expert systems. In some cases it has worked well. In others, it has mainly served to divert the students' time and efforts away from the more vital parts of the course.

- Inexpensive (in terms of purchase price, training costs, and supporting hardware requirements)
- Simple and easy to master (which, as a consequence, resulted in their employment by individuals *outside* the AI community)

The first factor in the most recent list was often considered as evidence of blatant heresy amongst the AI community, many of whom believed (and, some of whom still believe) that only AI languages should be employed. Oddly enough, the last two factors were also the basis for some criticism. The opinion was, "If something doesn't cost much and is easy to use, how can it be a *serious tool?*"

Despite these real and imagined limitations, rule-based shells became increasingly popular among practitioners and would-be practitioners of expert systems. Moreover, even with their early limitations, they found application in a number of important, real-world problems. This popularity, coupled with numerous actual implementations, served to accomplish two things. First, the shells became even more popular. Second, the capabilities of the shells were substantially enhanced. Consequently, the majority of the commercial shells that are available today, while they may no longer be quite as cheap or as easy to master, allow the user to solve rule bases with hundreds and even thousands of rules, to employ multiple modes of chaining, and to bridge to external programs (including those permitting the use of frames as an adjunct to the consultation process).

About the only *deficiency* that remains is the fact that most of today's expert systems shells do not employ an AI language. However, most of this criticism has been quite effectively muted as more and more expert system products have been developed or translated into a conventional language format, and as those firms that have tried to withstand this change have either modified their stance or fallen by the wayside.

Today, the majority of commercial, rule-based expert systems shells exhibit the following features:

- They may be implemented on inexpensive, general-purpose personal computers.
- They may be readily used to solve rule bases of small-to-modest sizes (i.e., up to a few hundred rules) and, in some cases, truly large-scale rule bases (i.e., many hundreds of rules)—on inexpensive, general-purpose personal computers.
- They exhibit computational efficiencies comparable to, and even superior to those available through packages implemented by means of an AI language (e.g., LISP)—even when these other packages employ expensive, dedicated computers.
- They (usually) permit multiple modes of chaining (including, quite often, mixed chaining).
- They provide bridging features that permit access to external programs (in particular, relatively efficient access to the most popular personal computer-based spreadsheets and databases).

Differences among Rule-Based Shells

While most existing rule-based shells feature the capabilities listed above, there are certain differences among packages. Ignoring the attribute of price, as most packages now sell in the general range of a few hundred to a few thousand dollars, some of the more important of these differences include the following:

* The language, or languages, by which extensions to the package may be developed (and, in some cases, through which the rules must be stated)
* The features of the rule editor (e.g., automatic prompting, checks for consistency and completeness, ordering of rules)
* The use of free form prompts versus menus
* The supporting method, or methods, for the consideration of uncertainty
* Provisions for rule-base development through a rule-induction process
* Restrictions on run-time versions of any expert systems developed
* Hardware requirements/restrictions

Let us consider each of these factors in turn.

Some shells require the knowledge engineer to have some degree of familiarity with a computer programming language. Some, in fact, require some exposure to one or more of the various AI languages. Personal Consultant, for example, uses ARL or SCHEME (as based on LISP) and the rule-base developer must, at the least, be familiar with the structure of the statements employed. For example, let us assume that you wish to include the following rule in a knowledge base to be run by Personal Consultant:

Rule: *If* distance to be traveled exceeds 500 miles
       Then best mode of transportation is an airplane

You would enter this rule, in response to the package's prompts, as

Premise: (DISTANCE>500)
 Action: (MODE=FLY)

Advocates of such an approach would point out its economy of words (and typing). We need only input the fundamentals (i.e., the attributes and values) of each rule. Critics would note that, through such a rigid structure, some degree of transparency of the rule base is lost. To counter this particular complaint, most shells employing this approach will include a facility that permits the rules to be viewed in either an English translation or in their LISP version.

Most shells allow the knowledge engineer to construct interfacing computer programs, in order to extend the developer's control over the consultation process. However, some shells restrict the computer programming language that may be employed. This may not be a problem if the developer, or developers, are already familiar with the particular programming language that is required. However, if they are not, then such extensions may involve a fair amount of time and effort.

Further, if the language that must be used is LISP (or a language based upon LISP), as it so often is, the amount of training can be truly substantial.

The differences between rule editors (and the overall user interface) can play a significant role in the acceptance or rejection of a shell. Some packages appear (and this is sometimes a function of individual developer preferences) to be well thought out and easy to use. Others are complex, confusing, and serve to irritate rather than assist the developer. And there are even some software packages that do not contain rule editors per se. Instead, the knowledge engineer must employ a word processor and construct the entire rule base outside the environment of the shell. For example, Level5/Macintosh requires the developer to use an outside text editor. The resulting text is then read in (or compiled) by the Level5/Macintosh shell that checks the syntax. Consequently, the final set of rules are not available until after a successful compilation. Other shells, such as EXSYS, permit the knowledge engineer to both develop and run the rule base, or any portion of the rule base, from within its own editing environment.

Shells also differ with respect to how they produce user prompts. Menu-based shells, such as EXSYS, provide automatic prompting as based upon the rule that has been input and the values associated with the attribute(s) of that rule. For example, consider the following rule:

Rule: *If* the color of the smoke is gray
        Then sound the alarm

Further, let us assume that the values associated with the premise clause attribute (*color of the smoke*) are black, gray, and white. Thus, whenever the user must be prompted for this attribute, the program will *automatically* produce a menu of the legal values of the attribute. For example, the user prompt might appear as:

```
Is the color of the smoke
1. Black
2. Gray
3. White
Input choice: _____
```

On the other hand, some shells require the knowledge engineer to input the prompt (i.e., in its complete form) associated with each attribute. This certainly provides the knowledge engineer with more flexibility over the format of the prompt but does impose some additional work.

Shells also differ according to the method, or methods, that they provide to support uncertainty. Some, like VP-Expert, permit the use of confidence factors for both rules and user prompts. Others, like EXSYS, only allow confidence factors for rules, as they are designed under the belief that any uncertainty on the part of user input may be best encompassed by proper design of the menu selection (recall our discussion in Chap. 7).

Another difference lies in the choice of specific algorithms for handling uncertainty. Some shells employ the Bayesian algorithm, others may use a MYCIN-type approach, others may utilize some form of fuzzy sets, and still others may permit the knowledge engineer to use his or her own algorithm (e.g., as permitted in the Professional version of EXSYS).

Certain shells include their own rule induction software. Recalling our discussion from Chap. 5, this type of facility permits the knowledge engineer to construct, for certain categories of problems (i.e., specifically for problems of classification), rules from examples. However, as was noted in our evaluation of VP-Expert and Xi Rule, the underlying algorithms that accomplish such induction may differ substantially.

Yet another factor that separates the various shells is that of the restrictions on run-time versions of the expert systems developed. In some instances, you pay an up-front fee to obtain the right to produce an unlimited number of copies of any expert system that you may develop. In others, you must pay a royalty (sometimes, a quite hefty royalty) to the shell developer for each copy that you either produce or sell. If you intend to distribute copies of your expert systems, this is likely a factor that you will most definitely want to investigate in depth prior to selecting the shell.

Shells differ according to the computers that they may be run on. Some run only on personal computers. Others (relatively few) run only on mainframes. While still others may be run on a variety of types and sizes of computing systems. Personal computers are a particularly convenient and inexpensive means through which expert systems may be developed and tested. However, if you are building a really large expert system, or if you have to directly access databases that are located on a mainframe, you may wish to select a shell that either runs on both personal computers and larger systems, or consider developing the entire system on the mainframe.

A further factor that serves to separate shells is their computational efficiency. Some shells may perform brilliantly on the simple, small problems found in textbooks or software manuals, while exhibiting poor-to-mediocre performance on problem sizes approaching those typically faced in practice. In the September 1988 issue of *AI Expert* [Press, 1988], a comparison of different expert system shells was described that serves to provide a snapshot of the computational efficiencies of 11 popular shells, exhibited by the versions available at the time of the evaluation. These shells were all tested on benchmark knowledge bases. Each knowledge base is characterized by certain factors, as listed below:

- S = a sequential rule set (i.e., only a single premise clause for each rule)
- C = a conjunctive rule set (i.e., where two or more premise clauses are connected by AND operators for each rule)
- D = a disjunctive rule set (i.e., wherein two or more premise clauses are connected by OR operators for each rule)
- B = the number of rules in the knowledge base
- W = the "width" of the rules in the knowledge base (i.e., the number of premise clauses in each rule)

TABLE 10.2
Execution time (in seconds)

Expert systems shell	C 20 × 120	S 1 × 100	C 2 × 100	D 2 × 100
Arity Compiler	71.0	NP	17.4	7.3
Arity Interpreter	71.0	NP	16.5	6.6
EXSYS Standard	**1.7**	**0.3**	**0.4**	**0.6**
EXSYS Professional	1.9	0.5	0.6	0.8
Guru	20.2	10.5	10.6	10.5
Level 5	6.2	NP	2.6	1.8
M.1	20.3	3.6	3.7	2.2
Nexpert	NR	9.3	7.3	NR
PC Easy	58.5	9.6	10.7	5.9
PC Plus	94.8	38.5	34.0	19.0
VP-Expert	27.7	NP	NP	NP

Source: From *AI Expert,* 500 Howard Street, San Francisco, California 94105, Copyright 1988. Reprinted with permission.

All tests were conducted on identical IBM PS-2 model-50 computers. Four knowledge bases were used:

	B	W
Sequential	100	1
Conjunctive	100	2
Disjunctive	100	2
Conjunctive	120	20

Just some of the results obtained in the evaluation are listed in Tables 10.2 and 10.3. Note that NP means that it was not possible to solve the associated

TABLE 10.3
Largest possible sequential knowledge base

Expert systems shell	Number of rules	Execution time in seconds	Rules per second
Arity Compiler	70	9.3	7.5
Arity Interpreter	70	8.8	8.0
EXSYS Standard	397	11.3	35.2
EXSYS Professional	**500**	26.4	18.9
Guru	188	5.7	33.0
Level 5	97	1.3	**77.6**
M.1	264	27.4	9.6
Nexpert	NR	NR	NR
PC Easy	211	22.9	9.2
PC Plus	225	45.5	4.9
VP-Expert	17	2.1	7.9

Source: From *AI Expert,* 500 Howard Street, San Francisco, California 94105, Copyright 1988. Reprinted with permission.

knowledge base with the shell while NR means that the shell was not run (i.e., not used) for the associated knowledge base. We have also set in boldface the best results for each evaluation. Thus, in terms of execution time, EXSYS (Standard version) was superior for all four benchmark knowledge bases. In terms of the largest possible sequential knowledge base, EXSYS (Professional version) was able to solve the largest number of rules while Level 5 had the fastest execution time in terms of rules per second. Unfortunately, however, Level 5 simultaneously had one of the lowest ratings in terms of the number of rules in its largest sequential knowledge base. While these results may have changed by the time you read this text (i.e., through the introduction of upgraded versions of the packages), they do serve to indicate the substantial differences that can exist among shells.

Finally, there are differences concerning certain features of the inference process. For example, there are two variations in how disjunctive clauses are treated. Some shells (including VP-Expert and M.1) evaluate all of the disjunctive premise clauses, whether or not one of the clauses has already been found to be satisfied. Consider the following rule:

Rule: *If* temperature is over 100 degrees Fahrenheit
        or throat is sore
        or glands are swollen
        Then the person is ill

Now, assume that we know that the person's temperature is over 100 degrees and that this fact has been made available to the expert system. Despite this, some shells would proceed to evaluate all of the disjunctive clauses in this rule. Other shells (e.g., EXSYS Professional version, Level5, Personal Consultant) would immediately trigger the above rule as soon as any single disjunctive clause is true. Of course, we can take care of this particular difference by simply avoiding the use of disjunctive clauses.

Other differences among shells include the manner in which the use of unknown (as a user response) is handled and the manner in which rules are selected for evaluation and firing. Sometimes these points are brought out in the user manuals for such packages; other times one only learns through actual experience with the shell. In order to minimize the number of surprises you may encounter, we recommend that you prepare a small knowledge base, or a number of such knowledge bases, for the sole purpose of determining the specific operating characteristics of the expert systems shell under consideration. This is probably the safest way to assess the manner in which any shell deals with inferencing decisions.

Advantages and Disadvantages of Rule-Based Shells

Some of the advantages of rule-based expert systems shells include the fact that

- The employment of rule bases by such shells results in
    A particularly natural and transparent mode of knowledge representation
    Ease of knowledge base modification
    Ease of knowledge base validation

- Such packages are inexpensive, in terms of both initial and overall costs
- Most commercial shells may be run on inexpensive personal computers, and a host of such shells are available to choose from
- Training requirements are minimal
- Experience in computer programming is, for most shells, unnecessary
- Rule-based expert systems can be employed to mimic most features of frame-based representation schemes
- Rule-based expert systems shells permit the knowledge engineer to focus his or her attention on the two most critical phases of the development of an expert system: knowledge acquisition and knowledge representation

There are two main disadvantages associated with shells. First, since one does not (normally) have access to the source code, control over the consultation/inferencing process is restricted to whatever provisions the developer has provided. Second, the lack of frames may make the development of some knowledge bases awkward. This second factor, however, is becoming less of a problem as more and more commercial shells provide for access to auxiliary frame-based representation packages.

Most other disadvantages of rule-based shells are a consequence of their traditionally low price (i.e., as compared with expert systems environments, the subject of the next section). Adding more "bells and whistles" to a rule-based shell will raise its price and, as a result, can have a negative impact on sales. Environments, on the other hand, have always been expensive (often, unreasonably so), and the audience for this mode of development is accustomed to paying higher prices. As a consequence, many firms think long and hard before including an enhancement to a shell if that enhancement will require a significant increase in price. One result of this has been the proliferation of rule-based shell "add-ons," available from a range of sources (including the individual entrepreneur).

EXPERT SYSTEMS ENVIRONMENTS

Only a few years ago, there seemed to be a clear distinction between expert systems *shells* and expert systems *environments*. A shell was typically a production rule-based system composed of a rule editor, a user/knowledge engineer interface, and an inference engine that was usually limited to one mode (i.e., forward or backward) of chaining. For the most part, the user needed only to insert his or her particular knowledge base (in production rule format) into the system and then run the consultation sessions. These shells usually ran on personal computers and the amount of training required to learn to use and implement them was relatively minimal. An environment, on the other hand, typically used several modes of knowledge representation (e.g., frames and production rules), employed multiple modes of chaining, and had various types of (directly linked) supporting routines (e.g., for graphics, simulation, and databases). Environments were also almost always limited (at least in development) to expensive, special-purpose LISP computers, and the amount of training necessary to run such systems represented, in itself, a major investment in time and money. Among the choices of environments are KEE, ART, S.1, Knowledge Craft, and Nexpert Object.[6]

Today, with the advent of powerful personal computers and low-cost workstations, it is now possible to implement an environment on a personal computer — or, alternately, to substantially enhance the performance and capabilities of expert systems shells. However, an unfortunate consequence of the enhancement of rule-based shells has been the resulting increase in their complexity, and the amount of time necessary for training. In this and other regards, it is becoming more and more difficult to draw any really meaningful distinction between the developmental capabilities of many present-day shells and environments. As a consequence, the main distinction today between shells and environments lies most usually in three areas: cost, complexity, and the emphasis on frame-based knowledge representation. Probably the easiest way in which to describe an environment, and to contrast environments with shells, is to discuss one specific expert systems environment: the KEE development package.

KEE: A Representative Expert Systems Environment

KEE is a product of IntelliCorp and is, at this time, representative of the high end of development packages, in terms of both cost and capabilities. In the late 1980s, it was claimed that KEE [IntelliCorp, 1987] "remains first in the industry, with more installed systems than all competing products *combined*."[7] Like most environments, KEE is primarily a frame-based development tool. In 1985, estimates of the total cost for the purchase, training, and implementation of KEE ranged from $250,000 to more than $1 million [Linden, 1985]. This was broken down as follows:

- Software and initial support cost, $60,000
- LISP machine, $50,000 to $100,000 (or more)
- Personnel commitment and training costs, upwards of several hundreds of thousands of dollars

---

[6] Nexpert Object is a relatively new expert system environment (or hybrid shell) that has recently managed to capture a large share (about 40 percent) of the high-end expert system market. This $5,000 package competes directly with such tools as KEE, ART, and Knowledge Craft. Nexpert Object runs on personal computers (either IBM compatibles or Macintosh) and interfaces with programs or functions written in either C or Pascal. The success of this package should most likely result in decreases in the prices of its competitors.

[7] It is not clear what is actually meant by "installed systems." Evidently, the comparison is being made between expert systems *environments*, rather than installed expert systems *shells*. Either VP-Expert or EXSYS, for example, certainly have sold far more copies of its software. In the 1986 Annual Report of IntelliCorp, it is cited, "As of the writing of this letter over 950 copies of the KEE system have been licensed to more than 300 customer sites, 621 copies in the past fiscal year alone."

However, only 3 years later, IntelliCorp offered a "full-blown" version of the KEE software for use on a 386 personal computer (i.e., a personal computer based on the 80386 chip) for $15,000.

Most rule-based shells have attempted to emphasize ease of use coupled with minimal training requirements. The basic idea has been to focus one's attention on small-to-medium-sized rule bases (e.g., up to about 200 rules) and to minimize the time and effort required on the part of the developer. KEE (as well as most other environments) seems to have been developed under a philosophy virtually 180 degrees out of phase with that of shells. Rather than providing customers with a straightforward, easy-to-use tool with which they may almost immediately begin to build (at least, relatively modest) expert systems, the developers of KEE evidently opted to develop a tool that stressed power first, and ease of use second (some might say a distant second). In their defense, it is clear that the developers of KEE took great pains to try to consider all the factors involved in the construction of expert systems, and to then incorporate the features that supported these aspects into KEE. The cost of this flexibility is, however, evident in the complexity of the system—and the quite substantial amount of training necessary for one to become even reasonably competent in its use.[8] However, the very complexity of the tool, coupled with the investment of time in training, has resulted in a dedicated group of KEE advocates who consider environments to be the tools of the serious professional, while shells are deemed to be but the toys of the novice. Because of the complexity of KEE, and its requirement for extensive training, a virtual "cottage industry" has developed, dedicated to the provision of support in its training and implementation.

While KEE can be criticized for its complexity, it should not be dismissed out of hand on that basis alone. If you are willing to invest the time and funds necessary to become well-versed in its usage and capabilities, you will find yourself in possession of a powerful, flexible tool for dealing with real-world problems, with virtually all of the support necessary for (effective) implementation immediately and readily available. In particular, if you have an affinity to frames and LISP, and can actually benefit from the power available in such an environment, you are likely to be attracted to KEE over most rule-based shells.

KEE is written in COMMON LISP and was, until recently, restricted to implementation on (expensive) LISP machines. However, as was mentioned, a *full-blown* version is now being sold for implementation on 80386 chip-based personal computers. As of 1990, the cost of KEE (software only) ranged from $15,000 to $98,000, depending on the machine employed. The KEE inference engine operates in either forward or backward chaining and incorporates a variety of search options (e.g., depth first, breadth first). Since the introduction of the basic KEE environment, IntelliCorp has also introduced a wide variety of options that will operate with KEE. These include

---

[8] IntelliCorp has recently made available a *training* package, titled KEEtutor, that covers the basics of KEE. Its price was $5,000.

- KEEconnection: a KEE-based expert system that provides a bridge between SQL® relational databases and KEE knowledge bases
- IntelliScope: using the KEEconnection software, this option accesses databases and formulates database queries, while displaying the analyses in graphic format
- SimKit: an option that facilitates the development and operation of graphics-based simulations
- PC-Host: provides the capability to develop applications on the LISP computer and then deliver them to PCs as linked to mainframe/mini-host computers.

KEE, like most other expert systems tools, possesses explanation facilities, and may thus state why it has requested certain information from the user or how it arrived at a specific conclusion. To do so, it displays the supporting rules and facts that led to the user prompt or consultation conclusion. The screen display available through KEE is divided into several windows. For example, user prompts and responses may be in one window, and explanations in another. In addition, the tool can generate graphics that indicate the hierarchical structure of the knowledge base, and may display objects and attributes in the form of graphical images known as icons. The user either types in information or selects options through a computer mouse.

KEE permits one to use multiple modes of representation, and is thus a multi-paradigm expert systems tool. Included among these paradigms are rules, frames, demons, object-oriented programming facilities, and LISP code. An important feature of KEE, as well as most other environments, is the use of a frame-based representation, and this concept is so fundamental to the use of KEE that it really needs to be addressed before going further. While frames were briefly touched upon in Chap. 4, the discussion to follow will focus on the nature of frames within the specific environment of KEE.

THE USE OF FRAMES IN KEE. Recall that a frame provides a type of structured representation of an object or class of objects. For example, one frame might be used to represent a specific automobile, while another is used to represent the entire class of automobiles. Consequently, the first frame would be a subclass of the second, while the second frame (automobiles in general) might well be a subclass of yet another frame (e.g., of the frame: surface vehicles). In KEE, each frame represents a *unit*.

The frame corresponding to automobiles in general will include a number of slots, such as slots corresponding to the group of attributes associated with automobiles in general. These might include number of doors, height, weight, length, color, horsepower, and so on. The frame associated with a specific automobile (*my* automobile, for example) would have corresponding slots, where such slots are said to be *inherited* from the frame for automobiles in general—and these slots may then be filled with the data specifically pertaining to my car. Since all member frames typically inherit the slots of the superframe, the frame-based representation process should make knowledge representation more ordered and consistent and, as well, reduce the amount of storage requirements. At this level, frames exhibit a useful, systematic manner in which the data associated with a knowledge base may be stored and accessed. For example, the number of spark plugs needed for an automotive tune-up is a direct function of the number of cylinders. Thus, if the expert system deals with automotive tune-ups, and *our* automobile is under consideration, we would access the number of cylinders directly from the frame associated with our car.

However, there is much more that may be accomplished with frames. For example, messages may be transmitted from frame to frame. The KEE system incorporates two facilities used in such transmission. The first is known as the *methods* facility. Methods are simply LISP procedures, attached to frames, that respond to messages sent to the frames. Such methods are themselves stored in the slots of the frame (special slots that are designated as *message responders*). Consequently, messages transmitted to a frame must identify the target message responder slot and should include any arguments needed by the particular method stored within the slot.

The second facility is known as the *active values* facility. Active values are procedures or collections of production rules attached to slots, and these are accessed whenever the slot's values are either accessed or stored. Such facilities are known as *demons* as they are typically used to monitor changes and activities in the system. Or they may be used to dynamically compute values on an as-needed basis.

The KEE environment also uses frames as a foundation for the development and storage of production rules. The advantage of this approach is that each rule is represented as a single frame, and thus rules may be conveniently grouped into classes. In KEE, a predicate logic language is used to represent each rule's premises and conclusions. The rule designer, in this way, has full access to the frame language through the predicates. Also, the predicate logic language permits any LISP function to be a predicate, and thus any arbitrary computation may be used to determine the truth value of a rule premise.

The developers of KEE maintain that, as a production rule system becomes larger, frames provide a means for managing and organizing these rules. Specifically, they assert that production rules should be organized into small, easily managed modules, and that frames provide a means of organizing and indexing modular collections of production rules according to their intended usage.[9]

Advantages and Disadvantages of Environments

The primary advantages to be found in the employment of an expert systems environment as a development tool are listed below:

---

[9] In rule-based shells such as EXSYS, such management—or partitioning of rules—may be achieved to some degree via the use of a blackboarding capability. Alternatively, the guidelines discussed in Chap. 4 may (and should) be used in rule-based expert systems so as to more easily organize and control rule sets.

- They are considered, by many, to represent professional tools for serious implementation (i.e., as compared to many expert systems shells that are, in some quarters, perceived as "toys").
- They generally employ LISP machines with all the accompanying support (e.g, superb graphics, supporting packages, and the flexibility to use multiple paradigms).
- They are, for the most part, based on LISP and, as such, are attractive to those convinced of the superiority of this language.
- They may be used for much more than just expert systems development, that is, they are full-fledged environments for general AI research and development.
- They generally employ frames (and/or objects) and are thus attractive to those who prefer this mode of knowledge representation, or who wish to take advantage of its flexibility.
- A number of environments (in some cases, even from a personal computer resident system) have the ability to access *mainframe* resident databases.[10]

It may also be noted that, once a firm is committed to the substantial investment for such a system, it often becomes difficult to conceive of moving to another system, even if that system is much lower in price.

Some of the disadvantages of environments include:

- High initial cost
- High overall costs (e.g., training, hardware support)
- Complexity and amount of training and experience required

We feel that the last factor represents the single, major disadvantage of environments. If one is to use an environment, and to draw any real benefit from its supporting facilities, then one must spend a considerable amount of time in training. Further, since such packages are so complex, you must, almost continually, work and practice with the package to not forget what you have previously learned. As a consequence, such packages drain the time and resources that we believe could be better spent on knowledge-base acquisition and representation.

Our personal opinion is that, unless the expert system to be developed is truly one of major scale (e.g., many hundreds or thousands of rules—plus the need for direct access to a considerable amount of additional support), the power of the approach provided by an environment is most definitely abrogated by its complexity. Such packages simply tend, all too often, to divert the knowledge

---

[10] Nexpert Object, an expert systems environment that runs on either IBM (or compatibles) or Macintosh personal computers, has the ability to access mainframe resident databases and was available for $5,000. However, like most other environments, the system is complex and training requirements are far from minimal.

engineer's attention away from the fundamental tasks of knowledge acquisition and representation and toward areas that are nice, impressive, but not always necessary. Thus, for expert systems of modest size (i.e., a few hundred rules—as is quite typical for those of most recent expert systems implementations), it would seem considerably simpler, as well as more cost effective, to employ a rule-based expert systems shell.

EXPERT SYSTEMS HARDWARE

The primary hardware for the support of expert systems development is, of course, the digital computer. Until relatively recently, few analysts voiced any hope for the development and implementation of expert systems on anything other than mainframe, or large-scale, dedicated computers. The development of powerful personal computers, along with the use of more conventional programming languages, has proved them wrong. So wrong, in fact, that several companies that were established solely to provide dedicated computers for expert systems have either failed or are, at the least, finding it necessary to redirect their efforts.

Early work in expert systems was performed on either mainframe computers (e.g., IBM 370 class) or super-minis (e.g., VAX computers), and efforts still exist on such systems. However, the use of the LISP language, in particular, motivated a shift to so-called LISP machines, dedicated computers for AI research and development. Since such computers can do far more than just implement LISP efficiently, a better description would be "AI workstations." Some of the typical characteristics of the AI workstation include:

- A single user
- Hardware specialization such as special high-speed LISP processors that permit the faster running of LISP programs
- Large memory (e.g., 30 MB of RAM and a 474 MB hard disk)
- Bit-mapped display (e.g., 1024 × 808 pixels) to enhance the use of graphics and permit the use of windows and other tools
- Specialized keyboards
- Mouse and windows
- AI-specific software such as LISP, PROLOG, sophisticated editors, and expert systems development tools such as KEE and ART

Representative of such AI workstations are

- LAMDA (LISP Machine, Inc.)
- Symbolics 3600 series (Symbolics, Inc.)
- Xerox 1100 series (Xerox, Inc.)
- Explorer (Texas Instruments)

There are three main difficulties with such AI workstations. First, they are simply considered too expensive by the typical customer (one of the few exceptions has been the military where, as evident by their $600 toilet seats and $5000 coffee makers, cost may be no object). The second is that such workstations, and the LISP language, were simply incompatible with the installed computer system of the typical customer. And third, the time and training required to become proficient on such workstations were considered excessive.

More recently, a number of firms (e.g., Apollo, Sun Microsystems) have begun to offer *general purpose* workstations with features comparable to those on AI workstations, and at a cost of only a fraction of that of the typical AI workstation (e.g., $10,000 to $20,000 versus $50,000 to $100,000 or more for the AI workstation). Further, the recent versions of personal computers (such as the Macintosh II, Macintosh IIx, and the 386 and 486 chip-based machines) look, for all practical purposes, very much like workstations themselves—and can be purchased for under $10,000. As an example of just how serious a competitor such microcomputers are, consider the following statement that appeared in *IEEE Expert* [IEEE, 1988]:

> The Macintosh II is replacing specialized AI hardware in many university, aerospace, and other industrial laboratories, according to Dan Shafer, editor of the *Intelligen: Systems Analyst* newsletter. "The Macintosh II is being taken very seriously indeed by AI researchers at many levels and at a broad range of institutions," Shafer says in a recent issue of the newsletter in which he examines Hypercard and its relationship to expert systems on the Mac.

Further, if one must have a workstation with LISP capability, it is now possible to purchase a card to install on your personal computer that will permit performance levels that rival that of many AI workstations. One such offering is the result of a joint venture between Apple Computer and Texas Instruments. This resulted in the microExplorer. The microExplorer incorporates a specially developed coprocessor board that includes Texas Instruments' LISP microprocessor and the TI Explorer software environment for the development and delivery of AI applications. Explorer and Macintosh II operating environments execute concurrently. The list price for this system starts (as of the summer of 1988) at $14,995 which includes a Macintosh II with 2 M bytes of memory, a 40 Mb disk, the microExplorer processor with 4 Mb of memory, and the Explorer runtime software environment.

SUMMARY

There are a number of alternatives to rule-based expert systems shells. At one level, the more courageous among us may decide to construct the entire expert system from scratch, using either a *conventional* computer programming language (e.g., FORTRAN, C, Pascal) or an *AI language* (e.g., LISP, PROLOG). However, unless you simply cannot find an existing package to encompass your problem (an increasingly remote likelihood), such an approach is akin to reinventing the wheel. Further, if you are employing such an approach simply in order to save money, you had better think again, that is, unless you place no value whatsoever on your own time.

At the opposite extreme, one may choose to employ an expert systems environment. Such a selection is reasonable only if you need, and can benefit from, the additional features available through such software. Further, you need to carefully evaluate the hidden costs of such an approach, that is, the costs associated with training and implementation. Even more important, be careful that, when using environments, you do not lose sight of the two most vital aspects of expert systems development: knowledge acquisition and knowledge representation. No matter how much money you spend, or how fancy your system, or how colorful and dazzling its graphics, its ultimate performance depends on the knowledge that has been acquired, and the manner in which it has been represented and organized.

We are by no means advocating that the reader avoid either the use of computer languages or environments. A truly top-flight knowledge engineer should have familiarity with all modes of expert systems development. At some time or another, you will most likely wish to extend your capabilities beyond those encompassed by rule-based shells, and environments offer such extensions.

Environments, through the employment of frames, can enhance knowledge organization. Further, such tools can deal with certain situations that might be difficult for rule-based shells to handle. In particular, frame-based knowledge representation permits one to deal readily and effectively with situations that might be best described as involving multiple objects or instances. That is, the typical rule-based expert system consultation session involves a single object. For example, consider an expert system for the determination of new-product marketing (i.e., such as PEORIA from Chap. 1). If we are only dealing with a single product at a time, a rule-based shell may suffice. But, when we are considering the introduction of several projects, and where each of these must compete for advertising dollars, production time, storage space, and so on, the complexity of the problem has been enormously increased—and we might find that frames serve to alleviate the situation.

As another example in which rule-based shells have difficulty, consider a problem whose structure may actually change as intermediate conclusions are reached. For example, we may wish to construct an expert system for use by the police in their search for missing persons. As evidence mounts, we may proceed along one line of reasoning. However, a single new piece of evidence may cause us to abandon that approach and direct our attention toward other possibilities. In such an instance, some facts must be forgotten and certain beliefs revised. We have approached problems somewhat akin to this in this text (i.e., the robot landscaper of Chap. 6) through the use of contrived rule bases (e.g., the use of partitions). An environment may offer an easier approach if we are willing to devote the initial time and effort necessary to understand how to employ such a facility.

In the not-too-distant future, we will undoubtedly look back at the expert systems software and hardware of the late 1980s and early 1990s as being crude, limited, and inefficient. The tools that exist today have, most certainly, considerable room for improvement. Despite this, we now have tools of sufficient power and flexibility to construct very important, very useful expert systems for real-world implementation. And this may be accomplished at a very reasonable cost—*if* one is familiar with the actual scope and limitations of existing software and hardware. Hopefully, this chapter has served to somewhat enhance the reader's familiarity with these areas.

EXERCISES

10.1. On noticing that you are employing an expert systems shell, a friend of yours states that, unless you have actually constructed an entire expert systems software package from the ground up, you will never really understand expert systems. What is your response? Is there any truth to your friend's statement?

10.2. The same (know-it-all) friend as encountered in Exercise 10.1 also notes that it is unwise to rely on commercial expert systems software as such packages come and go, while development languages (particularly LISP) have been with us for decades. What is your response? Is there any truth to your friend's statement?

10.3. List those situations, in the construction of an expert system, for which (*a*) a development language might be preferred, (*b*) an expert systems shell might be preferred, (*c*) an expert systems environment might be preferred.

10.4. You have been asked to review an expert systems shell. What attributes (e.g., computational speed, maximum rule-base size) do you think should be used in the evaluation? Why?

10.5. How does the choice of an expert systems development package influence the choice of hardware, and vice versa?

10.6. Develop a small rule base that may be used to determine the precise manner in which an expert systems shell deals with disjunctive clauses. That is, if the shell permits disjunctive clauses (remember that not all do), then does it evaluate all of the disjunctive clauses or stop as soon as one is satisfied?

10.7. Develop a small rule base that may be used to determine the precise manner in which an expert systems shell deals with the employment of *unknown* as a user response.

CHAPTER
11

IMPLEMENTATION

CHAPTER OVERVIEW

In this chapter, we shall attempt to address certain fundamental issues concerning the implementation of an expert system, as well as its potential impact on the organization. The discussion is based on the following assumptions:

- The use of the expert systems approach has been justified.
- A prototype expert system has been developed and its performance validated.

The specific implementation issues to be addressed include

- The development of a schedule for implementation (a milestone chart)
- Personnel issues
- The utilization of outside support
- Software and hardware considerations
- Provisions for monitoring and maintenance
- Documentation

Let us begin by examining the role played by the milestone chart.

THE MILESTONE CHART

Implementation of an expert system, in most respects, is really not much different (other than, perhaps, for the expectations that may be held) from the implementation of any other computer-based tool for decision analysis. Further, unless the expert system to be implemented is quite trivial, its incorporation into the day-to-day workings of the organization has the potential for being a traumatic endeavor—at least if it is not accomplished in a careful, systematic, and professional manner. As such, the development of a milestone chart in support of implementation can do much to help alleviate potential problems and to generally smooth the transition process.

A milestone chart serves to indicate both the schedule and relationship of all (pertinent) activities pertaining to implementation, and helps to isolate potential problem areas. It also serves as a means to list the goals that support implementation, and to note when these have been, or should be, accomplished. Also, in the event of problems (e.g., the need to finish the project earlier than expected, or a reduction in personnel and/or funding), one may use such a chart to determine how to best reallocate one's efforts.

Various forms of milestone charts and supporting methodologies have been developed including PERT (Program Evaluation Review Technique), CPM (Critical Path Method), GERT (Graphical Evaluation Review Technique), and VERT (Venture Evaluation and Review Technique). In essence, each of these approaches attempts to systematically represent the network of activities required to complete a project. Then, through the use of various analytical procedures, one may determine those activities whose performance is critical to the scheduled completion date, as well as those whose performance may be delayed without an impact on the schedule. In order to keep things relatively simple, we shall focus on the utilization of a deterministic PERT chart approach. It is not, however, our intent to provide anything approaching a comprehensive introduction to this technique. The reader who desires further details may consult the references [Ignizio and Gupta, 1975; Moeller and Digman, 1978; Moore and Clayton, 1976; Wiest and Levy, 1969].

The first step in the development of the PERT chart is to identify the following problem parameters:

- The primary goals or milestones associated with the implementation of the expert system (e.g., software and hardware to be installed and tested, personnel training to be accomplished, trial runs to be executed)
- The activities that must be completed to achieve each milestone and their estimated duration

Once these factors have been determined, one may use the PERT procedure to construct a PERT chart (i.e., a special type of network) and to determine the so-called *critical path* for that network. The length of the critical path serves to determine the minimum time in which it will be possible, without resorting to extraordinary measures, to complete the implementation. Further, any reduction in the scheduled completion date may only be achieved through a corresponding reduction in time for one or more activities on the critical path.

In order to demonstrate the development of a PERT chart in support of expert systems implementation, let us consider a simple, illustrative situation. In particular, let us assume that we have developed and validated the performance of a prototype expert system for use in the scheduling of processing tasks at a petrochemical refinery. Presently, the scheduling activities are accomplished by two human process schedulers, and these individuals have previously served as the domain experts during the development of the prototype. We are now ready to attempt the implementation of the expert system within the organization.

Let us utilize a parallel implementation procedure. That is, we shall implement the expert system in parallel with the existing, manual system. Initially, both systems will be used to develop a schedule, but only that one developed by the human schedulers will be used in the establishment of the actual schedule. Once we have gained confidence in the schedules developed by the expert system, we will then use it to generate the actual schedules, and the human schedulers will be used only as a backup in the event of problems (or to deal with situations beyond the expertise captured by the expert system's rule base). Finally, if all goes well and we are sure the expert system is working properly, we might even terminate the backup role of the human schedulers and rely exclusively on the expert system (with the exception, again, of those special situations that are not dealt with by the knowledge base). The advantages of such an approach to implementation should be obvious.

In Table 11.1, we have identified the primary components of the PERT chart for the implementation of our expert system. Each one of these components represents a specific implementation *activity*. In this table, each activity has been coded by a letter, as follows:

A = development of the milestone chart itself
B = installation and checkout of all software and hardware
C = personnel training and familiarization
D = establishment of all monitors and controls
E = parallel testing of the expert system
F = implementation phase satisfactorily completed

The associated PERT network diagram for this problem is presented in Fig. 11.1. Along each branch are the activity indicators (i.e., A, B, C, etc.) and

TABLE 11.1
Implementation activity matrix

Activity	Estimated duration (working days)	Immediate predecessor(s)
A	5	None
B	20	A
C	28	A
D	15	A
E	60	B,C,D
F	0	E

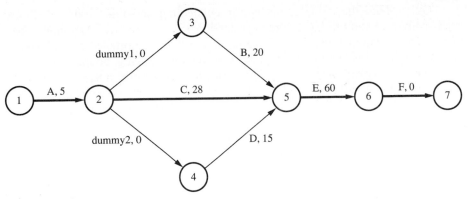

FIGURE 11.1
PERT network for example.

the required time for that activity. Thus, the branch from node 4 to node 5, for example, represents activity D—which requires 15 days to complete.

Note carefully that Fig. 11.1 includes two additional *activities*, the two *dummy* activities labeled as dummy1 and dummy2. These dummy activities simply serve to satisfy the requirement (i.e., of the PERT approach) that each activity must have a unique pair of originating and terminating nodes. As may be noted, if these two dummy activities were not used, activities B, C, and D would be represented by the same node pair (i.e., nodes 2 and 5).

The activities along the path from 1-2-5-6-7 (as denoted by the heavy lines) are those on the critical path. The length of this path, in turn, is 93 days.

The PERT network illustrated in Fig. 11.1 has been developed simply for the sake of illustration. However, it should be noted that even this simple network can provide immediate, visual information in support of our implementation effort. For example, we could reduce the time required for implementation by 5 days by simply developing the PERT chart *before* we complete the validation of the prototype. Also, personnel training could (and, in fact, *should*) be started during the prototype development phase, and this should also reduce the time required for implementation. Further, whenever one wishes to shorten the schedule duration, any additional resources should be committed only to those activities on the critical path. Consequently, we may wish to redirect some of the resources devoted to activities not on the critical path (i.e., B and D) to those on the critical path.

PERSONNEL ISSUES

The successful implementation of an expert system will ultimately depend on its acceptance by certain key personnel within the organization. This group includes those individuals who will be responsible for using the expert system, those whose duties will be directly or indirectly impacted by the expert system, and those who were originally responsible for the decision to acquire and implement such a system.

Quite often, the focus of discussion concerning personnel issues is almost exclusively centered on that of the organization's work force and, perhaps, labor-management relationships. One can hardly dismiss the importance of such issues; however, unless one has the complete, *unwavering* backing of management, it will be extremely difficult, if not impossible, to successfully introduce an expert system into the organization. *And the lack of such support will most definitely be sensed by the work force.* Consequently, first and foremost, you must have the support of management.

Equally as important as management's support is the support of the expert systems user(s). All too often, expert systems are constructed primarily to satisfy the needs of the domain expert and, possibly, the knowledge engineer. However, once the expert system is implemented, someone else must be able to access and use the package. And it is ultimately this class of user who must be satisfied. If not, they are going to complain and, if these complaints have a basis, management may well reconsider their support of the expert system. Both the prototype development and implementation phases simply must take the needs of the expert system user into consideration.

Another factor to be considered and dealt with is that of the handling of the organization's most adamant critics of the expert system. It would be a rare occurrence, indeed, to encounter only positive support of the introduction of an expert system (or, for that matter, the introduction of virtually any new approach). More likely, certain "adversaries" are going to surface within the organization; and one should be prepared to deal with such individuals (and, perhaps, groups) in such a way as to either mitigate their fears and suspicions or, failing this, making it clear that management intends to implement the expert system with or without their full support (if that is indeed the case). In any event, one must not be so naive as to think that everyone will rush to embrace such a system, regardless of how terrific you may believe it to be.

Oddly enough, the most vociferous critics of the introduction of an expert system are sometimes those within the organization's computer center and/or computer science group. Mixing artificial intelligence/expert systems with *more conventional* computer science is quite often like trying to mix oil and water. Specifically, the methodologies used are different (i.e., heuristics versus algorithms), the programming languages may be quite different (e.g., LISP, PROLOG, or an expert systems shell rather than COBOL or FORTRAN), and even the computers on which the system is implemented may be unfamiliar. And fear of the unknown should not be underestimated.

Certain groups within the organization may well view those implementing the expert system to be in competition with them for scarce company resources, as well as for the all so important attention of higher management. As one member of a firm's computer staff once told us, "These people (i.e., the individuals implementing an expert system) are stealing our thunder." Further, if the expert system is to be implemented on microcomputers or workstations, the organization's computer center may see this as a serious dilution of its responsibilities and influence. However, unless one wishes to continue doing business as usual, management must step in and make it crystal clear that the expert system (and the associated software and hardware) has their support.

Potential expert systems users as well as the general work force are also likely to view the introduction of an expert system with some degree of apprehension. Such misgivings may, on occasion, be well warranted. Automation of any form, despite the sometimes glib assurances of its advocates, is likely to have an impact on certain personnel. And, in some instances, this can even extend to the abolition of certain jobs. On the other hand, new jobs or, at the least, new and enhanced job responsibilities are more likely to be the result of the implementation of expert systems. Generally, these new situations are interesting and challenging. Most typically, the introduction of an expert system will result in the automation of routine decision making and permit the domain expert, as well as other users of the expert system, to address other, more pressing issues. In our experience, the best way to deal with the reservations of the work force is to take the time and effort to meet with them and explain just what the expert system will and won't do, and the manner in which the package will be implemented and used.

OUTSIDE SUPPORT

At this time, very few organizations have a staff of knowledge engineers. Even among those that do, all too often the training and level of expertise of such individuals leave much to be desired.[1] As such, the use of outside support, typically in the form of consultants, provides one possible means for the development, validation, and implementation of an expert system.

When one uses outside support, it is essential that such support extend through, and sometimes even beyond the implementation phase. Unfortunately, however, it is not uncommon for such outside support to be terminated just when it is most needed. In some cases this results from lack of interest (or is it perhaps the apprehension of actually being involved in such a "messy" phase as implementation?) on the part of the outside consultants. We must admit to being particularly skeptical of those individuals who shy away from the implementation phase. More often, however, it is the firm itself that terminates the outside support prior to implementation. For the most part, the only rationale for this seems to be either the perception that money can be saved by having the firm's own personnel proceed with implementation, or that the implementation phase is a predictable, smooth process. Unless one is extremely lucky, this view is shortsighted and dangerous. If the implementation phase runs into problems, as it likely will, one needs the support of those who understand the system and who have experience in previous, successful implementations.

---

[1] For example, in some of the organizations that we have dealt with, knowledge engineers have been "created" by assigning the responsibility of expert systems design to computer programmers. The results were, as one might expect, not particularly encouraging.

The utilization of outside support is not without its own set of problems and drawbacks. At this time, one company after another is being set up in an attempt to exploit the artificial intelligence/expert systems market. And, among existing consulting firms, we see an extensive redirection of effort toward what is seen to be a booming and lucrative area. As such, the level of competence of such firms and individuals covers an extremely wide range. In too many cases, there really is no competence per se. Thus, as the old saying goes, "one pays his money and takes his choice."

The best advice that we can think of, with regard to selecting outside support in expert systems, is to move carefully and deliberately. Further, this matter should be addressed well prior to the initiation of the development (or purchase) of an expert system within the organization. If no one within the organization has any reasonable level of expertise in expert systems, then the very first step that should be taken is to assign one or more employees the task of becoming educated in the area, at least to the degree that one has an understanding and appreciation of both the scope and limitations of expert systems. With such a foundation, any decisions on outside support may be made on a much more intelligent basis. In the following chapter, we shall deal with the training and education of knowledge engineers. This material should also contribute toward a more effective decision on external support, or the development of an internal knowledge engineering group.

SOFTWARE AND HARDWARE CONSIDERATIONS

It is not necessarily the case that the same type of software and/or hardware used in prototype development will be used in the final implementation of an expert system. In fact, those advocating the use of LISP machines often do so on the basis of using LISP, and the LISP machine, in (just) the development phase. Once this is complete, the software is transformed into some other, more conventional (and transportable) form, such as C or FORTRAN, for use in actual implementation. As a result, much of what has been learned in development may not be directly transferable to practice.

However, even when the software and hardware for development coincide with the choice for implementation, there are certain issues that still must be considered. These include

- The decision of whether to use a mainframe computer, minicomputers, micro-computers, dedicated workstations, or some combination of these
- The number and types of microcomputers or workstations to use
- The networking of such computers
- The provisions for direct access to supporting databases or other external programs
- The potential for using the computers for other tasks (i.e., either other expert systems or other, more conventional support)
- The useful life of the software and hardware, and their potential for upgrading

The last factor is one that should be given particularly serious consideration. The evolution of computer hardware has been quite astonishing. Only a few years ago, the use of personal computers in expert systems development was considered impractical if not impossible. Today, the performance of personal computers rivals that of the mainframe systems of only a decade ago and, in fact, the majority of the expert systems developed today are being done so on personal computers. Moreover, the rapid evolution of computer systems continues.

The evolution of expert systems development packages has been almost as dramatic as that seen in the hardware. In but a period of a few years, packages that once captured the majority of the expert systems audience have seen their market share virtually disappear. Moreover, new packages are being developed at a remarkable pace—and new features are being added to existing software (e.g., such as hypertext, multiple inference paradigms, enhanced database access, pull-down menus, and overlapping windows).[2]

Consequently, any decision to purchase a machine or software package today should be carefully made in light of one's best estimate of what will be available tomorrow, and how amenable the system we purchase now may be to upgrading. Actually, these changes in hardware and software support, and the uncertainty surrounding such support, are yet another reason for focusing the bulk of one's attention on the development of the knowledge base. It is the knowledge base, not the hardware or software, that serves to capture the specific area of expertise within the organization, and which serves to represent the most important tangible asset resulting from any expert systems development effort. As such, and in light of the rapidly changing status of supporting hardware and software, one should design the knowledge base with portability in mind. And this is just another way of saying that the knowledge engineer should, in general, avoid designing his or her rule base solely about any specific set of hardware and software.

PROVISIONS FOR MONITORING AND MAINTENANCE

All too often, the aspects of monitoring and maintenance of an expert system are either totally overlooked or are afterthoughts [Newquist, 1988b]. However, consideration of the monitoring and maintenance of an expert system should really start with the very initiation of the expert systems conceptualization and development phases. Specifically, *any expert system should be designed with monitoring and maintenance in mind*. The reasons for this include the following:

- The organization within which the expert system resides will change with time, resulting in the need to change the expert system accordingly.

---

[2] While in some cases these new features serve to enhance the expert systems development effort, in others they only seem to complicate the process. This is just another example of the fact that *more* is not always *better*.

- New procedures and concepts will develop over time, and these should be incorporated into the knowledge base of the expert system.
- New, improved implementation tools (i.e., both software and hardware) will emerge—and a move to such tools must be considered.
- Over time, both minor and major problems are likely to be encountered in the use of the expert system, necessitating changes to its design.

Simply put, an expert system for anything other than a trivial problem is not a static entity. Rather, it is (or should be) an evolving system and, as such, must be constantly monitored to determine when and how modifications to the system should be made.

Typically, the human expert from whom the expert system's knowledge base was developed is a good candidate for the monitoring and maintenance responsibility—if he or she has received sufficient training in expert systems. Whenever the performance of the expert system is in question, this is a signal that the system should undergo an off-line evaluation. If the results indicate that the expert system needs to be revised, a maintenance effort should then be initiated. Such an effort is often a relatively minor exercise, perhaps the addition of a few rules to the knowledge base (plus a thorough validation of the performance of the revised system). To assist in this effort, one should have a set of validation exercises through which a complete evaluation of the expert system may be accomplished. In fact, the development of a package of such exercises should be a required part of the expert systems development and implementation effort.

One can draw an analogy as well as a contrast between monitoring and maintenance of expert systems and that of databases. Virtually everyone realizes that a database is a dynamic, changing entity. The purpose of the database is to reflect the facts within the organization at any given time. Obviously, if this purpose is to be accomplished, constant, or at least frequent updating of the database is necessary. The same thing is true for an expert system. Its knowledge base should reflect two aspects in particular: (1) the most current set of heuristic rules for solving the problem under consideration, and (2) the present status and configuration of the organization in which the system is implemented.

There is, however, one big difference. It is an easy procedure to modify a database, but a much more difficult as well as delicate procedure to change a knowledge base. In some instances (as we discussed in Chap. 4 and illustrated at the conclusion to Chap. 6), the addition or deletion of only a single rule can change, dramatically, the performance of the expert system. This is particularly true for rule bases that are not well designed. Any such changes must therefore be dealt with in a careful, systematic manner, and then be subjected to a thorough validation.

DOCUMENTATION

A key factor in implementation, as well as continued maintenance of the expert system, is its documentation. Whether one purchases or develops an expert system, documentation should be a prime consideration. To be acceptable, documentation should be

- Clear
- Current
- Complete
- Concise
- Accompanied by illustrative examples
- Accompanied by validation exercises

Further, such documentation should itself be validated. That is, those who are expected to use the expert system should be provided the documentation for comments and criticism. If any ambiguities exist, if steps are missing, or if any other problems are noted, the documentation should be modified to eliminate the difficulties encountered.

Another good way to evaluate documentation is to give it to someone who has not been associated with the development of the expert system, but who understands what it is to accomplish. If this individual can both operate and understand the system, using only the documentation provided, the documentation is probably adequate.

However, in addition to documenting the operation of the expert system, the documents should also cover its installation, verification tests, and all aspects of monitoring and maintenance. Based on an examination of existing expert systems documentation, the attributes cited here are those that should be satisfied in theory, rather than those characteristic of most existing documentation.

SUMMARY

A considerable portion of the existing literature on expert systems is focused on topics of more interest to the academician than to the practitioner. Further, even when expert systems applications are discussed, it may be difficult to determine whether or not they were actually implemented. There is, in fact, a relative paucity of articles concerning the aspect of implementation. Among these, we once again recommend the article on DMES [Cochard and Yost, 1985] as previously discussed in Chap. 8. Cochard and Yost describe, in some detail, their decisions about the development language and hardware. Further, they describe the manner in which DMES was gradually introduced into operation, first as simply a backup during training exercises and then, ultimately, as the sole air-cargo loading system for the Grenada invasion.

However, when all is said and done, the implementation of an expert system is not much different from that of most any other sophisticated software package for automation—and the lessons learned in these other areas generally hold true for the expert systems area. In particular, one must consider such factors as the implementation schedule, the impact of the system on the organization and its personnel, the supporting personnel, software and hardware, and the inclusion of provisions for the monitoring, maintenance, and documentation of the package. If due consideration is given to these issues, and if the expert system itself has been properly selected and designed, the probability of its ultimate acceptance within the organization should be substantially increased.

To conclude this chapter, we have provided a short list of do's and don'ts with regard to expert systems implementation:

- Do consider the aspect of implementation (including documentation, monitoring, and maintenance) as early as possible in the expert systems development effort
- Do seek to obtain the full, unwavering support of management
- Do seek the support and involvement of those who will work with, or who will be impacted by, the expert system
- Do be candid and forthright in your estimate of the time and resources required for implementation
- Don't underestimate the importance of implementation
- Don't delude yourself into believing that the implementation phase is going to be trouble free
- Don't expect to be able to please everyone
- Don't expect the expert system to be able to handle every situation encountered, and don't feel that such a fact reflects any *failure* on the part of the expert system
- And, *in particular*, don't forget the user

EXERCISES

11.1. A firm has asked you to build an expert system for them that will provide decision support for a task presently being accomplished by a domain expert. You are satisfied that the problem is one that should be addressed by expert systems and that the domain expert who has been identified will most likely be an easy person to deal with and, in fact, you are confident that the system can be built. However, the firm has stipulated that you will not receive any compensation whatsoever unless the expert system developed meets its expectations. Would you consider taking on this job? Why or why not?

11.2. Discuss any ideas that you may have for obtaining the cooperation of the work force during expert systems implementation. Also note some of the things that should be avoided during this phase.

11.3. Can you think of some ways in which a student taking a formal course in expert systems might be able to, realistically, gain a better appreciation of expert systems implementation?

11.4. An expert system has been developed with the close cooperation of a domain expert. This particular expert exhibits a level of performance far above that of his colleagues, and seems to have an almost incredible insight into the problem under consideration. The domain expert has insisted that the expert system satisfy his expectations, and that it perform precisely in the manner he employs in his decision making. What problems do you foresee?

11.5. What factors play key roles in the maintenance of any expert system?

CHAPTER
12

STAFFING
AND TRAINING

CHAPTER OVERVIEW

In this chapter, an attempt will be made to address the issues of the staffing and training of a (hopefully, successful) knowledge engineering group—as well as consideration for its placement within the organization. Such a group may either be a part of a much larger organization or the main focus of, say, a consulting firm. In either case, our fundamental assumption is that the primary purpose of such a group is that of actually using the methodology of expert systems to find solutions to real-world problems (i.e., as opposed to a group that is strictly involved in expert systems research and/or the development of supporting software).

The material in this chapter may deviate somewhat from that found in other texts on expert systems. And this is a result of a difference in emphasis. As we have discussed, there are two distinctly different views of expert systems. One perspective tends to feature the role of the computer and computer programming and, in fact, sees expert systems as essentially an alternative to *conventional* computer programming.

The second perspective, and the one to which we have adhered within this text, emphasizes the development of the model of the expert systems knowledge base—where the computer plays a strictly supporting role. As such, we view the primary pursuit of the knowledge engineer to be that of the identification of those problems that might be approached by expert systems, the acquisition of the domain knowledge, the representation of this knowledge using a model of the heuristics employed by the domain expert, the validation of this model, and the successful implementation of the expert system within the organization.

In particular, we see the tasks of knowledge acquisition and representation as the two factors that are most vital to the construction of an expert system. And it is upon this perspective that the issues discussed in this chapter have been based.

APTITUDE, EDUCATION, AND TRAINING

Undergraduate education, mainly because of the often severe constraints imposed by curriculum accreditation requirements, seldom permits the inclusion of what we would consider to be sufficient, formal course work in support of knowledge engineering.[1] However, those with undergraduate degrees in engineering may hold some slight advantage in one respect. This is because of the long history of the use of heuristics in engineering, and of its focus on pragmatic, real-world problem solving. In addition, students in science and engineering usually, although not always, receive considerably more exposure to computers as a tool for problem solving.

We would not, however, advocate that one restrict the field of knowledge engineering to just those with a degree in science or engineering. Far more important than the discipline are such matters as

- A familiarity with a wide range of tools in support of decision analysis
- A personality compatible with that needed by a successful knowledge engineer
- A sincere interest in attempting to gain an understanding of how things work, and of problem solving in general—and the ability to construct models that serve to represent this understanding
- The desire and ability to work on a variety of diverse, and generally *unstructured* problems

We shall address the latter three aspects first.

A knowledge engineer must work closely with other people, including domain experts as well as other knowledge engineers. As such, a knowledge engineer may well be technically brilliant, and still fail miserably because of his or her inability to deal effectively with others. If one cannot gain the respect and confidence of the domain expert, there is little hope of drawing out the expert's knowledge base. As a result, those with degrees in engineering and science may (sometimes) actually be at some disadvantage. Such individuals are sometimes, by nature, introspective. As such, they may find it difficult to deal effectively on a one-to-one basis with others.

A knowledge engineer should, we believe, be curious by nature. When we say that he or she should be interested in how "things" work, we are speaking on the broadest level. These things may be machines, organizations, or even people.

---

[1] Our discussion here pertains primarily to students in the United States.

Most of all, the knowledge engineer must be interested in solving problems, where there are no restrictions placed on the particular type of problem. A good knowledge engineer simply wants to find a way to gain the necessary insight and understanding that will permit him or her to obtain an acceptable answer to a given problem. As a result, we do not feel that specialists, in general, make for very good knowledge engineers.

The ideal candidate for a knowledge engineer should also be able to deal effectively with ill-structured problems. Unfortunately, traditional schooling does little to prepare one for this aspect of real-world problem solving. It may even, in fact, hamper one's abilities in this area. Students typically encounter their problems as exercises at the end of each chapter (as in this text), or as examination questions periodically throughout the term. Because of the limited time for responding to such questions, few instructors would dare attempt to incorporate a problem that was not well-structured, clear, concise, and directly amenable to one of those methods most recently covered in the course. In the real world however, things are not nearly so tidy. One does not, for example, encounter a linear programming problem one day, a queuing problem the next, and a *lease versus buy* issue the day after. Rather, most often all that is known is that *something* is wrong, and that *something* is most likely to span several methodologies, if not several disciplines. Some individuals cope well with such problems, many others don't.

Given that the knowledge engineering candidate works harmoniously with others, communicates effectively, has a keen interest in problem solving, and has minimal difficulty in dealing with ill-structured problems, let us now consider the types of tools that such an individual should, ideally, possess. As mentioned earlier, these tools are seldom a part of an undergraduate program. Further, very few graduate programs effectively address the development of a knowledge engineer. Thus, by default, knowledge engineering is often accomplished through a program in artificial intelligence/expert systems—usually taught within a computer science program. While such programs may certainly provide the background necessary in the development of supporting expert systems software, they do not, in our opinion, adequately support the education and training that should form the foundation of a knowledge engineering program that is directed, primarily, toward expert systems *application*.

In terms of the tools that provide the foundation for knowledge engineering, we have attempted to categorize these into (1) those that we believe are absolutely essential, and (2) those that we believe are helpful, but not absolutely necessary. We have prepared a listing of these tools below, where we have assumed for the sake of discussion that the individual under consideration has an undergraduate degree in business, engineering, or science.

Essential Tools

We believe that any effective knowledge engineer *must* be familiar with the following concepts and methodologies:

- Expert systems (including a familiarity with shells and environments, but focused on the topics of knowledge acquisition, knowledge representation, search, inference, validation, and implementation)
- Heuristic programming
- Applied statistics
- Cluster analysis
- Discriminant analysis (i.e., pattern recognition or classification analysis)
- Neural networks
- Combinatorial optimization (and, in particular, problems of sequencing and scheduling)
- Mathematical programming (i.e., the techniques of mathematical optimization such as linear programming, integer programming, nonlinear programming, dynamic programming, goal programming, and fuzzy programming)
- Relational databases (both as a source of data for the expert system and as they relate to frames and objects in expert systems environments)
- Management information systems (MIS) and, in particular, decision support systems (DSS)
- One or more computer programming languages (e.g., C, FORTRAN, BASIC, PASCAL, LISP, PROLOG)
- Cost analysis (or engineering economy or economic analysis)

As such, there are at least four disciplines in which many of these essential courses may be found. These are (1) operations research/management science [OR/MS], (2) industrial engineering [IE], (3) computer science [CS], and (4) electrical engineering [EE].

Table 12.1 indicates just which of these four disciplines is most likely to include courses in support of the tools listed above. An enclosed x ($\otimes$) indicates that such a topic is very likely to be offered in the respective program. A triangle (Δ) represents a topic that is only sometimes a part of such a program of study. If neither of these symbols appear, it is unlikely that the topic will be found within the respective program of study (at this time). It should be noted that this list was compiled through the examination of 1987–1990 catalogs of course offerings, within accredited programs, from some 30 universities within the United States, including both private and state institutions.

It would then appear that one should take at least 12 different courses (i.e., about 36 semester hours of course work). However, it is quite likely that the topics of applied statistics, programming languages, and cost analysis were a part of the undergraduate preparation of most candidates for a knowledge engineering position. This leaves nine courses, or about 27 semester hours of work. Most master's degree programs in the United States involve about 24 semester hours of course work (plus a thesis). As such, the 27 semester hour figure seems plausible, and could be further reduced by combining certain of the topics listed above. For example, cluster analysis, discriminant analysis, and neural networks might well

TABLE 12.1
Tools versus program of study

Tool	OR/MS	IE	CS	EE
Expert systems	⊗	⊗	⊗	⊗
Heuristic programming*	Δ	⊗	Δ	
Applied statistics	⊗	⊗	Δ	Δ
Cluster analysis	Δ	Δ	Δ	Δ
Discriminant analysis	Δ	Δ	Δ	⊗
Neural networks	Δ	Δ	⊗	⊗
Combinatorial programming	⊗	⊗	⊗	
Math programming	⊗	⊗	Δ	Δ
Relational databases			⊗	
MIS/DSS	⊗	Δ	⊗	
Programming language(s)	Δ	⊗	⊗	⊗
Cost analysis	Δ	⊗		

* Heuristic programming is seldom taught under that designation. Rather, it most often appears as a part (often a major part) of courses in such topics as scheduling, nonlinear programming, facility location, and facility layout.

be forged into a single course since each methodology deals with the same general type of problem. And this would leave us with 21 semester credit hours.

　　We could even further reduce the number of courses by combining expert systems with heuristic programming and including the topic of relational databases within a course on MIS/DSS. And this would represent, in our opinion, an absolutely minimal plan of study for preparation as a knowledge engineer.

Supplemental Tools

In addition to those topics listed above, it would be nice, but not absolutely necessary, for the knowledge engineer to have at least some familiarity with

- Artificial intelligence (including such topics as robotics, machine vision, and speech comprehension)
- Frames and/or object-oriented programming
- An AI language such as LISP or PROLOG
- Computer-based spreadsheets

JUSTIFICATION

Obviously, the list of necessary and supplemental topics listed above indicates only one opinion. Even with our own opinion and biases, we can still appreciate the fact that one could conceive of several alternative topic listings that might be just as suitable for one's preparation in knowledge engineering. However, we feel just as certain that any program that focuses primarily (if not exclusively) on AI languages, AI workstations, and expert systems shells and/or environments does not adequately prepare one to be an effective knowledge engineer. Unfortunately, a number of existing programs have just such a focus.

In order to more fully appreciate the rationale for our selection of the necessary and supplemental topics, let us try to briefly explain our reasoning. Referring to the listing of necessary topics first, expert systems has been included as the first course on this list. Rather obviously, if one is to consider oneself to have any expertise in expert systems, one should be familiar with the topic. However, our concern is that far too many individuals are under the impression that they have an adequate knowledge of expert systems when all they really have is a familiarity with the use and manipulation of an expert systems shell, or environment, or an AI language. As such, we would emphasize an exposure to the topics of knowledge acquisition, knowledge representation, search methods (i.e., chaining), inference, validation, and implementation. In particular, and first and foremost, the candidate knowledge engineer must learn how to develop models in support of knowledge-base representation.

Listed next is the topic of heuristic programming. Since the field of expert systems is itself based on heuristics, and since there is considerable debate about whether or not expert systems is not itself just another format for heuristic programming, we believe that it is imperative that one have a comprehensive introduction to heuristics and heuristic programming. Unfortunately, few programs offer courses either designated as heuristic programming or focused solely on that topic. Instead, the methodology may most often be found within a number of courses including scheduling, nonlinear programming, facility location, facility layout, and combinatorial programming. Scheduling, in particular, is a topic in which one is usually exposed to a variety of heuristic programming approaches.

Applied statistics is another topic that we consider essential—for two reasons. First, applied statistics provides the support necessary to better understand stochastic processes. Second, it provides the foundation necessary to better understand and more fully appreciate the related topics of discriminant and cluster analysis, two topics that we consider absolutely essential to the knowledge engineer's tool kit.

A personal survey of existing expert systems applications (or proposed applications) has convinced us that the *majority* of the problems solved by expert systems have, thus far, been problems that may be categorized as either those of discriminant analysis (e.g., pattern recognition, classification analysis) or cluster analysis (i.e., grouping or partitioning problems). As such, some of these applications could have been more appropriately and effectively addressed through the tools of either conventional discriminant analysis, conventional cluster analysis, neural networks, or a mathematical programming approach to discriminant/cluster analysis. However, discussions with practicing knowledge engineers indicate that most of these individuals are only vaguely aware of the tools associated with statistically based discriminant or cluster analysis, and none was aware of the existence of mathematical programming approaches to such problems. Had they been familiar with these topics, they would have been far better prepared to more effectively model and solve the problems that they encountered, and are likely to encounter in the future.

Combinatorial optimization (or combinatorial programming) is yet another topic that is essential. The vast majority of problems solved through expert systems are problems of combinatorics. Further, many of these are combinatorially explosive, and thus one must resort to heuristic solution methods. Courses in combinatorial optimization typically focus on the use of such heuristic methods, and particularly so in solving problems of layout, location, and sequencing/scheduling. They also provide an excellent exposure to the development of quantitative models. The knowledge engineer who is familiar with such approaches may well find them preferable, in some cases, to that of the construction of an expert system.

Mentioned already is the belief that mathematical programming may be used, in some instances, in the solution of problems of discriminant analysis or clustering. In fact, the use of this approach may offer certain substantial advantages over more conventional, statistics-based approaches or neural networks. This is because the mathematical programming approaches are nonparametric and, in addition, amenable to the consideration of virtually any number of side conditions [Cavalier et al., 1989; Freed and Glover, 1981a, 1981b; Ignizio, 1987a]. However, there is another reason to have a familiarity with mathematical programming. Specifically, it may be used to find an optimal, or perhaps near optimal, solution to a wide variety of problem types. Again, in the survey that we mentioned earlier, we discovered several instances in which the problem being solved by an expert system could have been solved, more effectively, by means of mathematical programming. For those readers who are not familiar with mathematical programming, let us emphasize that it has nothing to do, per se, with computer programming. Rather, the term *program* is used to refer to the *solution* to a problem. Programming is then the act, or methodology used to find the solution. Finally, mathematical programming simply means that one seeks to find the solution to a problem that has been represented by a mathematical model. As with combinatorial optimization, mathematical programming provides an excellent introduction to problem modeling.

Also listed is the topic of relational databases. A familiarity with computerized databases, and particularly relational databases, is an essential skill. An expert system simply cannot operate without data, whether those data are provided by the user or through access to an existing database. Most likely, at least some of the necessary data will reside in a database. While most expert systems shells and environments have bridges to such databases, one cannot fully appreciate nor understand such a process without at least some familiarity with databases. In addition, many databases provide control languages through which one can actually write a self-contained expert system. Further, we believe that anyone who uses or will use a frame-based mode of knowledge representation will benefit from an exposure to relational databases. The approach and concepts employed in relational databases are, quite often, very similar to those used in frame representations. In our opinion, the topic of relational databases is simply too important (and practical) to be neglected.

Next, MIS (management information systems) and DSS (decision support systems) are listed. As mentioned before, an expert system is perceived as just another tool for decision analysis. MIS and, in particular, DSS represent companion philosophies and approaches to the support of the decision-making process. As such, we believe that reasonably informed knowledge engineers must be aware of these concepts. In particular, they should have a familiarity with DSS.

That one should have a familiarity with one or more computer programming languages has been noted. Frankly, what language one selects is not important, and it certainly need not be an AI language. Our reason for listing this as a requirement is that the development of a skill in the use of a computer language should provide one with at least two things. First, one has a medium through which one may develop computer programs to actually solve problems. Second, through the learning process, one gains a better appreciation of how to interface with a computer, as well as its capabilities and limitations.

The final topic on our required list is that of cost analysis (or engineering economy, or economic analysis). In some cases, early expert systems were developed and (occasionally) implemented simply for the sake of having an expert system. The economics of such development and implementation were either not considered or were not of much concern. However, as the field matures, an expert system will have to be justified in precisely the same manner necessary for the justification of any other product, that is, according to its *cost effectiveness*. Successful knowledge engineers must know how to do more than just develop and implement an expert system, they must know how to sell such a system, and to do so on a sound, systematic basis. Thus, knowledge engineers must, we believe, be able to understand, use, and intelligently discuss the tools of economic analysis.

SOURCES OF EDUCATION AND TRAINING

Before details on thoughts of the best sources of education and training are presented, let us first describe one of the worst possible ways that one may approach this issue. Some time ago, within a government agency, it was decided to establish an in-house knowledge engineering group. One of the key decision makers in this endeavor had read precisely one book on expert systems, a slim paperback touting the various wonders that may be achieved through the magic of expert systems. With this as his single point of reference, a decision was made to establish a knowledge engineering group in the following manner:

- The knowledge engineers would be formed by retraining a group of computer programmers. At the time of the decision, these individuals were a part of the computer programming service department within the agency. That is, if someone needed a computer program developed (or debugged, modified, revised, etc.), these individuals might be called on to perform such a task.
- The manager of the knowledge engineering group was to be an individual with some previous experience in the management of computer programmers. He, himself, was considered to be a real *computer whiz*. He, too, was to be given some (formal) training in expert systems.
- The retraining effort consisted, for the most part, of attendance at a variety of short courses on expert systems and artificial intelligence. These courses ranged

from 1-to-5 day offerings. Most were presented by AI/expert systems hardware and/or software vendors. As such, they focused almost exclusively on the use of the particular product that the vendor hoped to sell to the organization.

The last time we checked, this "knowledge engineering" group had purchased a number of quite expensive AI workstations, considerable software, and was contemplating the attendance of yet more short courses and seminars. In addition, they now saw a need for the attendance, by most of the staff, of formal courses in LISP as offered by a nearby university (the software and hardware they had previously purchased were based upon the LISP language). They had, however, yet to even begin to develop an expert system for the solution of any actual problem.

We frankly do not expect much to come from this group. And we truly feel sorry for these poor souls who have so suddenly been thrust into a role for which they have had so little preparation, or inclination. It would be far better to simply forego even the thought of the development of an internal knowledge engineering group if this were the only way it could be achieved.

Our personal suggestions for the education and training of knowledge engineers are based on the points cited earlier within this chapter. First, a candidate for a position as a knowledge engineer must have the appropriate psychological profile (e.g., works harmoniously with others, communicates effectively, has a keen interest in problem solving, and has an affinity for dealing with ill-structured problems). If so, we can then consider education and training.

Ideally, the knowledge engineer should be exposed to the topics listed in Table 12.1 through formal course work. Such courses typically extend over a 10-to-16 week period, permitting one the time to give some thought to just what it is that is being taught, and just how it might be applied. Further, most formal courses of study in the listed areas include projects, as well as periodic feedback, from the instructor—something that is just not possible (at least to any sufficient degree) in a seminar or short course.

However, short courses, seminars, and self-education may be used to learn some portion of that which is necessary in support of one's role as a knowledge engineer. But this will only be true if

- The individual attending the course is highly *motivated*
- Sufficient time can be allocated for additional *self-study*
- The *right* seminars and short courses are attended

The first two aspects are self-explanatory. Let us then concentrate on the third. What do we mean by the *right* seminars and short courses? Let us try to answer this question by first citing those types of seminars and short courses that are, in our opinion, precisely the *wrong* ones to attend.

First, we would be particularly wary of any seminars or short courses presented by someone who has a product to sell. This does not mean that we believe that all such presentations are worthless. Some are, in fact, quite good. However, most focus so exclusively on the product being touted that one simply does not get any sort of comprehensive, impartial perspective of the subject; nor any reasonably objective introduction to alternative products. Further, those presenting such material too often, for our taste, focus on how to "push the buttons" on a particular product, rather than on how the underlying methodology works, where it works, and where it does not.

Second, we feel that too many short courses on expert systems (and artificial intelligence, neural networks, etc.) are put together in a thinly veiled attempt to "exploit a hot topic." While we cannot fault someone for wanting to make a few dollars, we do believe that they should at least attempt to provide reasonable value for those dollars. Unfortunately, the majority of short courses that we have attended have been rather hastily put together, and have had no real theme, focus, or evident coordination. Most, in fact, are assembled through the collection of a number (often, a very large number) of experts and pseudoexperts. Some of the speakers may be from academia, some from consulting firms, and others from government agencies and contractors. Unfortunately, they are not always selected on the basis of their knowledge, experience, or speaking skills.

A series of 1- or perhaps 2-hour presentations by each speaker is then considered to comprise the program. In those seminars and short courses that we have attended, all too often none of the speakers is even vaguely aware of the material to be presented by the others. As a result, in one rather nightmarish 2-day seminar that we attended, no less than six of the speakers provided rather extensive, and redundant histories of AI and expert systems. Four gave rather long-winded definitions of expert systems, and several of these definitions were in direct contradiction to others. The presentation of two of the speakers consisted solely of 35 mm slides of products that they were all too obviously hoping to sell to the audience.

A third type of seminar or short course that we would not recommend, or at least not recommend until one has achieved some level of proficiency as a knowledge engineer, is one devoted to a programming language (e.g., a seminar on LISP or PROLOG). LISP, in particular, is difficult enough without having to be faced by the prospect of attempting to learn it in a few days. This type of material can only, in our opinion, be learned through formal courses or rather extensive self-study. We might also remark that a seminar or short course on a computer language is likely to be deadly dull for most participants.

Recommendation for the type of short course or seminar *to attend* would be one that

- Has a clear theme and precise focus
- Describes, objectively and realistically, both the advantages and disadvantages of the expert systems approach
- Focuses on the methodology, as opposed to a product (i.e., one which focuses on expert systems rather than on the use of a particular shell, environment, or workstation)

- Utilizes a minimal number of speakers (e.g., anywhere from one person up to no more than three or four)
- Utilizes live speakers rather than videotaped presentations
- Utilizes speakers with actual experience in the development, implementation, and validation of expert systems
- Avoids a dependence on any single expert systems development package or supporting hardware
- Avoids speakers whose main interest is in selling themselves or their services

Yet another way to augment one's training in expert systems is to initially obtain the services of an external consultant (or consultants). Such a consultant may be used to help establish the internal knowledge engineering group, in the screening of potential knowledge engineering candidates, and in the identification of problems that the group might initially address. Also, and until the firm's own personnel gain some confidence, such a consultant should be retained to assist in the conduct of the group's initial development efforts.

ORGANIZATIONAL CONSIDERATIONS

Up until now, we have focused exclusively on the psychological profile, education, and training of the individual knowledge engineer. Let us now address the problem of precisely where to place the knowledge engineering group within the organization.

Again, let us begin by noting just where *not* to put a knowledge engineering group. First, a knowledge engineering group should most definitely not, in our opinion, be placed within the organization's computer center, or in its computer science or computer engineering department. The role of a knowledge engineering group is (or certainly should be) *far* broader than that of those departments that focus on conventional, computational support. And the preconceived notion of a computer center, as held by most organizations, is simply not supportive of the type of image that a knowledge engineering group should strive to achieve. The emerging knowledge engineering group will find the going tough enough without having to deal with such perceptions.

Second, a knowledge engineering group should not be established as a line-item group. Rather, the purpose of such a group should be associated with the broadest definition of decision support—for the entire organization.

Our recommendation is that a knowledge engineering group have a staff function. If an existing DSS (or MIS) department exists, the knowledge engineering group should probably be placed within that department.[2] Typically, this will

---

[2] Unfortunately, some organizations have placed the MIS and, on occasion, the DSS group within the computer department. For the same reasons as noted above, neither should, in our opinion, be so located.

require some level of initial education of both the DSS (or MIS) staff as well as the knowledge engineers. That is, pains should be taken so that each is made aware of what the others do. It should be made clear that knowledge engineering represents simply another, albeit quite powerful, tool for decision support, and that the best interests of all can be served only through mutual cooperation and mutual respect.

In the (unusual) event that the organization does not have a DSS or MIS department in place, we can only suggest that they form the structure of a DSS department. They may place the knowledge engineering group within this structure and then strive to develop and nurture the entire department. One possible structure for such a DSS department is presented in Fig. 12.1.

Actually, in Fig. 12.1, both the knowledge engineering staff and analytical methods staff could (and possibly should) be placed in a single group, that is, a "decision modeling and methods" group. The expertise of these individuals will be in those methodologies listed previously in Table 12.1.

The computational support group's purpose is precisely that indicated by its name. That is, this group will provide the expertise in, and support of, the tools (both software and hardware) used by the analytical methods and knowledge engineering staff. The members of this group must keep abreast of developments in the hardware and software that serve to support the development and implementation of expert systems (as well as that related to analytical methods). They should also provide the support necessary to bridge between expert systems packages and external programs (and, in particular, databases). Through the provision of such support, the knowledge engineers should be able to focus the bulk of their attention toward the identification of potential applications of expert systems, and the tasks of knowledge acquisition, knowledge representation, and product implementation.

Finally, the projects group, which serves as staff to the DSS department manager, consists of individuals familiar with both analytical methods and ex-

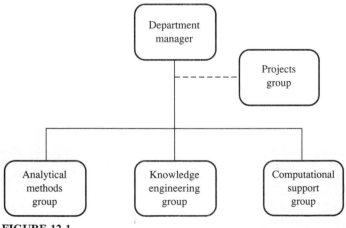

FIGURE 12.1
Decision support systems department.

pert systems. However, in addition, these individuals must have a good, working knowledge of the organization as a whole. It is their responsibility to seek out areas in which either the analytical or expert systems-based tools may be profitably employed. Such opportunities should then be matched to the capabilities and experience level of the DSS personnel. In particular, one should generally avoid attempting to address the development of massive expert systems until the group has first exhibited some success in more modest endeavors. Once a project with a realistic potential for success has been identified and approved, these individuals may then serve as project managers for the efforts. As such, the role of those within the projects group is multifaceted. They must be troubleshooters, salespersons, and, ultimately, project managers for the efforts that they have identified for investigation.

OVERSIGHT AND EVALUATION

As in the case of any other group within an organization, provisions must be made for the oversight and evaluation of the knowledge engineering group. However, since the role of such a group is relatively new to most organizations, and since both the potential and risk of the establishment of such a group is high, care should be taken to provide the group with both the time and resources deemed necessary to sufficiently support its initial growth and development. Simply put, one can hope for, but should not expect immediate, far-reaching results.

It is particularly vital, to the growth and ultimate success of such a group, to provide it with the right kind of manager. Specifically, the attributes of such a manager, in addition to those normally deemed essential for any management position, should include,

- A familiarity with the history, philosophy, and tools for the support of decision analysis in general
- Sufficient familiarity with both expert systems and decision support systems to appreciate both the scope and limitations of these tools, and to effectively communicate such to the rest of the organization
- The confidence and respect of top management
- Access to top management

In the past, there has been some tendency to place specialists, particularly computer specialists, in charge of the DSS, MIS, or knowledge engineering groups. We feel that this is, most usually, a mistake. A specialist of any sort, whether from operations research/management science, engineering, business, or computer science, is generally not the type of person who should be given such a responsibility. Rather, the ideal candidate should be one who is interested in the very broad area of real-world decision analysis, without a bias toward one approach over another.

Finally, let us consider the problem of how to evaluate the performance of a knowledge engineering group. This is a difficult aspect which does not seem amenable to any quick and easy answers. Ideally, the knowledge engineering group could be evaluated by means of simply comparing the costs saved (or profits increased) through their developments versus the costs associated with the existence of the group. If the savings (or profits) exceeded these costs, the group is clearly a success according to any measure of performance based solely on dollars.

However, if a knowledge engineering group addresses any nontrivial type of decision-analysis problem, it will be difficult to impossible to measure the outcome according to dollars alone. Improved decision making should result in numerous direct as well as indirect enhancements throughout the organization. And it is extremely difficult to place a monetary value on *enhancements* to decision making. However, for such intangible benefits, it may be possible to set aspiration levels for various performance measures in the organization. For example, one might list the following goals for the achievements of the knowledge engineering group:

- Reduce paperwork (e.g., by some percent)
- Reduce time required to make and implement a decision
- Increase the "throughput" of the organization (e.g., number of tasks accomplished per given time)
- Reduce the number of "late" decisions (e.g., those resulting in lost opportunities)
- Improve worker morale

Moreover, associated with each goal should be some aspiration level. For example, we might cite a target of a 20 percent reduction in paperwork through the efforts of the knowledge engineering group. Of those goals cited in the above list, we feel that the second is usually of the most importance. Any reduction in the time required to make and implement a decision should serve to enhance the position of an organization. This single factor can, in fact, even make the difference between an organization that survives and one that fails.

SUMMARY

In this chapter, we have attempted to present some ideas about the position and training of the knowledge engineer, the purpose of the knowledge engineering group, and its possible role and placement within the organization. The notions presented reflect our personal views and it is most likely that numerous alternatives exist that would provide results of equal benefit. However, we still believe that most knowledge engineers in this early period of the discipline are ill-prepared for the tasks that they will face. Experience is a great teacher, but perhaps some of the ideas and concepts introduced in this chapter may serve to reduce the number of errors made in a trial-and-error approach to knowledge engineering.

EXERCISES

12.1. A colleague remarks that the easiest way to develop knowledge engineers would be to simply augment the training of DSS (decision support systems) analysts with education and experience in expert systems. What do you think of such a concept?

12.2. Examine the various expert systems-related journals (e.g., *AI Expert, IEEE Expert, AI Magazine*) for advertisements for knowledge engineering positions. In particular, try to determine the percentage of employment advertisements that stress computational support skills rather than knowledge acquisition and representation skills. How might this situation be changed?

12.3. A fair number of people see no reason for the establishment of a formal program in knowledge engineering. They note that the field is evolving too rapidly, touches on a diversity of areas, and is highly dependent upon each individual project encountered. Do you think that they just might be right? If so, just who should perform the function of knowledge engineer?

12.4. A firm is considering the establishment of a knowledge engineering group. The decision about whether or not to set up such a group will be based primarily on the estimated cost effectiveness of the group to the organization. What would you advise?

CHAPTER
13

SUMMARY AND CONCLUSIONS

CHAPTER OVERVIEW

The purpose of this, the final chapter, is three-fold. First, we would like to share thoughts and feelings on an issue that has been raised both in this text as well as elsewhere, that is, the ongoing debate concerning the distinction between expert systems and heuristic programming. Second, we shall attempt to address the future of expert systems as well as the trends that are foreseen in this area. Finally, we will try to "put it all together," and endeavor to summarize the primary points made in this text.

EXPERT SYSTEMS
AND HEURISTIC PROGRAMMING

In earlier chapters, we have either raised or alluded to the question of whether or not expert systems is really nothing more than just another name for heuristic programming. As we have mentioned, the bulk of the AI community traces expert systems back only to the late 1960s, and specifically to the development of DENDRAL. Others see expert systems as only another name or, at best, simply another format for heuristic programming—a tool with a considerably longer, albeit a far less publicized history.

334

Should We Really Care?

Many readers may view such bickering as pointless—or even as sour grapes. Further, we are sure that some readers may wonder why anyone should even care about such seemingly unimportant topics. We feel, however, that this is a matter that must be addressed. We simply do not believe that one can truly appreciate a field until one has an appreciation of the history of that field, and an accurate picture of that history. Further, without an appreciation and understanding of heuristic programming, we do not believe that one can consider oneself truly schooled in expert systems—or knowledge engineering.

Whether or not there is any meaningful distinction between expert systems and heuristic programming, there is at least one very practical reason for being aware of both areas. Simply put, those intent on the development of expert systems, and who are simultaneously unaware of heuristic programming, run the very real risk of reinventing the wheel—or of paying someone else to do so. This is not a matter of conjecture—it has happened, is happening, and most likely will continue to happen. In some cases, such reinvention is due to ignorance. In others, the reason is less benign. We have already discussed the matter of "rebranding," that is, the marketing of methods developed originally as heuristic programs and ' now promoted as expert systems (or AI). As we noted, in some cases an existing heuristic program is simply renamed and sold as an expert system. In others, existing software is translated into an *acceptable* AI language. If the surveys that have been conducted are any indication, this practice is not uncommon.

In most cases, however, we do not believe that those who engage in such practices feel that it is really wrong. One argument that has been presented to us, by those who would defend rebranding, is that it is really a positive endeavor that actually encourages the growth and acceptance of expert systems. Specifically, as they point out, the methodology provides the customer with the capabilities to enhance their existing decision-making process—regardless of the name applied to the technique. Moreover, if they did not call it expert systems (or artificial intelligence), the customer would most likely not consider purchasing the methodology. So, as the argument goes, the customer and supplier both win—and what could possibly be wrong with that?

However, there are losers. In many cases, the heuristic programs that have been or are being rebranded were developed by others—and these individuals receive neither recognition nor compensation for their contributions. In some cases, and evidently particularly so in the military sector, the government (i.e., the taxpayer) is being double-billed—first for the development of a *heuristic program* and later for development of an *expert system*.

Such practices are bound to create a backlash. As the customer becomes more aware of these practices (and this is already happening), the ultimate result may be the development of a distrust for *anything* termed an expert system— whether legitimate or not. This only serves to hurt the field as well as the honest practitioners of the methodology. Finally, through such practices, the perspective that there really are no meaningful differences between expert systems and heuristic programming will only be reinforced. And, as we shall discuss, this is a perception that is not completely valid.

Heuristic Programming Revisited

As was discussed in Chap. 2, a heuristic is a "rule of thumb," developed through experience, judgment, intuition, and insight. When we combine several heuristics together, in some evidently logical format, the result is a heuristic program. If one is to truly understand expert systems, one must (we believe) first understand heuristics and heuristic programming—and be aware of previous contributions in this allied area. In particular, one must be familiar with heuristic programming as it exists, and has existed, both within *and* outside of the AI community. We would recommend that the reader consult the references for a more detailed perspective of heuristic programming [Fisher, 1980; Fuller, 1978; Ignizio, 1980; Koen, 1985; Müller-Merbach, 1981; Polya, 1957; Simon and Newell, 1958; Tonge, 1961; Wiest, 1966]. Here, we will attempt to simply review heuristic programming in order to compare it with expert systems.

A heuristic program provides a solution to a problem; however A heuristic program provides a solution to a problem; however the solution cannot be guaranteed to be optimal. Heuristic programs are (or should be) employed whenever

- Exact (algorithmic) methods cannot be efficiently used
- Acceptable, rather than optimal solutions are sought[1]

In addition, heuristic programs may offer certain advantages whenever a problem is difficult to quantify. As a result, heuristic programs have been developed to solve a host of complex problems, particularly those of a combinatoric nature. Sometimes they are called heuristic programs; at other times they are designated as *approximate methods, suboptimal procedures*, or even *numerical search*. While widely employed and respected by the practitioner, heuristic programming has not found corresponding acceptance in the scientific literature.[2] This is so even when such methods solve (i.e., provide acceptable solutions) to problems that could not otherwise be dealt with in an effective manner.

The three most typical ways of deriving the heuristic rules that comprise a heuristic program are to

- Acquire these rules through interviews with a human who is expert in obtaining acceptable solutions to the problem under consideration (rule acquisition via the domain expert)

---

[1] While one would always prefer an optimal solution, there are times (many times) when it is obvious that the search for an optimal solution would be impractical, if not impossible.

[2] This should not be particularly surprising. Such methods are intuitive in nature, lack supporting proof of convergence, and are generally considered to offer "nothing new" to the literature.

- Hypothesize the rules based on a review of past decisions (rule induction)
- Develop the rules by oneself, generally by means of drawing analogies among the problem under consideration and others that have previously been dealt with . . . successfully (become one's own expert)

Once the rule set has been developed, it is then represented as a step-by-step procedure, and finally by a computer program.

So far, the description of the development of a heuristic program should sound very much (if not exactly) like a description of the development of an expert system. This is most likely why those with a familiarity and experience in heuristic programming perceive the methodologies to be basically identical. However, there are certain quite important differences.

Some Differences

One difference between a heuristic program and an expert system lies in the documentation of the heuristic rule set (or, as it is known in expert systems, the knowledge base). This is known as the knowledge representation phase in expert systems, wherein a relatively formal procedure exists regardless of the ultimate choice of representation (e.g., frames, rules). No such formal scheme exists in heuristic programming, and the ultimate representation is a matter of personal preference. While such a free-form schema provides flexibility, it can lead to unstructured and unwieldy representations.

Another difference lies in the separation of the knowledge base from the inference engine within an expert system—as contrasted with the comparably unstructured manner that this is dealt with in a heuristic program. In a typical heuristic program, the heuristic rules and the inference procedures are all part of a single unit within the associated computer program. And this comingling tends to complicate changes to either the heuristic rules or the solution strategy, as well as to obscure the differences between the two elements of knowledge.

Yet another difference, and certainly the most obvious one, lies in the development of the computer program for the implementation of the heuristic program. Again, the translation of a heuristic programming procedure into a computer program has been, for the most part, an exercise in free form. Emphasis has, in general, been placed almost exclusively upon the *results* obtained as opposed to the structure or characteristics of the computer program, computer language, or the hardware used. And this most likely reflects the pragmatism of those who implement heuristic programming methods.

In the case of expert systems, the emphasis often seems to be in an opposite direction. That is, a presentation of an expert system will, in many instances, appear to focus almost exclusively upon the development package, programming language, and hardware. In one recent, and not atypical exposition, the bulk of the discussion concerning an expert system centered mainly about computer implementation of the expert system, including such aspects as

- The number of rules in the knowledge base[3]
- A defense of the use of PROLOG (i.e., rather than LISP) in development
- The cost and features of the hardware on which the expert system was implemented (a relatively expensive workstation)

There was virtually no discussion as to the effectiveness of the system, of its impact on the organization, or of its acceptance within the organization. With such emphasis given to the single aspect of the computer implementation of expert systems, it should not be surprising that the relationship between heuristic programming and expert systems has become so obscured.

An Illustration

In an attempt to further clarify the similarities and differences between expert systems and heuristic programming, consider the following example. This example represents a somewhat simplified version of an actual implementation of both a heuristic program and an expert system—to the *same* problem.[4] Specifically, when first faced with the problem to be described, the sponsoring organization approved the development of a heuristic program by an outside consulting firm. The development of this particular effort occurred in the late 1960s. Less than two decades later, the same procedure (i.e., collection of heuristic rules) was "converted" into an expert system by yet another outside firm.[5]

The particular problem faced by this organization was that of the selection of candidates, from among a very large pool of individuals, for appointment to advanced training classes. Since the costs incurred for such training are considerable, it was deemed important to carry out this process in a logical, efficient manner. Initially, as mentioned, an outside firm was employed to provide an "automated" procedure for candidate selection. In essence, the desire was to convert the process employed by a group of human experts (a so-called applicant screening board) into a computerized procedure. Since this was the late 1960s, consideration of the development of an expert system was not a factor. And since the organization was impatient for results, the development of a heuristic program was decided on.

To develop the heuristic program, the consultants first interviewed the human experts who were then performing the selection task. They translated these discussions into what they termed a *decision-flow schematic*. The reader should recognize this as a simple decision tree. A version of this tree is presented in Fig. 13.1. As may be seen, based on certain attributes, the subject under consideration is either accepted, placed on a waiting list, or rejected. Further, rather than attempting to immediately evaluate every possible piece of data with regard to each

---

[3] And there is an unfortunate tendency to believe that, the more, the better.

[4] Another example of this would be the DMES and AALPS aircraft cargo loading efforts, as discussed previously.

[5] We are not aware of any attempt to, or any interest in, comparing the results of the two procedures; however, our guess is that the solutions would be virtually identical.

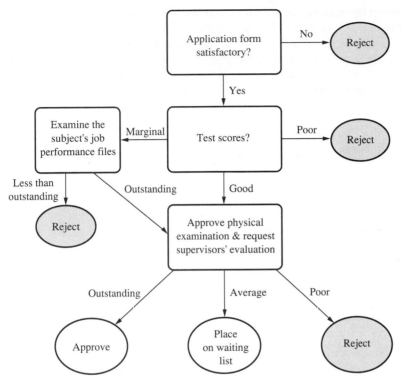

FIGURE 13.1
Decision-flow schematic.

subject, certain subsets of the data are used to determine whether or not to proceed further with the evaluation, that is, to serve to periodically screen out subjects during the procedure. The analogous component of an expert system would be, of course, the set of intermediate conclusions.

As may be noted from Fig. 13.1, certain questions are considered at the various decision points in the process. For example, one such point occurs with respect to test scores. That is, if the test scores (on a certain subset of examinations) of the subject are *poor*, then he or she is rejected. If *marginal*, the process continues and the job files of the subject are carefully scrutinized. Finally, if the test scores are *good*, the subject is sent directly to a thorough physical examination while, simultaneously, the subject's supervisor is asked to provide detailed performance evaluations. The consulting team decided to transform such qualitative concepts into equivalent, quantitative responses.[6]

To accomplish this, they assigned numerical values to each response. As just one example, given the test scores of the subject under consideration, they

---

[6] Although not considered by the team, this aspect of the problem would appear to be a logical place in which to apply either conventional statistically based discriminant analysis or, perhaps, a mathematical programming approach to classification.

computed a weighted sum of the scores. If this weighted sum (say WS) exceeded a certain figure (say A), the candidate was considered to have *good* scores. If the sum was less than A but greater than, say B, the candidate was considered to have *marginal* scores. Finally, if the weighted sum was less than B, the candidate was assumed to have *poor* scores. A small portion of a computer program to accomplish this, written in a rather unsophisticated set of BASIC statements, follows:

```
LET WS = W1*S1 + W2*S2 + W3*S3: 'Compute the
weighted score.
IF WS ≥ A THEN GOTO 100: 'Subject's scores are
considered good.
IF B ≤ WS < A THEN GOTO 110: 'Subject's scores
are considered marginal.
PRINT ''Subject should be rejected.": GOTO 10:
'Reject this subject and evaluate next sub-
ject.
```

The completed software package was then evaluated, mainly through the comparison of the classifications it derived versus those of the human experts. When the results were considered essentially the same, the package was put into actual operation.

Essentially the same approach was taken nearly 20 years later, in the development of an expert system to accomplish the same task. However, rather than constructing the decision-flow diagram, the consulting team developed a set of production rules—similar to the illustrative subset listed below. As should be obvious, these are identical with those implied by the decision tree of Fig. 13.1.

Rule 1:   *If* the application form is satisfactory
          and test scores are good
          and physical exam and evaluations are outstanding
          Then decision is to approve subject for advanced training

Rule 2:   *If* the application form is satisfactory
          and test scores are good
          and physical exam and evaluations are average
          Then decision is to place subject on wait list

Rule 3:   *If* application form is not satisfactory
          Then decision is to reject subject

Rule 4:   *If* application form is satisfactory
          and test scores are poor
          Then decision is to reject subject

Rule 5:   *If* application form is satisfactory
          and test scores are good
          and physical exam and evaluations are poor
          Then decision is to reject subject

Rule 6: *If* application form is satisfactory
and test scores are marginal
and job performance files are less than outstanding
Then decision is to reject subject

Rule 7: *If* application is satisfactory
and test scores are marginal
and job performance files are outstanding
and physical exam and evaluations are outstanding
Then decision is to accept subject for advanced training

Rule 8: *If* application is satisfactory
and test scores are marginal
and job performance files are outstanding
and physical exam and evaluations are average
Then decision is to place subject on wait list

Rule 9: *If* application is satisfactory
and test scores are marginal
and job performance files are outstanding
and physical exam and evaluations are poor
Then decision is to reject subject

Placing these rules into the knowledge base of an expert systems shell, which was essentially the approach used, will result in decisions that are virtually the same as those developed by the original heuristic programming approach. The only difference (and likely a rather minor one) is that the qualitative responses (e.g., satisfactory, marginal, outstanding, poor) were replaced by numerical scores in the heuristic program.

Comparing the two approaches, it is our opinion that they are essentially different representations of precisely the same procedure. However, employment of the expert systems approach does achieve at least six things:

- Through separation of the knowledge base (heuristic rules) and inference engine (solution procedure), the expert system approach allows for easier maintenance of the package (e.g., changes to rules).
- While *explanation* is certainly possible in a heuristic program (i.e., via a trace of the procedure), it is easier to achieve and comprehend through the use of an expert system.
- The expert systems approach permits us to more easily include such features as uncertainty and the use of *unknown* in response to user queries.
- Through the separation of the rule base, and by means of the steps outlined in Chap. 8, we may easily check the rule base for completeness and consistency.
- The implementation of an *expert system*, as opposed to a *heuristic program*, is certainly more attractive from a general, public relations point of view.
- If we use a rule-based expert systems shell, we may focus the bulk of our time and attention toward the construction of the model of the domain expert's rule

base whereas, in the development of an equivalent heuristic program, a major portion of that effort will likely have to be devoted to the development of each step of the process, and the proper combination of these steps.

Based on this and a substantial number of similar examples, our personal opinion is that expert systems and heuristic programming are, *philosophically*, indeed one and the same thing—despite certain differences in emphasis, terminology, and software. However, we do believe that the additional rigor and sophistication provided by the expert systems approach (e.g., schemes for knowledge representation, the separation of the knowledge base and inference engine, and sophisticated chaining schemes) serve to enhance, considerably, the solution approach. As such, we see expert systems as an evolutionary step in the heuristic programming approach—not unlike the evolution from manually implemented trial-and-error approaches into computer-based simulation. However, in our opinion, the most important benefit of the expert systems approach over that of heuristic programming is that cited in the last factor of the above list. That is, the expert system permits the knowledge engineer to address the development of the rule-based model, without being burdened by the need to construct the purely supporting components of the package. Yet another way to state this advantage is to note that, through the employment of an expert system, we may concentrate our full attention on *what to know*, rather than *what to do*.

FUTURE TRENDS IN EXPERT SYSTEMS

Previous projections made for the use and growth of expert systems have varied widely, ranging from estimates that indicate dramatic increases to those of truly stupendous proportions. While we personally would hope that even the most tempered of these prove true, we believe that it is more reasonable to expect a relatively gradual period of growth in the near future with a substantial increase over the long run.

We believe that it is clear that, over the decade of the 1980s, we have witnessed a rather phenomenal level of growth of interest in this methodology—but this has been achieved primarily through the efforts of only a relatively limited number of individuals, within an even more limited number of disciplines. In the 1980s, those who have been the *movers and shakers* in expert systems have been composed, for the most part, of individuals from computer science, electrical engineering, and (although to a much more limited degree) psychology and operations research/management science. However, in the decade of the 1990s, and beyond, this situation can be expected to change considerably.

Specifically, interest in expert systems has transcended the artificial boundaries of computer science. Today, the most rapid growth in the introduction of formal courses in expert systems is within the colleges of business, engineering, and the social sciences. Whereas, in the recent past, those who wished to learn about expert systems had to, for the most part, either do so on their own—through books espousing more or less an exclusively computer science-based perspective, or by attending formal courses offered through a computer science department. As a consequence, those who previously wished to be able to understand and actually apply expert systems had to first encounter and surmount the topics of AI languages, LISP and LISP machines, along with a host of other details far more pertinent to the development and support of software than of the implementation of the expert systems methodology. An analogy to such an approach would be to force students to first learn how to design a computer, and then how to program it, before they were permitted to learn such topics as accounting, mathematical optimization, or economics.

In the 1990s and beyond, we expect to see at least one introductory level course in expert systems in virtually every program in business and engineering—and quite likely, within a large portion of the programs in the social sciences. In fact, within most major universities in this country, we may already note this development.[7] The introduction of such courses should provide the graduates of these disciplines with a powerful, exceptionally robust, and relatively straightforward tool for dealing with a host of real-world problems. And one should observe a concomitant growth of actual implementations.

The 1990s should also be a period in which the proliferation of many small expert systems (i.e., roughly 50 to 200 rules) should continue. Although counter to conventional wisdom, which holds that the only expert systems worth building are those of large size, this has been the area of most growth during the mid- to late 1980s. Small expert systems are attractive in that they may be built in a relatively short time and can be run on inexpensive personal computers—while simultaneously providing benefits that *far* outweigh their (minimal) costs. They also reduce, considerably, the risk involved in expert systems development. That is, rather than expending all of one's funds and time on one mammoth program, one may address a host of smaller efforts. Quite likely, most of the small expert systems will be successful. Even if a few are failures, any disappointment should be mitigated by the success of the others. However, focus on one major development effort, if a failure, can quite effectively curtail any further interest in expert systems on the part of the organization concerned.

Another trend that should continue is that of the embedding of expert systems within other methods and software. While the end user may be unaware of even the existence of the expert system, the performance of the overall tool may be vastly enhanced. Expert systems have been introduced within database software, to diagnose the data for anomalies and otherwise unnoticed data clusters and abnormalities. They have been used to develop so-called intelligent front ends to software programs, in order to ease user training and implementation requirements. It is simply a matter of imagination and observation as to where else such systems may enhance existing tools and techniques.

---

[7] For example, just within the University of Houston, there are two courses in expert systems in the industrial engineering department, two in electrical engineering, one in mechanical engineering, and at least one each in computer science, business, and psychology.

We should also observe a dramatic increase in the growth of hybrid expert systems. In fact, embedded expert systems may be thought of as one rather broad class of hybrid expert systems. Particular interest in hybrid expert systems (i.e., as described in Chap. 9) has occurred in the area of operations research/management science where one sees various combinations of expert systems and *more conventional* decision support methods. Hybrid systems combining expert systems with neural networks should, in particular, flourish.

We also expect to see fewer expert systems development packages (i.e., shells and environments) as purely a result of the *survival of the fittest*. Specifically, the majority of those development tools that will exist in the 1990s and beyond will not only be more powerful, they should be easier to use. This too should motivate a wider interest in expert systems, as well as an increase in actual applications.

A recent projection by the Market Intelligence Research Company (MIRC) serves to point out a number of factors that should be of vital interest to the knowledge engineer of the 1990s and beyond. Specifically, MIRC predicts that [AI Week, 1990]:

> ... the U.S. expert systems market currently stands at $820 million per year, and will climb to at least $6 billion per year in total revenues by 1995. The trend away from LISP and mainframe workstations and toward PCs will continue; ... by the mid-1990s, virtually all expert system products will be capable of functioning in a PC environment.

PUTTING IT ALL TOGETHER

Until now, we have resisted the temptation to provide the reader with a comprehensive, step-by-step procedure for expert systems development and implementation. Instead, we have focused on the individual elements of such a process. We have done so because the development and implementation of an expert system is not normally (if ever) an orderly, predictable process—and we did not want to lead the reader to think so. However, in this section we would like to describe an *ideal* procedure—one that may be used as a rough guide in actual practice.

In Fig. 13.2, we have attempted to summarize the elements and their relationships, as they pertain to the phases of expert systems development and implementation. Initially, we must identify the problem at hand. For example, just what is the problem? Where does it exist? Whom, or what, does it affect? Once this has been accomplished, our next step is to document the problem through a detailed problem statement—including the goals, objectives, constraints, and variables associated with the problem. If possible, the problem statement should be placed into the form of a mathematical model. Otherwise, a schematic of the elements of the problem might be constructed.

With a complete problem statement in hand, we should next try to match the problem with an appropriate solution methodology. If the most appropriate and cost-effective methodology is that of expert systems, we have found a match and

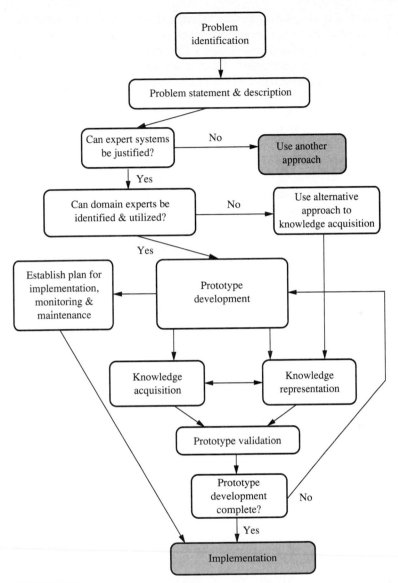

FIGURE 13.2
Expert systems development flow diagram.

justified the use of the expert systems approach. Otherwise, we should employ some other approach to solving the problem. The importance of this particular decision should not be underestimated.

If we are able to justify the use of an expert systems approach, we should next determine if a human expert (i.e., the domain expert) exists and, if so, whether or not he or she may be utilized in the knowledge acquisition phase. If not, we will have to resort to some other way to acquire, or induce, the knowledge base. If so, we may proceed to the prototype development phase. This phase, in essence, consists of two interrelated and virtually simultaneous efforts: knowledge acquisition and knowledge representation. Further, we should, at this time, begin to develop a firm plan for the ultimate validation and implementation of the expert system.

In Fig. 13.2, the tasks of knowledge acquisition and knowledge representation are depicted as being conducted in parallel. While knowledge acquisition and rule-base development may be neatly partitioned in theory, in practice these two activities form the basic elements of that composite activity known as prototype development. That is, as soon as a small set of heuristics has been identified, it is normally represented by a set of production rules. Further, as soon as there are *any* production rules, one may conduct an evaluation of the completeness and consistency of this set. Next, new knowledge is acquired and the process repeats itself.

Finally, when we are satisfied that prototype development has been completed, we may proceed to the implementation phase, using the plan as derived earlier in the effort. While implementation is shown as the terminal phase of expert systems development, it may well be necessary to perform some revisions even during this phase. Ultimately, if conducted properly, these series of activities should lead to the implementation of a working, successful expert system.

RECOMMENDATIONS
FOR FUTURE STUDY

We have attempted, to the best of our ability, to design this textbook to support those who wish to understand, appreciate, and implement rule-based expert systems—in real-world situations and where no more appropriate methods might be employed. Those who have found the topic of interest should now, in our opinion, strive to continue their education in this area—as well as in related, supporting areas. Chapter 12 provides thoughts about those tools that are intrinsic to such support.

However, in addition, it is absolutely essential to develop experience in expert systems development. Specifically (if one has not yet done so), the reader should start to actually build, evaluate, and implement expert systems. We would advise that one begin with expert systems of modest size (e.g., less than 100 rules), directed toward real problems within a particularly narrow domain of expertise. In this way, one should build experience and confidence in one's abilities, and be prepared to deal later with increasingly more difficult situations.

Later, when readers have obtained some confidence in rule-base development and the implementation of small-to-modest sized expert systems, they will likely wish to consider the use of more sophisticated development software, the so-called expert system environment development packages. When dealing with larger problems, the additional features (particularly, frames) of these packages may prove worthwhile.

CONCLUSION

An expert system is a representation of mankind's attempt to replicate the processes involved in human decision making. Typically, the *ideal* toward which we strive is to provide a methodology that incorporates the most desirable characteristics of human decision making. These include the ability to

- Identify the most relevant data concerning a problem
- Prune away all unnecessary paths
- Sequentially filter out possibilities
- Ask only the most pertinent questions
- Backtrack (in the search process) whenever necessary
- Recognize the limits of our skills, and the impact of uncertainty
- Provide conclusions even in the absence of certain facts or data
- Break down large, complex problems into smaller, simpler ones
- Look for analogies
- Seek a solution that is acceptable, rather than attempt to obtain optimal results

At this time, it should be rather obvious that we have a way to go. However, despite the limitations of expert systems, they have already demonstrated an ability to effectively deal with an incredibly diverse array of real-world problems. As with any other tool, to be most effective, they should be applied by the right individual, to the right problem, in the right manner.

APPENDIX

EXSYS USER'S GUIDE

EXSYS: AN OVERVIEW

There are, at this time, two versions of the EXSYS expert systems shell. The first is denoted as the *Standard* version while the second is termed the *Professional* version. Both versions operate in a similar fashion and employ a somewhat similar format. Some of the features that the versions share in common include

- Support of IF-THEN-ELSE production rules
- Support of both forward and backward chaining (backward is the default mode)
- Automatic menu-driven user prompts
- Several choices for *rule* confidence factors[1]
- Consultation sessions may be run, at any time, from *within* the rule editor (i.e., it is not necessary to compile the rule base)
- Bridges to external programs for either data acquisition or external program execution (including access to programs supporting frame-based knowledge representation)
- Report generator and command language
- Numeric and string variables
- Support of various mathematical functions

---

[1] Neither package supports confidence factors for user input. Instead, any uncertainty associated with user input should be handled by the design of the menu associated with the respective attribute. The reader may wish to refer to our discussion in Chap. 7.

- Blackboarding
- Written in C

The two versions differ mainly with respect to the fact that the Professional version incorporates a number of additional features. In particular, the Professional version includes:

- An enhanced command language
- A rule compiler that permits word-processor editing of the knowledge bases plus ease of movement of the rule bases of other shells into the EXSYS environment
- Development of custom screens
- Direct access to dBase files
- Custom formulas for confidence factors
- A redesigned user interface
- Hypertext

The software that has been provided with this text is an educational, or demonstration, version of Standard EXSYS. This package is intended for use on IBM personal computers or fully compatible systems. The only difference between the demonstration package and the Standard EXSYS shell is that the demonstration package restricts the user to the development and saving of no more than 25 rules per knowledge base. However, since the demonstration version of EXSYS supports blackboarding, one may actually, although laboriously, construct expert systems consisting of more than 25 rules by dividing the complete rule base into a number of smaller ones (i.e., where each one contains 25 rules or less and is connected through the EXSYS blackboard).

We should note that the Standard (*non*demonstration) version of EXSYS will permit the solution of rule bases of up to roughly 5000 rules on an IBM personal computer (or compatible) with 640K RAM. Other size restrictions associated with this version include:

- No more than 126 clauses in each of the premise and conclusion (and ELSE) portions of a rule
- No more than 30 values per attribute

Other performance factors associated with both the Standard and Professional versions of EXSYS have already been summarized in Tables 10.2 and 10.3 of Chap. 10. Rather obviously, either version represents a top-of-the-line expert systems shell.

This appendix has been designed to be used in conjunction with the EXSYS tutorial lessons that are a part of the demonstration package software. Together, this appendix and the tutorial should provide the reader with a good overview of the more routine capabilities of Standard EXSYS, along with basic guidelines for the development of rule bases in the EXSYS environment. However, the appendix is most definitely not intended as a complete software manual in support of all

EXSYS functions and capabilities. Those readers who wish to learn more about EXSYS have at least three choices. First, the EXSYS software provided with this text includes a five-part, on-screen tutorial. Readers may access this tutorial once they have installed the EXSYS demonstration package on their hard disk (installation procedures will be described in the following section). Our appendix will deal, primarily, with the capabilities of EXSYS that are described in the first three lessons of the tutorial (lesson1, lesson2, and lesson3), and in lesser detail with those of the fourth and fifth lessons (lesson4 and lesson5).

A second choice is available through the purchase of the book, *Developing Expert Systems Using EXSYS*, by K. G. Sprague and S. R. Ruth [Sprague and Ruth, 1988]. This 157 page guide to EXSYS provides an in-depth look at the basic capabilities of the shell—but does not delve into the more advanced features of the package.

Alternatively, the reader may purchase the complete EXSYS software manual (either for the Standard or Professional versions) directly from EXSYS, Inc. And it should be noted that copies of the demonstration software are also included with each manual. Those readers who wish to employ the more advanced features of EXSYS (e.g., command language, report generator, calls to external programs, blackboarding, and spreadsheet/database access) should most definitely consider this option. Orders should be placed directly with EXSYS, Inc. at the following address:

EXSYS, Inc.
P.O. Box 11247
Albuquerque, New Mexico 87192
Phone: (505) 256-8356

INSTALLATION

Before proceeding with the installation procedure, make sure that you have made a copy of the disk provided. Insert the copy of this disk into drive A of your computer. Installation of the EXSYS package on an IBM personal computer, or compatible, with a *hard disk,* may then be accomplished by typing:

A: README

and by then following the directions that appear on the computer screen.

TERMINOLOGY AND RULE FORMAT

Before proceeding to the tutorial sessions, we should discuss the vocabulary employed by the EXSYS package. EXSYS uses its own rather unique terminology and, unfortunately, this terminology does not correspond with that presented in either this text or, in fact, in most other texts on expert systems. This terminology includes

- *Condition*. A condition is simply a *complete clause,* and may be either a premise or intermediate conclusion clause.
- *Qualifiers and values*. A condition (i.e., clause) is made up of two parts, a qualifier and one or more values. The qualifier is the portion of the condition up to and including the verb. Specifically, a qualifier is an *attribute plus a verb*. The values are those values that are acceptable for assignment to the attribute described in the qualifier (i.e., the set of legal values). For example, if the condition is IF THE COLOR IS RED, then the qualifier is THE COLOR IS, while the value is RED.
- *Choice*. The choices of a knowledge base are *all of the possible conclusions* (or solutions) to the problem under investigation. That is, a choice is (normally) a *final* conclusion. For example, if the expert system is one to be used to diagnose reasons why an automobile will not start, then some possible choices (i.e., conclusions) might include BATTERY FAILURE, VOLTAGE REGULATOR FAILURE, OUT OF FUEL, and so on. Notice, in particular, that an EXSYS *choice* is a conclusion clause attribute *plus* a specific value.

Rules in EXSYS are constructed according to the following format:

```
RULE NUMBER:   IF (conditions)
               THEN (conditions or choices)
               ELSE (conditions or choices)
               NOTE (a note to the user)
               REFERENCES (a reference that is available to
               the user on demand)
```

The ELSE, NOTE, and REFERENCES portions of the rule are all optional. Again, we advise against the use of ELSE. The note and reference facilities are, however, ones that should definitely be employed to document any real-world rule base.

There is another aspect of EXSYS rule formatting that needs to be mentioned. Specifically, the Standard version of EXSYS does not (at least at this time) permit the employment of *true* disjunctive *clauses*. It does, however, allow for disjunctive *values*. To illustrate, the following rule is permitted:

Rule disjunctive values: *If* the day is a Saturday *or* Sunday
                           Then the establishment is closed

That is, since the values (Saturday and Sunday) for the *same* attribute (the day) are connected by the logical OR, we may use such a rule.

However, a rule having true (i.e., completely general) disjunctive clauses, as shown below, is not supported (such rules are, however, permitted in the Professional version).

Rule disjunctive clauses: *If* the day is a holiday
                            or the weather is rainy
                            Then the establishment is closed

Of course, all one has to do to circumvent this particular restriction is to form two equivalent rules.

Despite the capabilities of the EXSYS package, we advise the reader, in general, to

- Avoid the use of disjunctive clauses (of any type)
- Avoid the use of the ELSE condition
- Avoid the use of the negation operator (i.e., NOT) in rule *construction*

With this brief exposure to terminology and rule format, we may now proceed to the tutorial sessions. At this time, it is assumed that the readers have installed the EXSYS demonstration package on their computer and are ready to proceed.

TUTORIAL SESSIONS

While it is intended that this appendix be used in conjunction with the tutorial lessons provided on the software disks, it is actually possible to skip those lessons if you so wish. Those readers who choose to do so may proceed directly to the next section. However, if you find the going difficult, you should consider returning to this section and consulting the tutorials.

If you have installed EXSYS on your hard disk drive, you need to access the hard drive and the directory in which the demonstration version of EXSYS has been stored. This should result in a screen prompt that appears as C:\EXSYS>. Once this prompt, or its equivalent, appears, type LEARN if you have a color monitor or LEARN NOCOLOR if you have a monochrome monitor. You will then be presented with a screen that introduces you to the tutorial lessons that are provided. [If you are using floppy drives, place the floppy disk with the file EDITDEMO.EXE (the EDITOR disk) in drive A and the second EXSYS disk in drive B. Next, check the directories of both disks to determine which one contains the file LEARN.BAT, and access that file (i.e., type LEARN or, if necessary, B: LEARN).]

We now advise the reader to finish the first three lessons of the tutorial before proceeding further. These lessons deal with the most basic fundamentals of EXSYS (in particular, the general use of the rule editor, the development of a rule base, and the editing of a rule base). Upon completion of these lessons, the reader should be prepared to construct and run relatively simple rule bases. We shall deal, in fact, with the step-by-step construction of such a rule base in the section to follow.

ILLUSTRATIVE RULE BASE

Using the material of the first three lessons of the tutorial as our basis, we shall now employ the demonstration version of EXSYS to actually develop and run a simple knowledge base. Let us assume that this knowledge base is to be used to determine the career choices for the graduates of the Incessant Patter School for Television News Broadcasting. The rules have been acquired from the school's job placement director. The portion of the knowledge base that shall be used in the demonstration is listed below. Note that the object, for every clause in the rule base, is the job candidate.

Rule 1: *If* physical abilities include reads from a teleprompter
      and physical appearance is attractive
      Then candidate's career choice is anchorperson

Rule 2: *If* physical abilities include reads from a teleprompter
      and physical abilities include correctly pronounces Navratilova
      and physical appearance is athletic
      Then candidate's career choice is sportscaster

Rule 3: *If* physical abilities include reads from a teleprompter
      and physical appearance is not attractive
      and dress habits include wears funny hats
      Then candidate's career choice is weatherperson

Upon closer examination of this rule base, it should be apparent that we are assuming multivalued premise clause attributes (e.g., for physical abilities in rule 2) as well as multivalued conclusion attributes (i.e., for career choices). This will present no problem as EXSYS allows the use of multivalued attributes. Now, and suppressing the desire to immediately enter these rules in the knowledge base of the expert systems shell, let us first develop the AV-pair table for this rule set. The results are depicted in Table A.1.

We may note, from the table, that there are no intermediate conclusions. Further, we have listed four AV pairs that do not explicitly appear in the original rule base. These are presented in italics in Table A.1 and listed as *"implied* AV pairs." These four values have been included because EXSYS is menu driven and, had they not been made available in the menu, users might be forced to reply with a value that they know is incorrect—simply to proceed with the consultation.

To illustrate, consider the menu that would be presented to the user for the attribute *physical abilities*, had not the implied values been added:

```
Physical abilities include
1. Can read from a teleprompter
2. Can pronounce Navratilova
Input selection(s): _____
```

Further, let us assume that the candidate can neither read from a teleprompter nor pronounce Navratilova. One cannot proceed any further with the consultation unless at least one number is input—but neither are appropriate in this instance. Thus, the user either types in a wrong value or is stuck. A similar situation could

TABLE A.1
Attributes and values for knowledge base

Rule	Clause	Attribute	Value
1	1	Physical abilities	Reads from teleprompter
2	1	Physical abilities	Reads from teleprompter
3	1	Physical abilities	Reads from teleprompter
Implied	*Implied*	Physical abilities	*Cannot read from teleprompter*
2	3	Physical abilities	Correctly pronounces Navratilova
Implied	*Implied*	Physical abilities	*Does not correctly pronounce Navratilova*
1	2	Physical appearance	Attractive
2	2	Physical appearance	Athletic
Implied	*Implied*	Physical appearance	*Not athletic*
3	2	Physical appearance	Not attractive
3	3	Dress habits	Wears funny hats
Implied	*Implied*	Dress habits	*Does not wear funny hats*
1	3	Career choice	Anchorperson
2	4	Career choice	Sportscaster
3	4	Career choice	Weatherperson

occur if the attribute *dress habits* is provided with only a single value. To correct this situation for *physical abilities,* we have added two implied values.

Specifically, we need to provide the user a choice if the candidate can neither read from a teleprompter nor correctly pronounce Navratilova. As a result, the associated menu will appear as[2]

---

[2] Actually, there are at least two other alternatives that we might consider. First, we could have added a third value to the previous menu: *None of the above.* Second, we might consider a refinement of the attribute *physical abilities.* That is, we could (and probably should) divide this attribute into two separate attributes: *reading ability* and *speaking ability.* And this second approach would serve to mitigate a number of the problems that we shall be faced with in a subsequent rule base entry. In actual practice, we would use this latter approach. However, for the purpose of illustration, we shall continue to employ the single attribute *physical abilities.*

```
Physical abilities include
1. Can read from a teleprompter
2. Cannot read from a teleprompter
3. Can pronounce ''Navratilova''
4. Cannot pronounce ''Navratilova''
Input selection(s): _____
```

Consequently, to use EXSYS, we will include the four implied values listed in Table A.1. However, the observant reader may note yet another apparent problem. That is, the only career choices are anchorperson, sportscaster, and weatherperson. What, you may wonder, will happen if a candidate is not suited for any of these occupations? We could, if we so desired, modify the rule base to handle such a situation. However, it is easier and faster to leave the rule base as it is and simply note that, whenever a choice cannot be concluded, the package will indicate this by the following message: NO CHOICE RECEIVED A VALUE GREATER THAN THE THRESHOLD VALUE. In essence, this simply means that there is no appropriate career match.

ENTRY OF ILLUSTRATIVE RULE BASE

Before proceeding further, let us establish a convention for listing the computer keyboard actions that the user must follow in entering data or commands into the system. Specifically, whenever the user must type something, we shall enclose whatever is to be typed within special brackets. For example, if the user must type the letter H, this will be denoted as type <H>.

Let us now employ the rule editor to enter the television careers rule set into the EXSYS knowledge base. Assuming that you are in the directory in which the EXSYS files exist, you may simply type <EDITDEMO> to enter the rule editor (again, if your monitor does not support color, type <EDITDEMO NOCOLOR>). This not only takes you into the expert system development mode, but it also permits you to test run any expert systems that you develop (or have already developed).

Before proceeding further, let us note that the entry of any rule base will require the knowledge engineers to first answer a number of questions before they are permitted to enter the components (i.e., final conclusions, attributes, values, and clauses) of each rule. Specifically, after typing <EDITDEMO>, you will be asked the following questions, in the order shown:

- Rule-base filename
- If the rule base is new, you will be asked if you wish to start a new file
- Subject of the knowledge base
- Author of the knowledge base
- Choice of confidence factors

EDITXS
Expert System Rule Editor
© Copyright 1983, 84, 85, EXSYS, Inc.
Ver 3.2.1

Expert System filename: **CAREERS**

FIGURE A.1
EXSYS logo window.

- Should you select something other than the 0 to 1 confidence factor system, you will be asked if you wish to change the threshold limit when the final results are presented
- Number of rules to use in data derivation
- Explanatory text to be shown to user at start of consultation
- Explanatory text to be shown to user at termination of consultation
- Whether or not you wish the rules displayed to the user during consultation
- Whether or not you wish to call an external program at the start of the run
- List of final conclusions (the choices) as derived by the inference process

    Once you type <EDITDEMO>, an initial screen will appear (see Fig. A.1) and you will be asked to provide the filename for your expert system's rule base. For the sake of illustration, let us name our knowledge base CAREERS (i.e., type <CAREERS> and then press the ENTER key). If you type in a filename that is not in the EXSYS files, you will be asked if this is a new knowledge base. Assuming that our knowledge base is new, you would respond in the affirmative. You will then be asked to enter some narrative on the "subject of the knowledge base." Here, you may type in any material that you wish to be used to describe the specific knowledge base and its purpose to the user. When you finish this input and press the ENTER key, you will be asked to input the author's name—and you may respond by typing in your own name.

    Pressing ENTER will now lead to a new screen in which you will be asked for your choice of confidence factors. Actually, the rather cryptic question asked is, "How do you wish the data on the available choices structured?"—and you are given three choices. Recall from Chap. 7 that (the Standard version of) EXSYS permits the use of three types of confidence factors (and, in the more recent Professional version, user-defined formulas for combining confidence factors).

It is once again stressed that these are *confidence factors for rules only*, and that EXSYS does not provide a facility for confidence factors associated with user responses to prompts. Typically, EXSYS will permit the knowledge base developer to change and modify most any input. However, this is *not* the case with the choice of confidence factors. Once this choice is made, it cannot be changed. The three rule confidence factor choices available are (1) simple yes or no (i.e., confidence 1 or confidence 0), (2) rule confidence factors on a 0 to 10 scale, and (3) rule confidence factors on a -100 to $+100$ scale. To keep things simple, let us assume that we are dealing with a deterministic situation (i.e., you have complete confidence in the rules), and thus we may select the first choice.

You will then be asked for the "number of rules to be used in data derivation"—all or just the first. This refers to the fact that several rules may reach the same (intermediate) conclusion and, should one of these be fired, we may not wish to attempt to fire the others. In particular, we may want to avoid asking unnecessary questions of the user. However, to be on the safe side, select ALL (the default selection) initially (this can be changed later, if you so desire). You will next be asked to input text for display to the user on how to employ the expert system and, after that, any text that you wish to be displayed at the conclusion of the consultation session. You may input any text you wish and, if you so desire, change this input at any time through the editor. The next screen you encounter will ask if the rules should be displayed (i.e., during consultation), and we suggest that you initially use the default (NO) response. Following this, you will be asked if external programs are to be called at the *start* of the consultation process. This refers to just one of the external call capabilities of EXSYS, and one that we need not employ in this simple demonstration—so respond with a NO.

The next screen that you see should now ask you to input the CHOICES (i.e., final conclusions) to be considered. There will be only three final conclusions for our rule base.[3] Examining our knowledge base (or Table A.1), it should be clear that the final choices are

1. CAREER CHOICE IS ANCHORPERSON
2. CAREER CHOICE IS SPORTSCASTER
3. CAREER CHOICE IS WEATHERPERSON

Each of these must be typed in at the EXSYS prompt (be sure to press the ENTER key after each choice is typed in) and, at the fourth prompt, you must press the ENTER key to move to the next step in the process.

Following the input of the choices, you will enter the primary rule editing environment of the rule editor. Specifically, the editing window will be divided

---

[3] These three choices correspond to a single conclusion clause attribute: *career choice*, plus three associated values (anchorperson, sportscaster, and weatherperson).

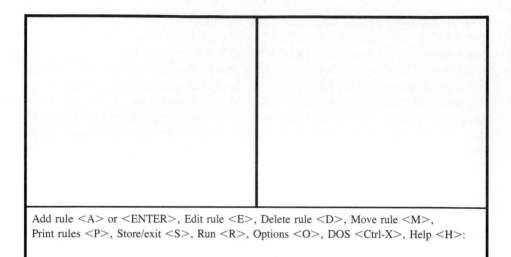

Add rule <A> or <ENTER>, Edit rule <E>, Delete rule <D>, Move rule <M>,
Print rules <P>, Store/exit <S>, Run <R>, Options <O>, DOS <Ctrl-X>, Help <H>:

FIGURE A.2
Main menu.

into two main areas by a vertical line. All actions available through this window
(or any other rule editing window) will be listed below the horizontal line at the
bottom of the screen. And, should you desire any information on how to use
any of the features of the editor, simply type <H> (i.e., type the letter H, for
HELP) and all the operations possible for the particular screen shown will be
explained. The first rule editing window is shown in Fig. A.2. This particular
screen represents the *main menu* of the rule editor.

Entry of Rule 1

Since we are ready to start constructing our rule base, we need to first "add a rule"
and thus we press either <A> or the ENTER key. This moves us to yet another
screen (the *rule component entry menu* as illustrated by Fig. A.3) and permits us
to start entering the premise, or IF clauses (i.e., the EXSYS conditions). When
dealing with the entry of the premise clauses, it is particularly helpful to refer to
a table of rule attributes and values, as listed for this rule set in Table A.1. Since
we are entering a new qualifier, we type <N>. This results in a new prompt at
the bottom of the screen: "Input text of qualifier (ending in verb):." The text for
this qualifier will be

PHYSICAL ABILITIES INCLUDE

which refers to the candidate's physical ability. The associated screen is depicted
in Fig. A.4.

Thus, type in <PHYSICAL ABILITIES INCLUDE> and press the ENTER
key. You can then type in the values for the associated premise clause attribute.
The new prompt at the bottom of the screen is now "Input acceptable values.

```
┌─────────────────────────────────┬──────────────────────────┐
│ RULE NUMBER: 1                  │                          │
│                                 │                          │
│ IF:                             │                          │
│                                 │                          │
│                                 │                          │
│                                 │                          │
│                                 │                          │
│                                 │                          │
│                                 │                          │
├─────────────────────────────────┴──────────────────────────┤
│ Enter a qualifier #, New qual. <N>, Find <F>, Last qual.    │
│ <L>, Copy cond <K>                                          │
│ Repeat cond <R>, Choice <C>, Math/Variable <M>, Help<H> or  │
│ <ENTER> when done                                           │
└─────────────────────────────────────────────────────────────┘
```

FIGURE A.3
Rule component entry menu.

Input just <ENTER> when done." These acceptable values are simply the values associated, in Table A.1, with the attribute *physical abilities*. These entries are listed below:

1. READS FROM A TELEPROMPTER
2. CANNOT READ FROM A TELEPROMPTER
3. PRONOUNCES NAVRATILOVA
4. CANNOT PRONOUNCE NAVRATILOVA

```
┌─────────────────────────────────┬──────────────────────────┐
│ RULE NUMBER: 1                  │ physical abilities include│
│                                 │                          │
│ IF:                             │                          │
│                                 │                          │
│                                 │                          │
│                                 │                          │
│                                 │                          │
│                                 │                          │
│                                 │                          │
├─────────────────────────────────┴──────────────────────────┤
│ Input text of qualifier (ending in verb):                   │
│                                                             │
└─────────────────────────────────────────────────────────────┘
```

FIGURE A.4
Entry of qualifier for rule one.

RULE NUMBER: 1	Qualifier #1
IF:	physical abilities include 1. reads from teleprompter 2. cannot read from teleprompter 3. correctly pronounces "Navratilova" 4. cannot pronounce "Navratilova" **1**

Enter number(s), NOT + number (s), New value <N>, Type correction <T>,
Delete <D>, ↑ or ↓ to scroll, Help <H> or just <ENTER> to cancel

FIGURE A.5
Selection of first value of first premise for rule one.

After typing <CANNOT PRONOUNCE NAVRATILOVA> we simply press the ENTER key when 5. appears on the screen (i.e., there are, from Table A.1, only four values associated with this attribute). This allows us to finish the construction of the first premise clause. Since our first premise clause should be "If physical abilities include reads from a teleprompter," we type <1> (i.e., type the number 1 to assign the first value to the first qualifier). This action is shown in Fig. A.5. The first premise clause, in its complete and correct form (i.e., IF PHYSICAL ABILITIES INCLUDE READS FROM A TELEPROMPTER) should then appear on the left of the screen (see Fig. A.6). We next need to enter the second premise clause, and since this also involves a new qualifier (i.e., new attribute), we type <N> to develop the second qualifier. This qualifier is associated, as may be noted in Table A.1, with the attribute *physical appearance*. The text for this second qualifier is

   PHYSICAL APPEARANCE IS

and the values are[4]

---

[4] Actually, should we so wish, we could employ one of rule editor's facilities to reduce our typing with regard to the four values listed above. That is, it is possible to simply type in *two* values, ATTRACTIVE and ATHLETIC. Should you then, for example, wish to employ the value NOT ATTRACTIVE in a premise clause (i.e., as in rule 3), you may form such a clause by typing <NOT 1> when prompted by the rule editor. We, however, tend to avoid this particular facility. The specific reason for this is that, when the consultation session is run, the only values that will be presented on the menu associated with the attribute PHYSICAL APPEARANCE will be ATTRACTIVE and ATHLETIC. And, if the individual happens to be neither attractive nor athletic in appearance, we will be stuck.

```
┌─────────────────────────────────────────────────────────────────────┐
│                                                                       │
│  RULE NUMBER: 1                     │                                 │
│                                     │                                 │
│  IF:                                │                                 │
│  physical abilities include reads from                                │
│  teleprompter                       │                                 │
│                                     │                                 │
│                                     │                                 │
│                                     │                                 │
│                                     │                                 │
│                                     │                                 │
│                                     │                                 │
│                                     │                                 │
│                                     │                                 │
│                                     │                                 │
│─────────────────────────────────────┴────────────────────────────────│
│  Enter a qualifier #, New qual. <N>, Find <F>, Last qual. <L>, Copy cond <K> │
│  Repeat cond <R>, Choice <C>, Math/Variable <M>, Help<H> or <ENTER> when done │
│                                                                       │
└─────────────────────────────────────────────────────────────────────┘
```

FIGURE A.6
First premise of first rule.

1. ATTRACTIVE
2. NOT ATTRACTIVE
3. ATHLETIC
4. NOT ATHLETIC

Simply press the ENTER key when asked to provide the fifth value. If we respond to the next prompt by typing <1>, the second premise clause will appear on the left of the screen as PHYSICAL APPEARANCE IS ATTRACTIVE. Since the complete set of premise clauses for rule 1 have been entered, we should next press the ENTER key. This should result in the word THEN being displayed on the left of the screen (see Fig. A.7). We are now ready to develop the conclusion clause for our first rule.

Since the conclusion clause of rule 1 is a final conclusion (or, in the terminology of EXSYS, a *choice*), we next type <C> (for choices). The three choices we initially specified will then appear on the right-hand side of the screen (see Fig. A.8). Type <1> (to select the first choice). A new prompt shall appear at the bottom of the screen:

```
Enter value 0 (false) or 1 (true)
Help <H>, <ENTER> or <ESC> to cancel: _____
```

Here, simply type <1> to assign a *probability* of 1 to this choice. At this point, you should have both the IF and THEN conditions for the first rule of the knowledge base (see Fig. A.9).

RULE NUMBER: 1

IF:

physical abilities include reads from
teleprompter
physical appearance is attractive

THEN:

Enter a qualifier #, New qual. <N>, Find <F>, Last qual. <L>, Copy cond <K>
Repeat cond <R>, Choice <C>, Math/Variable <M>, Help<H> or <ENTER> when done

FIGURE A.7
Ready for entry of conclusion clause (choice).

Next, press the ENTER key again and the ELSE condition should appear in the left-hand side of the screen. Since we are not employing ELSE conditions in our knowledge base, simply press the ENTER key again. You may now enter a NOTE and/or REFERENCE if you so choose. Next, you will be asked if the rule is correct. If it is, respond with a YES to return to the main editing screen. However, even if the rule is not correct, it is advisable to respond with a YES as a NO response *will erase the entire rule*. Consequently, if the rule is all right,

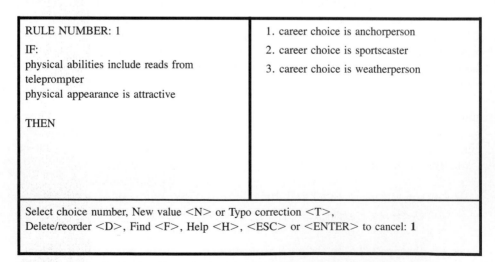

FIGURE A.8
Selection of choice 1.

```
┌─────────────────────────────────────────────────────────────────────┐
│ ┌──────────────────────────────────────────┬────────────────────────┐│
│ │ RULE NUMBER: 1                            │                        ││
│ │                                           │                        ││
│ │ IF:                                       │                        ││
│ │ physical abilities include reads from     │                        ││
│ │ teleprompter                              │                        ││
│ │ physical appearance is attractive         │                        ││
│ │                                           │                        ││
│ │ THEN:                                     │                        ││
│ │ career choice is anchorperson-Probability=1│                       ││
│ │                                           │                        ││
│ │                                           │                        ││
│ │                                           │                        ││
│ ├───────────────────────────────────────────┴───────────────────────┤│
│ │ Enter a qualifier #, New qual. <N>, Find <F>, Last qual. <L>, Copy cond <K> ││
│ │ Repeat cond <R>, Choice <C>, Math/Variable <M>, Help<H> or <ENTER> when done││
│ └────────────────────────────────────────────────────────────────────┘│
└─────────────────────────────────────────────────────────────────────┘
```

FIGURE A.9
Premise and conclusion clauses for rule one.

or if it has but minor errors, we suggest that you respond with a YES. Later on, you may make any necessary changes to the rule through the editing facilities.

Entry of Subsequent Rules

We are now ready to enter rule 2. However, before doing this it is important to note, once again, that EXSYS is menu driven. As such, it stores all the qualifiers, choices, and values by both name and number. Consequently, once these are initially entered, *you must not enter them again* (i.e., type them out in full). Rather, you simply call these rule components *by number* when needed. Not only does this approach save typing, it serves to reduce errors in input. For example, in constructing rule 2, the first two premise clauses contain qualifier 1 while the third premise clause contains qualifier 2 (this may also be noted by an examination of Table A.1); and these *numbers* must be called rather than requesting a new qualifier.

There is yet another factor of critical importance in the entry of rule 2. Note that, when we begin to enter this rule (by first typing <A> at the main menu), we must call up qualifier 1 (*physical abilities*) from the rule component entry menu—as shown in Fig. A.10. Observe carefully that rule 2 consists of *two* premise clauses associated with the same attribute: *physical abilities*, and that these clauses are to be connected by the AND operator (*not* the OR operator). However, if you are not careful, and type <1,3> when asked to "Enter number(s)," you will produce a disjunctive clause.

The correct manner in which to enter the two premises containing the attribute *physical abilities* is to enter them *one at a time*. That is, first call up

RULE NUMBER: 2	1
IF:	
Enter a qualifier #, New qual. <N>, Find <F>, Last qual. <L>, Copy cond <K> Repeat cond <R>, Choice <C>, Math/Variable <M>, Help<H> or <ENTER> when done	

FIGURE A.10
Selection of qualifier one—rule two.

qualifier 1 and type <1>, as illustrated in Fig. A.11. Then, call up qualifier 1 again and this time type <3>.

Using the general procedure described above, we may continue to enter the complete rule set. Once all the rules have been entered, and you are satisfied that your rule base is correct, you may run a consultation session by simply typing <R> (for RUN) from the main menu. We shall discuss the running of this expert system in the section to follow.

RULE NUMBER: 2	Qualifier #1
IF:	physical abilities include 1. reads from teleprompter 2. cannot read from teleprompter 3. correctly pronounces "Navratilova" 4. cannot pronounce "Navratilova" 1
Enter number(s), NOT + number(s), New value <N>, Type correction <T>, Delete <D>, ↑ or ↓ to scroll, Help <H> or just <ENTER> to cancel	

FIGURE A.11
Selection of value one for qualifier one.

A CONSULTATION SESSION

A particularly valuable feature of EXSYS is its ability to run a consultation session directly from the rule editor. Thus, if you are confident that you have properly entered the rule base for our television careers example, you should now return to the primary rule editing window (i.e., the main menu, as depicted in Fig. A.2). The consultation session may then be initiated by simply typing <R> (for RUN).

For the sake of illustration, let us assume that we wish to determine the proper career choice for a candidate who

- Can read from a teleprompter *and* can correctly pronounce Navratilova
- Has an attractive physical appearance *and* has an athletic appearance
- Does not wear funny hats

Consequently, whenever prompted by EXSYS for an attribute value, enter the appropriate selection from the list provided above.

Once you have typed <R>, you will be presented with the title screen for the television careers rule base. The information shown (title and author's name) will be whatever you entered in response to the prompts during rule-base construction. You may simply press any key to move to the next screen.

The next screen provides the user with the text you entered for presentation prior to the consultation session. Press any key and the consultation session begins.

The screen you should now see should list the values for the attribute *physical abilities* (see Fig. A.12). And you need to type in the appropriate response, or responses. Since our candidate can *both* read from a teleprompter *and* correctly pronounce Navratilova, you should type <1,3>. Note that, should you wonder

---

physical abilities include
    1 reads from teleprompter
    2 cannot read from teleprompter
    3 correctly pronounces "Navratilova"
    4 cannot pronounce "Navratilova"
1,3

Enter number(s) of value(s), WHY for information on the rule,
<?> for more details, QUIT to save data entered or <H> for help

---

FIGURE A.12
User response to first prompt.

```
physical appearance is
    1 attractive
    2 not attractive
    3 athletic
    4 not athletic
1,3




Enter number(s) of value(s), WHY for information on the rule,
<?> for more details, QUIT to save data entered or <H> for help
```

FIGURE A.13
User response to second prompt.

just why this question has been presented, you may type <WHY> rather than a numerical response. This will result in the display of the rule under consideration in the inference process.

You will now be presented with a screen listing the values for the attribute *physical appearance* (see Fig. A.13). The response here is to type <1,3>. Once this has been done, the consultation session for this candidate is finished (there is no need to ask about the candidate's dress habits) and you should be presented with a screen listing whatever text you input for the user to see upon completion of the consultation session. Pressing any key will result in the display of the final results screen.

The results screen for this example is presented in Fig. A.14. As may be noted, this particular candidate should be considered for either an anchorperson or a sportscaster. If you wonder just how each of these conclusions were reached, you may, at this time, type in the line number associated with a choice (e.g., type <1> and press the ENTER key to determine how the first choice was derived). The rule, or rules, associated with the determination of this choice will then be displayed.

Once you are finished with the consultation session, simply type <D> (for DONE) from the results screen. You will then be asked if you wish to run the consultation again. Type <N> for NO (or just press the ENTER key) to return to the primary rule editing window. At this point, we are ready to discuss the editing facilities of the rule editor.

Values based on 0/1 system	VALUE
1 career choice is anchorperson	1
2 career choice is sportscaster	1

All choices <A>, only if value>1 <G>, Print <P>, Change and rerun <C>, rules used <line number>, Quit/save <Q>, Help <H>, Done <D>:

FIGURE A.14
Results screen.

EDITING THE RULE BASE

Entry of any rule base without some error or future need for revision would be a remarkable event indeed. Consequently, any good expert systems shell should have some means for editing the rule base (and associated information). EXSYS provides a particularly good editing facility.

For the purpose of illustration, let us assume that we have found yet another attribute that serves to determine a career choice for our television job candidates. This attribute is the score that each student receives in a comprehensive examination taken at the completion of his or her schooling. Using this new attribute, the revised set of rules are

Rule 1: *If* physical abilities include reads from a teleprompter
and physical appearance is attractive
and [TEST SCORE] > 50
and [TEST SCORE] < 80
Then candidate's career choice is anchorperson

Rule 2: *If* physical abilities include reads from a teleprompter
and physical abilities include correctly pronounces Navratilova
and physical appearance is athletic
and [TEST SCORE] < 70
Then candidate's career choice is sportscaster

Rule 3: *If* physical abilities include reads from a teleprompter
and physical appearance is not attractive
and dress habits include wears funny hats

RULE NUMBER: 1

IF:

     (1) physical abilities include reads from teleprompter

and   (2) physical appearance is attractive

THEN:

     (1) career choice is anchorperson-Probability = 1

CHANGE: If <I>, Then <T>, Else <E>, Note <N>, Reference <R>, Done <ENTER>
↑ for previous rule ↓ for next rule

FIGURE A.15
Change screen for rule one.

   and [TEST SCORE] > 85
   Then candidate's career choice is weatherperson

Note carefully that this new attribute is termed a mathematical variable. The EXSYS shell requires that we enclose such variables in square brackets. And we shall explore the use of variables (of various types) in more detail later on in the appendix.

Obviously, we need to add some additional premise clauses to each of our three rules. Further, this requires the addition of a new attribute, the mathematical variable associated with the candidate's test score.

To accomplish our revisions, we must return to the primary rule editing window (the main menu as depicted earlier in Fig. A.2). From this window, we simply type <E> to begin the rule editing process. You will then be asked for the rule number of the rule to be changed. Simply type <1> and press the ENTER key so that we may begin our revisions with rule 1 of the knowledge base.

Rule 1 of the knowledge base should now be displayed (see Fig. A.15). We need to add two new premise clauses to this rule so type <I> to indicate that we need to modify the IF portion of the rule. Once we have typed <I>, the next screen will present the list of premise clauses for rule 1 (see Fig. A.16). Since we wish to add premise clauses, we must type <A> at this point in time.

We should now be presented with the window that permits us to add a new premise clause (see Fig. A.17). Since we need to add a mathematical variable, we must type <M>, which leads us to a new screen. At this time, we need to enter the formula associated with one of the new premise clauses for rule 1. Thus, simply type [TEST SCORE]>50 and then press the ENTER key.

Again, we have a new screen. At the bottom of the screen, we are asked to enter the text to be used to describe this new variable. Type <THE NUMERICAL

---

1. physical abilities include reads from
teleprompter
2. physical appearance is attractive

Delete <D>, Add <A>, Change <C>, Reorder <R>, Help <H>, Done <ENTER>

FIGURE A.16
Premise clauses for revision.

TEST SCORE FOR THE CANDIDATE'S COMPREHENSIVE EXAM> and press the ENTER key. Next, you will be asked if you want this variable displayed at the end of the consultation session. There is no need for this, so respond with <N>. You will then be asked if this variable (i.e., test score) is a numeric or string variable. Since it is numeric, type <N>. You will be asked yet another question: "Do you wish this variable to be initialized to a value at the start of an EXSYS run?" Respond with <N>. You will then be asked if you wish to place limits around this variable (e.g., a test score cannot be less than 0 or greater than 100).

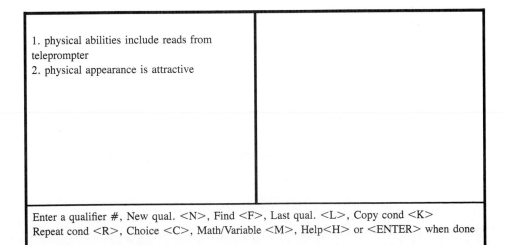

1. physical abilities include reads from
teleprompter
2. physical appearance is attractive

Enter a qualifier #, New qual. <N>, Find <F>, Last qual. <L>, Copy cond <K>
Repeat cond <R>, Choice <C>, Math/Variable <M>, Help<H> or <ENTER> when done

FIGURE A.17
Ready for entry of new premise clause (math/variable).

Usually, this is a good idea but, for this case, let us simply type <N>. Once this entry is made, you will be presented with a screen that includes the new premise clause.

Since we need to add yet another premise clause to this rule, we need to repeat the above process for the clause: [TEST SCORE]<80. The procedure should now be apparent so go ahead and add this new clause to rule 1. When both clauses have been entered, simply press the ENTER key to see the revised version of rule 1.

We should then repeat this process for the new premise clauses for rules 2 and 3. We may reach these rules by use of the arrow, or scroll keys, as available from the editing window. Once this has been accomplished, the rule base in EXSYS should look like the rule base presented at the beginning of this section.

Obviously, from the commands listed at the bottom of the supporting editing screens, we may make such changes as

- Additions, deletions, or corrections to premise clauses
- Additions, deletions, or corrections to conclusion clauses
- Additions, deletions, or corrections to ELSE clauses
- Revisions of notes and references
- Reordering of premises and/or rules

However, we may also make additional changes from the primary rule editing window (i.e., the main menu of Fig. A.2). Here, we may simply type <O> to proceed to the options window. The changes possible from this window include

- Subject line
- Author
- Beginning text
- Ending text
- External program
- Rules used to derive
- Variables
- Display threshold
- Display rules
- Check new rules

The majority of these options should be self-explanatory and we shall not address them here.

ALTERNATIVE MODES OF CHAINING

The default mode of chaining in EXSYS is backward chaining. However, as will be explained, the backward chaining method that is used is somewhat different from that presented in Chap. 6 of the text. In addition, the shell permits forward chaining. In this section, we shall briefly cover the following chaining options:

- Backward chaining (the default mode)
- Finalpass option
- Forward chaining
- Nobackward option (and forward chaining with nobackward)

Backward Chaining through EXSYS

EXSYS begins backward chaining by finding the first rule with the first choice in its conclusion clause. It then employs conventional backward chaining in an attempt to determine whether or not it can conclude that choice. Once it has searched all the rules relevant to the first choice, it will proceed to determine whether or not the next choice (i.e., the second choice) can be concluded. Again, it proceeds to the first (active) rule that includes the second choice in its conclusion and then backward chains from that rule. The process is repeated until *all* of the choices have been examined.

The user need not do anything special to invoke the backward chaining procedure. This is the default mode in EXSYS and thus, unless commanded otherwise, backward chaining will be employed on any rule base that is run.

Finalpass Option

When backward chaining is employed, it may be (and generally will be) the case that the consultation session ends and not all of the rules have been fired or discarded. We can invoke the FINALPASS option to perform a final pass through all active rules once backward chaining has been terminated. If any active rules are found, they will be tested and either fired or discarded, depending on the data provided. Whether such results are of interest will depend, of course, strictly on the interests of the user.

To invoke the FINALPASS option, we use the FINALPASS command when calling EXSYS. For example, if we wish to run a rule base named CAREERS and use backward chaining with FINALPASS, we would type

EDITDEMO CAREERS FINALPASS

Forward Chaining

The approach used by EXSYS to perform forward chaining also differs somewhat from that presented in Chap. 6. Specifically, when this option is used, the program will examine each rule in order of its occurrence (i.e., rule 1 first). *Backward chaining* will then be employed to derive any information required by the premise of the rule under consideration. Once the rule has been evaluated, we then move to the next active rule and repeat the process.

To invoke forward chaining, we simply type <FORWARD> in the command line when the rule base is called. Thus, for the rule base named CAREERS, the command line would be

EDITDEMO CAREERS FORWARD

Nobackward Option

If the NOBACKWARD option is employed by itself, it will force EXSYS to not use backward chaining to derive any information needed by the premise clauses. The process will start by moving to the first rule with the first choice in its conclusion. Since backward chaining is not employed, the user will be asked to provide any necessary input. Once the first choice has been determined, we move to the first rule with the second choice in its conclusion and repeat the process. The entire procedure is repeated until all the choices have been evaluated. The command line for a nobackward run of the CAREERS rule base would look like

EDITDEMO CAREERS NOBACKWARD

Usually, the NOBACKWARD option is used in conjunction with forward chaining. When so employed, a *pure* forward chaining process results. That is, the program will process each rule in order (as in the case with forward chaining). However, backward chaining will not be employed to derive any data needed by the rule premise. In essence, we simply have a single forward pass through all the rules (without using backward chaining to derive any supporting facts). *If we combine forward and nobackward, we must make sure that the rules are properly ordered.* Specifically, any rules needed to derive any facts needed by some other rule (or rules) must appear in the rule base before the rule (or rules) that require these facts. The command line for a forward, nobackward run of the CAREERS rule base is

EDITDEMO CAREERS FORWARD NOBACKWARD

An Example

If you are still having trouble in determining the differences between the various modes of chaining, we suggest that you enter the following (unordered and not particularly well-formed) rule base into EXSYS and try out each option. This should clarify the concepts that have been presented. Notice that there are two final conclusions, or choices, listed as choice 1 and choice 2 (e.g., choice 1 may be the recommendation that the extra wide tires be fitted to the car while choice 2 may be the recommendation that the car be serviced every 5000 miles).

Rule 1:   *If* make of car is German
           Then choice 2

Rule 2:   *If* owner of car is Joe
           Then make of car is German

Rule 3:   *If* owner of car is Harry
           Then make of car is Japanese

Rule 4: *If* owner of car is Harry
        Then choice 2

Rule 5: *If* owner of car is Harry
        Then choice 1

Rule 6: *If* owner of car is Joe
        Then price is expensive

In order to tell the differences between each mode of chaining, you will need to request a *trace* of the inference process. This provides you with a summary of the firing sequence of the rules, and should serve to help clarify the operation of each option. To request a trace, simply type the following command line when accessing the rule base:

EDITDEMO [FILENAME] [OPTIONS] TRACE = PRN

Here, the filename is the name given to the rule base and the options are the chaining options used (if any). The command TRACE = PRN, will cause the trace to be printed out (make sure that a printer is attached to your computer and that it is turned on). Should you wish to print the trace to a file (for examination at some other time), you can simply use TRACE = XXX.TRC where XXX is any filename you wish to use for storage of the trace.

ADVANCED TOPICS

At this time, we would suggest that the reader access the fourth and fifth lessons of the tutorial. Here, we shall simply provide a very brief overview of some of the more advanced features of EXSYS, and conclude the material with an introduction to an illustrative example which serves to depict some of these capabilities. Before proceeding, be advised that some portions of the material to follow will require some knowledge, albeit at an elementary level, of the use of the computer and computer programming.

The particular topics that we shall deal with in this section include

- Variables (mathematical, string, and text only)
- External programs
- Report generator
- Blackboarding

Each of these topics will be briefly discussed and their use summarized by means of an illustrative example of the use of blackboarding. We shall begin our discussion with an overview of the employment of variables in EXSYS.

Variables

There are, in EXSYS, two types of variables: mathematical variables and string variables. A variable is denoted by any string of alphanumeric characters, including spaces, enclosed by square brackets. Up to 100 characters may be used but only the first 18 of these characters are significant. Numeric variables are stored as floating-point numbers and may be operated upon by most algebraic expressions. String variables are text strings and may be used for comparison or concatenation.

We have already presented, in the discussion of rule-base editing, an example of the use of a mathematical variable. In that example, the variable *test scores* was used in the premise clause of all three rules of the knowledge base. Further, it was assumed that the user, when prompted, would provide a *numerical* value for this variable. Yet another example of the use of mathematical variables is presented in the rule shown below, where such variables appear in both the premise and conclusion clauses. This rule simply serves to compute an individual's weekly salary if he or she works overtime (i.e., more than 40 hours). All overtime hours are paid at a rate of one and one-half that of the normal rate.

Rule math variables: *If* [HOURS] > 40
Then [SALARY] = ([HOURS]−40)*([RATE]*1.5)
+(40*[RATE])

The above rule contains three mathematical variables: *hours, salary,* and *rate*. The entry of these variables follows the same basic procedure as described previously for *test scores*. Notice carefully that variables, like any other rule attributes, are treated differently according to whether they appear in a premise or conclusion clause. For example, in the premise of the above rule, we are testing the variable *hours* to determine if it exceeds a value of 40. If so, the premise is true. However, in a conclusion clause we are simply assigning a value to the variable *salary* when, and if, the rule fires. That is, in the above rule, *salary* is given the value shown to the right of the equality sign if the rule is fired.

String variables are also supported. A rule employing a string variable is listed below. Here, if the person's last name happens to be Jones, his first name evidently must be Paul.

Rule string variables: *If* [LAST NAME] = "JONES"
Then [FIRST NAME] = "PAUL"

As may be seen, string variables may appear in either premise or conclusion clauses. If a text string is used, as in the premise clause of the above rule, the text must be enclosed by quotes. In a premise clause, a string variable may be compared (by means of the operators <, >, = , <= , >= , or <>) with another string variable or with a text string. For example, [SMITH] < "DAVIS" is used to determine if the string SMITH is *alphabetically* less than the string DAVIS. Entry of a string variable is similar to that of a mathematical variable except, when you are asked if the variable is numeric or string, you obviously must answer with *string* (i.e., type <S>).

String variables may also be used to display messages at the conclusion of the consultation session. In such an instance, they are termed *text only variables*.

The rule listed below contains such a variable. Here, the (text only) variable [NOTE] is simply employed to provide a reminder to the user at the end of the consultation session. To input this variable as a string variable, type <M> when presented with the rule component entry menu (i.e., the menu used to support the entry of a qualifier). This response indicates that either a math or string variable is to be input. Then, when you are prompted to enter the text associated with this variable, simply type <MAKE SURE THAT YOU SAVE ALL DATA BEFORE SHUTTING THE SYSTEM DOWN>. When asked if the variable has a value with it, *make sure that you type <N>, for no*. Finally, when asked if it is to be displayed at the end of the consultation, type <Y> for yes. If this rule is fired, the text associated with [NOTE] will be printed out along with the conclusions of the consultation session.

Rule text only variable: *If* number of test samples remaining is none
                           Then testing is finished
                           and [NOTE]

To conclude, we should emphasize that variables in premise clauses *must* have values assigned to them. Those in conclusion clauses need not have a value. This permits the rule base developer to display notes at the finish of a consultation session, as was the case with the last rule presented, or to access the internal commands available through EXSYS (as shall be discussed in the section to follow).

External Programs

As should be the case with any good expert systems shell, EXSYS allows the developer to access external programs. The approaches used to call such programs are covered in some detail in the fourth and fifth lessons of the tutorial and, consequently, only the basics of these operations shall be presented in this section.

There are four different ways in which EXSYS may be used to call external programs. In this section, we shall briefly summarize the first three means of external program calls. In the section to follow (i.e., the section on the report generator), the fourth approach shall be described. All four approaches are listed below:

- Programs may be called prior to the initiation of a consultation
- Programs may be called from a qualifier or variable (mathematical or string) in the premise of a rule
- Programs may be called from a text only variable in the conclusion of a rule
- Programs may be called from RUN commands in the report generator

Essential to the support of calls to external programs is the use of the RUN command. Before proceeding further, let us examine the format of this command, as listed below:

> RUN (filename <data1> <data2> ... <options>)

where   filename = the name of the external program
        <data#> = the list of parameters passed to the external program
        <options> = the options which serve to control the manner in which
              parameters are passed between the external program(s)
              and the rule base

We can also use this command to call programs written in interpretative BASIC. To accomplish this, we simply insert the word BASIC before the filename. For example, to call a BASIC program named DECODE, we might use the following command:

> RUN (BASIC DECODE /C)

Note that the /C is an option that serves to control the manner in which parameters are passed to the external program (the EXSYS manual should be consulted for a listing and description of all such options). We shall, for the most part, restrict our discussions to the use of this particular form of the RUN command.

PROGRAM CALLS PRIOR TO CONSULTATION SESSION. You may recall that one of the questions asked of the rule-base developer, prior to the entry of the rules, was whether or not the rule-base developer wished to call an external program at the start of the run. Should we respond by typing <Y>, for yes, we may use this option to call an external program immediately prior to the consultation session. For example, we may wish to access an external database prior to our run.

Any data provided by the external program will be passed back to EXSYS by means of the default file RETURN.DAT. This data must be in the following form:

> For qualifiers: Q<number> <value(s)>
> For variables: V<number> <value>
> For choices: C<number> <value>

The "number" is the number of the qualifier, variable, or choice (e.g., Q2, V7, C3). The value, or values, are the values associated with the respective qualifier, variable, or choice.

An illustration of a simple rule base that employs this external call option is listed below. Following the rule base is a BASIC program that will be called *before* the consultation session. The name of this BASIC program is EXTER.BAS.

Rule 1: *If* color is red   Rule 2: *If* color is white
           Then choice A   *Then* choice B

```
10  '
20  'Program EXTER.BAS
30  'Used to demonstrate external program calls
40  '
50  CLS
60  PRINT "YOU ARE IN THE EXTERNAL PROGRAM"
70  PRINT: PRINT: PRINT
80  INPUT "ENTER COLOR (RED/WHITE) ", FLOWER$
90  IF FLOWER$ = "RED" THEN VALUE = 1
100 IF FLOWER$ = "WHITE" THEN VALUE = 2
110 OPEN "RETURN.DAT" FOR OUTPUT AS #1
120 PRINT #1, "Q1 "; VALUE
130 CLOSE #1
140 SYSTEM
```

Notice that the input to the external program is placed into the file RE-TURN.DAT, where it may be accessed by EXSYS immediately prior to the consultation session. The last statement of this program serves to return control to the operating system, and thus return to the EXSYS operating environment.

When entering the above rule base into EXSYS, first respond with a "yes" when asked if an external program is to be called at the start of the run. You will then be asked for the name of that program. In response, type <BASIC EXTER /C>, and note that there is a space between EXTER and / in this command line. Then go ahead and enter the rules as usual. Next, construct the BASIC program EXTER.BAS, as listed above. Make sure that EXSYS has a path to a BASIC applications package and the EXTER.BAS file.[5] Once you have done this, go ahead and run the rule base.

You will notice, during the consultation session, that you will be presented with a screen display of the printout associated with line 60 and of the prompt associated with line 80 of the BASIC program. That is, EXSYS has called the external program and your response will be accepted by *that* program. Whatever color you input will then be placed in the RETURN.DAT file (as a numerical value, by means of step 90 or 100) and associated with qualifier 1 (i.e., THE COLOR IS), as may be noted in line 120 of the program. It should be obvious

---

[5] Alternatively, you may compile the program EXTER.BAS and respond with <EXTER /C> when asked for the name of the external program.

that you can use the RETURN.DAT file to return any data you so choose to the rule base. Consider, for example, the following entries in the RETURN.DAT file:

V2   1.7
Q7   1, 3

These entries simply mean that variable 2 is assigned a value of 1.7 and the first two values of qualifier 7 are selected.

PROGRAM CALLS FROM A QUALIFIER OR VARIABLE IN A PREMISE CLAUSE. The second type of call is one associated with a specific qualifier or variable in a rule premise. In this instance, the external program associated with that qualifier or variable will only be called when a value is needed for the qualifier/variable—and when that value cannot be derived from the rule base.

A provision for such a call is accomplished by means of the text associated with the qualifier or variable. Specifically, you must enter the RUN command at the start of the text. This may be illustrated by the rule base below wherein the qualifier *color is* shall be used to call the external program:

Rule 1:   *If* weight is heavy   Rule 2:   *If* weight is light
          Then size is large   *Then* size is small

Rule 3:   *If* size is large   Rule 4:   *If* size is small
          and color is red   *and* color is white
          Then choice A   *Then* choice B

When entering this rule base, be sure that you respond with a "no" when you are asked if an external program is to be called at the start of the run. Make sure that your first qualifier is *weight is* (with values of *heavy* and *light*) and that the second is *size is* (with values of *large* and *small*). Then, when entering the text for the third qualifier (i.e., *color is*), type the following:

<RUN(BASIC EXTER1 /C) The color is>

Where the values for this qualifier are *red* and *white*.

The external program to be called from qualifier 3 will be identical, with the one exception of line 120, to program EXTER.BAS as listed previously. That is, the external program must now provide the value for the *third* qualifier. To construct this new program, simply change line 120 of EXTER.BAS to

120 PRINT #1, VALUE

Then name the program EXTER1.BAS. Once this has been done, you may go ahead and consult the rule base. You will discover that EXSYS will ask you for the value of weight. However, when you are asked for the value of the color, that request will come from the external program that has been called by the third qualifier in your rule base.

Calls to external programs from variables in premise clauses follow the same procedure just described for qualifiers. In this case, you simply enter the RUN command at the start of the text associated with the variable. For example, if you have the mathematical variable [TOP SPEED] in a premise clause within your rule base, you can have it call an external program from which the value of *top speed* may be provided (or computed). For example, this external program might be named SPEED. Thus, [TOP SPEED] will be entered as a numeric variable with the following text:

RUN (BASIC SPEED /C) The top speed of the aircraft =

Further, make sure that you respond with a yes when asked if the variable [TOP SPEED] has a value associated with it. As before, the data provided by the external program SPEED will be transmitted to EXSYS by means of the RETURN.DAT file, using the same formatting as discussed previously.

PROGRAM CALLS FROM A TEXT ONLY VARIABLE IN THE CONCLUSION OF A RULE. The third approach to external program calls is by means of a text only variable in the conclusion of a rule. Unlike the two previous approaches, this type of call cannot be used to return data (i.e., to EXSYS) from the external program. Actually, this approach is very much like the concept that we previously used to demonstrate the use of the text only variable [NOTE], as discussed at the conclusion of the section on variables. However, before we input any text for the variable, we need to input a run command.

To demonstrate, let us assume that we have an external program named ASSIST. This program is called from the text only variable [ASSISTANCE]. The program ASSIST might be an external package that serves to provide the user with a graphics display. Once that is done, control is returned to EXSYS. The illustrative rule base is listed below:

Rule 1: *If* house style is Georgian
Then choice A
and [ASSISTANCE]

In turn, [ASSISTANCE] must be a text only variable. The specific text associated with [ASSISTANCE] is

RUN (ASSIST /C) The graphics program ASSIST is called

Now, when you enter the above rule by means of the rule editor, you will notice that the rule exhibited by EXSYS does not look precisely like rule 1. Rather, the conclusion clause associated with the variable [ASSISTANCE] will appear as RUN(ASSIST /C). And this is precisely as it should be.

Report Generator

EXSYS also provides the developer with a particularly versatile and useful report generator. This capability permits the developer to control and enhance the output provided to the user as well as to restart the program, exit to the DOS, support blackboarding, and so on. However, if one is to really achieve the full potential of this facility, it is necessary to have a fair degree of familiarity with the operating system of the personal computer. Consequently, we shall only touch on some of the more fundamental uses of the report generator here.

All report commands are entered in a "report specification" file. Such files are ASCII files and thus may be created by any text editor (e.g., the EDLIN text editor supplied with most operating systems) or word processor that permits storage in the text or ASCII mode. The filename of a report specification file is identical to that of the associated rule-base name, plus an .OUT extension. For example, if your rule base is named CAREERS, then the associated report specification file will be named CAREERS.OUT. The report generator may be called in one of two ways:

- Use the EXSYS internal command REPORT (<filename>) in association with a text only variable in a conclusion clause.
- Use the fact that EXSYS automatically looks for a report specification file at the end of the execution of any knowledge base.

There are a host of report generator commands and we do not intend to cover these in this appendix. The reader who wishes to take full advantage of the report generator is referred to the EXSYS manual. However, to illustrate just a few of these commands, consider the following report specification file. The contents of this file, named CAREERS.OUT (and intended for use with our earlier illustrative rule base of the same name), are listed below:

```
FILE RESULTS.DAT
"Conclusions from single consultation session"
"-------------------------------------" /L
" " /L
"The career choice(s) is (are):"
C
```

This simple report specification file will simply print out all the career choices concluded by the expert system for any given consultation. The first line (FILE RESULTS.DAT) is used to open the file RESULTS.DAT and send the results listed directly below this line to that file. The first line of every report specification file must have such a line (i.e., FILE <filename>). The second line simply prints a string that provides a title for the report. The third line prints another string and, by means of the /L command, adds a blank line after that string. The fourth line simply serves to add two more blank lines. The fifth line is another string. The final line (i.e., C) causes the text and values of all choices to be printed out.

Another useful report generator command is that of RESTART. This command resets all choices, qualifiers, and variables to their initial (i.e., free) state at the start of a new EXSYS consultation session. For example, in the television careers rule base, we could automatically restart the consultation session and print out the results for a number of candidates. An example of such a report output specification file for the TV careers rule base is provided below. Note the /A option at the end of the first line. This serves to append the results of the most recent consultation session to the existing contents of the file. Once the choices for a given consultation session have been printed, the command RESTART is encountered—and a new session is initiated.

```
FILE RESULTS.DAT /A
"Conclusions from multiple sessions"
"----------------------------------" /L
" " /L
"The career choice(s) is (are):"
C
RESTART
```

As mentioned earlier, we may call external programs from the report generator. This is accomplished in essentially the same way that external programs are called from text only variables in the conclusion clause. The only differences are that

- The command is inserted in the report specification file
- The parentheses around the filename, data, and options are omitted

For example, should we wish to call an external program from the report generator, we would place the following command in the report specification file:

RUN filename <data1> <data2> ... <options>

Thus, should we wish to call a BASIC program named DECODE from the report generator, we would include the command line RUN BASIC DECODE /C in the report-specification file for the rule base.

Blackboarding

The EXSYS shell supports blackboarding. Such a facility is generally useful for one of two reasons. First, you may simply want to divide very large rule bases into smaller, more convenient ones. Second, you can maintain a separation between multiple sources of knowledge that serve to support a single expert system. Either way, the multiple rule bases are overseen and coordinated by the blackboarding facility.

The blackboarding concept involves a bit more than just building a number of different rule bases. Certain guidelines must be followed if the process is to operate properly (or at all). In particular, care must be taken that a file exists which contains the values of any qualifiers, variables, or choices that are to be *shared among several rule bases*. In order to illustrate the development and implementation of a blackboarding system, we shall employ an extremely simple example.

AN EXPERT SYSTEM FOR PROCESS MONITORING. This expert system represents a simplified version of one for the monitoring of a production process. In essence, the pressure and temperature of the process will be monitored. A warning message will be presented to the user if the process appears to be on the verge of going out of control, while an alarm will sound if it actually has gone out of control.

Our expert system will employ three different knowledge bases. One will be used to monitor the pressure of the process. The second will be used to monitor the temperature (but will require knowledge of the pressure as derived in the first knowledge base). Finally, the third knowledge base will be employed to determine if the session is to be terminated.

Our expert system will also access two external programs, each of which has been written in BASIC. Consequently, to run this example, you must have a BASIC applications file that may be directly accessed by the program.

The three rule bases, and the names they should be filed under, are listed below. Following this, listings of the associated BASIC programs for the alarm process and termination program are presented.

Rule base 1:   PRESSURE

Rule 1:   *If* [PRESSURE] > 200
        Then [DISPLAY1]
        and CHOICE = CHECK TEMPERATURE

Rule base 2:   TEMPERTR

Rule 1:   *If* [TEMPERATURE] > 150
        and [PRESSURE] > 220
        Then [ALARM]
        and [DISPLAY2]
        and CHOICE = SHUT DOWN

Rule base 3:   TERMINAT

Rule 1:   *If* termination is yes
        Then [QUIT]
        and CHOICE = TERMINATE

The associated text for each of the variables employed in these three rule bases is summarized in Table A.2, below. Note carefully that PRESSURE is a variable that is shared between the first and second rule base, and has thus been placed in italics.

TABLE A.2
Variable description and text

Variable name	Variable type	Associated text
[ALARM]	Text only	RUN (BASIC ALARM /C)
[QUIT]	Text only	RUN(BASIC QUIT /C)
[PRESSURE]	Numeric	Current pressure in the processing system
[TEMPERATURE]	Numeric	Current temperature in system
[DISPLAY1]	Text only	WARNING: Pressure too high-check temperature
[DISPLAY2]	Text only	WARNING: Shut down system immediately!

```
100 CLS
110 '
120 ' PROGRAM ALARM.BAS
130 '
140 FOR I = 1000 TO 2000 STEP 100
150 SOUND I,1
160 NEXT I
170 SYSTEM
```

```
100 CLS
110 '
120 'PROGRAM QUIT.BAS
130 '
140 OPEN "QUIT.RPT" FOR OUTPUT AS #1
150 PRINT "TERMINATION REQUESTED"
160 CLOSE #1
170 SYSTEM
```

To run this blackboarded expert system, we still need to construct two more files. The first is a report specification file for the rule-base pressure. This file should be named PRESSURE.OUT and its contents are listed below:

```
FILE BBFILE.DAT
V1 /D2
```

The first line serves to open a file named BBFILE.DAT (the blackboard file for our expert system). The contents of this file will be the name and value of the first variable in the rule base PRESSURE, that is, the value of the pressure of the system. Following this, we see the /D2 option. The D indicates that the *datalist* option is being employed—a requirement when passing data by means of the blackboard. The 2 indicates that this data will be the value for the *second* variable of the file to which the data is passed. That is, in the rule-base TEMPERTR, pressure is the second variable. The last file that must be developed is a BAT (i.e., batch) file that serves to coordinate the operation of the complete system.[6] The contents of this file, named START.BAT, are listed on the next page.

The first two lines of this file serve to delete any existing files that are named BBFILE.DAT or QUIT.RPT. The third line is simply a label that serves to note the start of a sequence of commands. The consultation continues to cycle through the three rule bases until a termination is requested (as indicated by the existence of the file QUIT.RPT).

---

[6] As an alternative, we may avoid the need to build such a file by having each rule base call the next one. This approach is described in the EXSYS manual and shall not be discussed here.

```
IF EXIST BBFILE.DAT DEL BBFILE.DAT
IF EXIST QUIT.RPT DEL QUIT.RPT
:BEGINNING
   EDITDEMO PRESSURE RUNONLY
   EDITDEMO TEMPERTR RUNONLY DATALIST=BBFILE.DAT
   EDITDEMO TERMINAT RUNONLY
   IF EXIST QUIT.RPT GOTO EXIT
   GOTO BEGINNING
:EXIT
```

Each of the calls to the rule editor (EDITDEMO) may be explained as follows:

- EDITDEMO PRESSURE RUNONLY. This command serves to call the rule editor, load the rule base PRESSURE, and treat it as a *run only* rule base (i.e., the rule base is simply run, not edited).
- EDITDEMO TEMPERTR RUNONLY DATALIST = BBFILE.DAT. This command calls the rule editor, loads the rule base TEMPERTR, and treats it as a *run only* rule base. The option DATALIST = BBFILE.DAT serves to provide the rule base TEMPERTR with the contents of the file BBFILE.DAT *prior* to its initiation. That is, the value of pressure, as derived from the preceding rule base, is made available to this rule base.
- EDITDEMO TERMINAT RUNONLY. This command calls the rule editor, loads the rule base TERMINAT, and treats it as a *run only* rule base. If the user decides to terminate the process, the file QUIT.RPT is created—which causes the termination of the entire run. Otherwise, we cycle back to the rule base PRESSURE. However, since PRESSURE is not presented any data from the blackboard, we will once again have to provide it with a value for the variable pressure.

Figure A.18 serves to indicate the architecture of this particular expert system. The reader is advised to construct the three rule bases, the various supporting files (i.e., START.BAT, PRESSURE.OUT), and the BASIC files. To run the complete blackboarded system, simply type <START>. While this blackboarding example is extremely simplified, it should serve as a guide for the development of more realistic rule bases that utilize the blackboarding facility.

SUMMARY

We have attempted to present, in this appendix, a relatively complete description of rule-base entry and editing. In addition, we have briefly covered some of the

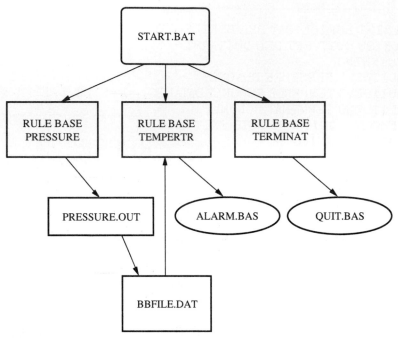

FIGURE A.18
Architecture of blackboarding example.

more advanced concepts available through the EXSYS shell. This material should be more than sufficient to give the reader an acquaintance with this tool. And for those who wish to pursue the matter further, we recommend the EXSYS user's manual available from EXSYS, Inc.

Most other expert systems shells operate in a more or less similar fashion. Consequently, much of what has been covered here is transferable to the utilization of alternative tools.

Finally, it may be noted that exercises appear within the text that may be used to explore the more basic features of the EXSYS package. However, they may also be addressed, if the reader so wishes, by virtually any other expert systems shell.

REFERENCES

AI WEEK, "Big Eight Accounting Firm Develops Tax ES," *AI WEEK*, July 1, 1988*a*, p. 4.

AI WEEK, "Chemical Bank Develops Foreign Trading ES," *AI WEEK*, September 15, 1988*b*, pp. 8–9.

AI WEEK, "ES Market Could Climb to $6 Billion by 1995," *AI WEEK*, February 1, 1990, p. 2.

Aikens, J. S., J. C. Kung, E. H. Shortliffe, and R. J. Fallat, "PUFF: An Expert System for Interpretation of Pulmonary Function Data," in B. C. Clancey and E. H. Shortliffe, (eds.), readings in *Medical Artificial Intelligence: The First Decade*, Addison-Wesley, Reading, Mass.,1984.

Anderson, D., and C. Ortiz, "AALPS: A Knowledge-Based System for Aircraft Loading," *IEEE Expert*, Winter 1987, pp. 71–79.

Baker, S. "Nexpert Object: Mainstreaming AI Applications," *IEEE Expert*, Winter 1988, p. 82.

Benchimol, G., P. Levine, and J. C. Pomerol, *Developing Expert Systems for Business*, North Oxford Academic, London, 1987.

Bonissone, P. P., and H. E. Johnson, "Expert System for Diesel Electric Locomotive Repair," Knowledge-Based System Report, General Electric Company, Schenectady, N.Y., 1983.

Brazile, R. P., and K. M. Swigger, "GATES: An Airline Gate Assignment and Tracking Expert System," *IEEE Expert*, Summer 1988, pp. 33–39.

Carter, C., and J. Catlett, "Assessing Credit Card Applications Using Machine Learning," *IEEE Expert*, vol. 2, no. 3, Fall 1987, pp. 71–79.

Casey, J., "Picking the Right Expert System Application," *AI Expert*, September 1989, pp. 44–47.

Cavalier, T. M., J. P. Ignizio, and A. L. Soyster, "Discriminant Analysis via Mathematical Programming: On Certain Problems and Their Causes," *Computers and Operations Research*, vol. 16, no. 4, 1989, pp. 353–362.

Cleland, D. I., and W. R. King, *Systems Analysis and Project Management*, McGraw-Hill, New York, 1968.

Cochard, D. D., and K. A. Yost, "Improving Utilization of Air Force Cargo Aircraft," *Interfaces*, vol. 15, no. 1, January–February 1985, pp. 53–68.

Cohon, J. L., *Multiobjective Programming and Planning*, Academic Press, New York, 1978.

Davidson, K., "Compound Is Vital to Stealth Bomber," *San Francisco Examiner*, June 26, 1988, p. 2.

Davis, R., "Applications of Meta-Level Knowledge to the Construction, Maintenance, and Use of Large Knowledge Bases," Ph.D. dissertation, Department of Computer Science, Stanford University, 1976.

Deal, D. E., J. G. Chen, J. P. Ignizio, and V. Jeyakumar, "An Expert System Scheduler: Some Reflections on Expert System Development," *Journal of Computers and Operations Research*, forthcoming.

Eskow, D., "Wendy's Turns to PCs to Beef Up Operations," *PC Week*, February 12, 1990, pp. 121–124.

Feigenbaum, E. A., Interview in *Knowledge-Based Systems: A Step-by-Step Guide to Getting Started*, Second Annual AI Satellite Symposium, Texas Instruments, 1986.

Feigenbaum, E. A., P. McCorduck, and H. P. Nii, *The Rise of the Expert Company*, Times Books, New York, 1988.

Fikes, R., and T. Kehler, "The Role of Frame-Based Representation in Reasoning," *Communications of the ACM*, vol. 28, no. 9, September 1985, pp. 904–920.

Fisher, M. L., "Worst Care Analysis of Heuristic Algorithms," *Management Science*, vol. 26, 1980, pp. 1–17.

Freed, N., and F. Glover, "A Linear Programming Approach to the Discriminant Problem," *Decision Sciences*, vol. 12, 1981a, pp. 68–73.

Freed, N., and F. Glover, "Simple but Powerful Goal Programming Models for Discriminant Problems," *European Journal of Operational Research*, vol. 7, 1981b, pp. 44–60.

Fuller, J. A., "Optimal Solutions Versus Good' Solutions: An Analysis of Heuristic Decision Making," *Omega*, vol. 6, 1978, pp. 479–484.

Gear, C. W., *Introduction to Computer Science*, SRA, Chicago, 1973.

Harmon, P., and D. King, *Expert Systems*, Wiley, New York, 1985.

Hayes-Roth, F., "Rule-Based Systems," *Communications of the ACM*, vol. 28, no. 9, September 1985, pp. 921–932.

Huntington, D., *EXSYS Expert Systems Development Package*, EXSYS Manual, Albuquerque, New Mexico, 1985.

IEEE, "Mac II: A Serious Contender for AI Applications?" *IEEE Expert*, Summer 1988, p. 84.

Ignizio, J. P., "Saturn SII Telemetry Antennas: Ground Test Procedures," *North American Aviation*, S&ID, Downey, Calif., 1962.

Ignizio, J. P., *A Heuristic Solution to Generalized Covering Problems*, Ph.D. dissertation, Virginia Polytechnic Institute, Blacksburg, Va., 1971.

Ignizio, J. P., R. M. Wyskida, and M. Wilhelm, "A Rationale for Heuristic Programing Selection and Validation," *Industrial Engineering*, January 1972, pp. 16–19.

Ignizio, J. P., and R. M. Harnett, "Heuristically Aided Set Covering Algorithms," *International Journal of Computers and Information Systems*, vol. 3, no. 1, 1974, pp. 59–70.

Ignizio, J. P., and J. N. D. Gupta, *Operations Research in Decision Making*, Crane Russak, New York, 1975.

Ignizio, J. P., "A Multicriteria Army Battalion Training Model: Refinements and Extensions," Final Report for Army Research Office, DAAG29-76-D-0100, 1978.

Ignizio, J. P., "Solving Large Scale Problems: A Venture Into a New Dimension," *Journal of Operational Research Society*, vol. 31, 1980, pp. 217–225.

Ignizio, J. P., "Antenna Array Beam Pattern Synthesis via Goal Programming," *European Journal of Operational Research*, vol. 6, 1981, pp. 286–290.

Ignizio, J. P., *Linear Programming in Single and Multiple Objective Systems*, Prentice-Hall, Englewood Cliffs, N.J., 1982.

Ignizio, J. P., *Introduction to Linear Goal Programming*, Sage Books, Beverly Hills, Calif., 1985.

Ignizio, J. P., "Discriminant Analysis via Goal Programming: On the Elimination of Certain Problems," *Proceedings S.W. IDS*, March 1987a, pp. 125–126.

Ignizio, J. P., "Identification of Incompleteness in Knowledge Bases via a Rule Dependency Matrix," technical paper, Department of Industrial Engineering, University of Houston, July 1987b.

Ignizio, J. P., K. W. Wiemann, and W. J. Hughes, "Sonar Array Element Location: A Hybrid Expert Systems Application," *European Journal of Operational Research*, vol. 32, 1987, pp. 76–85.

Ignizio, J. P., "Attribute-Value Pair Tables and Rule Base Architecture," technical paper, University of Houston, Houston, Tex., 1988.

IntelliCorp, *KEE: The Knowledge Engineering Environment*, a brochure distributed by IntelliCorp, 1987.

Jackson, P., *Introduction to Expert Systems*, Addison-Wesley, Wokingham, England, 1986.

James, M., *Classification Algorithms*, Wiley, New York, 1985.

Johnson, R. A., and D. W. Wichern, *Applied Multivariate Statistical Analysis,* Prentice-Hall, Englewood Cliffs, N.J., 1988.

Kane, B., and D. W. Rucker, "AI in Medicine," *AI Expert*, November 1988, pp. 48–55.

Karp, R. M., "On the Computational Complexity of Combinatorial Problems," *Networks*, vol. 5, 1975, pp. 45–68.

Kaufmann, A., and A. Henry-Labordère, *Integer and Mixed Programming*, Academic, New York, 1977.

Kendig, P. M., "A Methodology of Selecting Element Positions in a Hydrophone Array to Reduce Minor Lobes," *Journal of Acoustical Society of America*, vol. 50, 1971, pp. 310–313.

Kickert, W. J. M., *Fuzzy Theories on Decision Making*, Leiden, Boston, Mass., 1978.

Kim, J. W., Homework assignment for INDE-6369, Department of Industrial Engineering, University of Houston, Houston, Tex., May 8, 1989.

Kinnucan, P., "Software Tools Speed Expert System Development," *High Technology*, March 1985, pp. 16–21.

Koen, B. V., "Toward a Definition of the Engineering Method," *The Bent of Tau Bata Pi*, Spring 1985, pp. 28–33.

Lachenbruch, P. A., *Discriminant Analysis*, Hafner, New York, 1975.

Langston, M. A., "A Study of Composite Heuristic Algorithms," *Journal of Operational Research Society*, vol. 38, no. 6, June 1987, pp. 539–544.

Lin, S., and B. W. Kernighan, "An Effective Heuristic Algorithm for the Traveling-Salesman Problem," *Operations Research*, vol. 21, 1973, pp. 498–516.

Linden, E., "IntelliCorp; The Selling of Artificial Intelligence," *High Technology*, March 1985, pp. 22–27.

Lindsay, R. K., B. G. Buchanan, E. A. Feigenbaum, and J. Lederberg, *Applications of Artificial Intelligence for Chemical Inference: The DENDRAL Project*, McGraw-Hill, New York, 1980.

Lippman, R. P., "An Introduction to Computing with Neural Nets," *IEEE ASSP Magazine*, April 1987, pp. 4–22.

McDermott, J., "R1: A Rule-Based Configurer of Computer Systems," *Artificial Intelligence*, vol. 19, 1982, pp. 39–88.

Mingers, J., "Expert Systems—Experiments with Rule Induction," *Journal of Operational Research Society*, vol. 37, no. 11, 1986, pp. 1031–1037.

Mingers, J., "Expert Systems—Rule Induction with Statistical Data," *Journal of Operational Research Society*, vol. 38, no. 1, 1987, pp. 39–47.

Minsky, M., and S. Papert, *Perceptrons*, M.I.T., Cambridge, Mass., 1969.

Mishkoff, H. C., *Understanding Artificial Intelligence*, Sams, Indianapolis, Ind., 1985.

Moeller, G. L. and L. A. Digman, "VERT: A Technique to Assess Risks," *Proceedings of the American Institute for Decision Sciences*, vol. 2, 10th Annual Conference, St. Louis, Mo., October 1978.

Moore, L. J., and E. R. Clayton, *GERT Modeling and Simulation*, Petrocelli/Charter, New York, 1976.

Müller-Merbach, H., "Heuristics and Their Design: A Survey," *European Journal of Operational Research*, vol. 8, no. 1, September 1981, pp. 1–23.

Murty, K. G., *Linear and Combinatorial Programming*, Wiley, New York, 1976.

Newquist, H. P., III, "Somewhere Over The Rainbow: Using AI to Get to Wall Street," *AI Expert*, July 1988*a*, pp. 65–67.

Newquist, H. P., III, "Struggling to Maintain," *AI Expert*, August 1988*b*, pp. 69–71.

Newquist, H. P., III, "Tales from the Hearth of AI," *AI Expert*, December 1988*c*, pp. 61–64.

Newquist, H. P., III, "Expert Systems and Their Vendors: A Tale of Two Cities," *AI Expert*, May 1989, pp. 67–69.

Ng, K. Y. K., "A Knowledge Based Optimization (Goal Programming) Approach to Aircraft Loading," technical paper, Operational Research and Analysis Establishment, Department of National Defence, Ottawa, Canada, 1989.

Nguyen, T. A., W. A. Perkins, and D. Pecora, "Checking an Expert System's Knowledge Base for Consistency and Completeness," *Proceedings of the Ninth International Joint Conference on Artificial Intelligence*, American Association for Artificial Intelligence, Menlo Park, Calif., 1985, pp. 374–378.

Nguyen, T. A., W. A. Perkins, T. J. Laffey, and D. Pecora, "Knowledge Base Verification," *AI Magazine*, Summer 1987, pp. 69–75.

Nicholson, T. A. J., *Optimization in Industry*, Aldine-Atherton, Chicago, 1971.

Nii, H. P., E. A. Feigenbaum, J. J. Anton, and A. J. Rockmore, "Signal-to-Symbol Transformation: HASP/SIAP Case Study," *AI Magazine*, Spring 1982, pp. 22–35.

O'Connor, R., and E. W. Henry, *Input-Output Analysis and its Application*, Hafner, New York, 1975.

Peat, F. D., *Artificial Intelligence*, Simon & Schuster, New York, 1985.

Pederson, K., "Well-Structured Knowledge Bases—Part I," *AI Expert*, vol. 4, no. 4, April 1989*a*, pp. 44–55.

Pederson, K., "Well-Structured Knowledge Bases—Part II," *AI Expert*, vol. 4, no. 7, July 1989*b*, pp. 45–48.

Pederson, K., "Well-Structured Knowledge Bases—Part III," *AI Expert*, vol. 4, no. 11, November 1989*c*, pp. 37–41.

Pederson, K., *Expert System Programming*, Wiley, New York, 1989*d*.

Polya, G., *How to Solve It*, Doubleday, Garden City, N.Y., 1957.

Pople, H. E., Jr., "CADUCEUS: An Experimental Expert System for Medical Diagnosis," in P. H. Winston and K. A. Prendergast (eds.), *The AI Business*, M.I.T., Cambridge, Mass., 1984 pp. 67–80.

Prerau, D. S., "Knowledge Acquisition in Expert System Development," *AI Magazine*, vol. 8, no. 2, Summer 1987, pp. 43–51.

Press, L., "Eight-Product Wrap-up: PC Shells," *AI Expert*, September 1988, pp. 61–65.

Quinlan, J. R., "Learning Efficient Classification Procedures and Their Application to Chess End Games," in R. S. Michalski et al. (eds.), *Machine Learning: An Artificial Intelligence Approach*, Tioga, Palo Alto, Calif., 1983.

Quinlan, J. R, "Decision Trees and Multivalued Attributes," in J. E. Hayes and D. Michie (eds.), *Machine Intelligence II*, Oxford, London, 1987*a*.

Quinlan, J. R., "Simplifying Decision Trees," *Journal of Man-Machine Studies*, 1987*b*.

Reddy, D. R., L. D. Erman, R. D. Fennell, and R. B. Neely, "The HEARSAY Speech Understanding System: An Example of the Recognition Process," *IJCAI-3*, 1973, pp. 185–193.

Rich, E., *Artificial Intelligence*, McGraw-Hill, New York, 1983.

Rose, F., *Into the Heart of the Mind*, Harper & Row, New York, 1984.

Rose, F., "Thinking Machine: An 'Electronic Clone' of a Skilled Engineer Is Very Hard to Create," *The Wall Street Journal*, vol. CXIX, no. 30, August 12, 1988, p. 1.

Sage, A. P., *Methodology for Large-Scale Systems*, McGraw-Hill, New York, 1977.

Sanders, D. H., *Computers and Management*, McGraw-Hill, New York, 1970.

Schwartz, T., "DARPA Studies Neural Network Potential Military Application in All Areas," *AI WEEK*, 15 September 1988, p. 1.

Shafer, D., "Ask the Expert," *PC AI*, July–August 1989, pp. 22–23.

Sherer, P. M., "Abraxas Expert System Checks C Source Code," *PC Week*, August 21, 1989, p. 63.

Shinners, S. M., *Techniques of System Engineering*, McGraw-Hill, New York, 1967.

Shortliffe, E. H., *Computer-Based Medical Consultations: MYCIN*, Elsevier, New York, 1976.

Simon, H. A., *Administrative Behavior*, Macmillan, New York, 1957.

Simon, H. A., and A. Newell, "Heuristic Problem Solving: The Next Advance in Operations Research," *Operation Research*, January–February 1958, pp. 1–10.

Spitzer, M., "The Computer Art of Schedule Making," *Datamation*, April 1969, pp. 84–86.

Sprague, K. G., and S. R. Ruth, *Developing Expert Systems Using EXSYS*, Mitchell Publishing, Santa Cruz, Calif., 1988.

Stefik, M., "Strategic Computing at DARPA: Overview and Assessment," *Communications of the ACM*, vol. 28, no. 7, July 1985, pp. 690–703.

Surko, P., "Tips for Knowledge Acquisition," *PC AI*, May–June 1989, pp. 14–18.

Taha, H. A., *Integer Programming*, Academic Press, New York, 1975.

Ten Dyke, R. P., "A Taxonomy of AI Applications," *AI WEEK*, February 1, 1990, pp. 6–7.

Thompson, B., and W. Thompson, "Finding Rules in Data," *BYTE*, vol. 11, no. 12, November 1986, pp. 149–158.

Tonge, F. M., "The Use of Heuristic Programming in Management Science," *Management Science*, vol. 7, April 1961, pp. 231–237.

Townsend, C., *Mastering Expert Systems with Turbo Prolog*, Sams, Indianapolis, Ind., 1986.

Van Horn, M., *Understanding Expert Systems*, Bantam, New York, 1986.

Walker, L. L., and P. Gerkey, "The Seven Veils of AI: The Second Veil," *PC AI*, November–December 1988, p. 25.

Weiss, S. M., and C. A. Kulikowski, *Designing Expert Systems*, Rowman & Allanheld, Totowa, N.J., 1984.

Wiest, J. D., "Heuristic Programs for Decision Making," *Harvard Business Review*, September–October 1966, pp. 129–143.

Wiest, J. D., and F. K. Levy, *A Management Guide to PERT/CPM*, Prentice-Hall, Englewood Cliffs, N.J., 1969.

Yang, T., and J. P. Ignizio, "An Algorithm for the Scheduling of Army Battalion Training Exercises," *Computers and Operations Research*, vol. 14, no. 6, 1987, pp. 479–491.

Yang, T., and J. P. Ignizio, "An Exchange Heuristic Algorithm for Project Scheduling with Limited Resources," *Engineering Optimization*, vol. 14, 1989, pp. 189–205.

Zadeh, L. A., "Fuzzy Sets," *Information and Control*, vol. 8, 1965, pp. 338–353.

Zimmermann, H. J., *Fuzzy Set Theory and Its Applications*, Kluwer-Nijhoff, Boston, Mass., 1985.

INDEX

INDEX